Krystyna Jackowska, Paweł Krysiński

Applied Electrochemistry

Also of Interest

X-Ray Studies on Electrochemical Systems.
Synchrotron Methods for Energy Materials
Braun, 2024
ISBN 978-3-11-079400-7, e-ISBN 978-3-11-079403-8

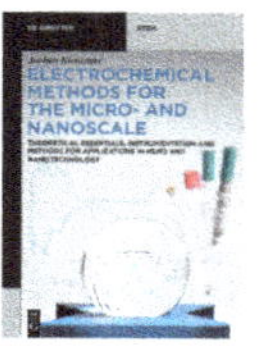

Electrochemical Methods for the Micro- and Nanoscale.
Theoretical Essentials, Instrumentation and Methods for Applications in MEMS and Nanotechnology
Kieninger, 2022
ISBN 978-3-11-064974-1, e-ISBN 978-3-11-064975-8

Electrochemical Energy Storage.
Physics and Chemistry of Batteries
Job, 2020
ISBN 978-3-11-048437-3, e-ISBN 978-3-11-048442-7

Maths in Chemistry.
Numerical Methods for Physical and Analytical Chemistry
Bansal, 2020
ISBN 978-3-11-069531-1, e-ISBN 978-3-11-069532-8

Solid-State Chemistry
Hofmann, 2023
ISBN 978-3-11-065724-1, e-ISBN 978-3-11-065729-6

Krystyna Jackowska, Paweł Krysiński

Applied Electrochemistry

Aspects in Material and Environmental Science

2nd Edition, expanded and amended

DE GRUYTER

Authors
Prof. Krystyna Jackowska
Professor Emeritus
University of Warsaw
Faculty of Chemistry
Laboratory of Electrochemistry
Pasteur 1
02-093 Warsaw
Poland

Prof. Paweł Krysiński
University of Warsaw
Faculty of Chemistry
Laboratory of Electrochemistry
Pasteur 1
02-093 Warsaw
Poland

ISBN 978-3-11-116034-4
e-ISBN (PDF) 978-3-11-116098-6
e-ISBN (EPUB) 978-3-11-116344-4

Library of Congress Control Number: 2024933394

Bibliographic information published by the Deutsche Nationalbibliothek
The Deutsche Nationalbibliothek lists this publication in the Deutsche Nationalbibliografie; detailed bibliographic data are available on the Internet at http://dnb.dnb.de.

Cover image: The template-assisted electrodeposited Pd nanowires. Courtesy of Dr. Magdalena Tagowska, Faculty of Chemistry, University of Warsaw
Typesetting: Integra Software Services Pvt. Ltd.

www.degruyter.com

Preface

A multiplicity of definitions exists for applied electrochemistry as a discipline. The most general, attempting to embrace all aspects, can be formulated as science that aims at improving our life, taking advantage of the phenomena occurring at the interfaces between metallic or semiconducting electrodes and electrolyte solutions (so-called electrodics) as well as those occurring in the bulk of the electrolyte solutions (so-called ionics). Such improvement can come through the understanding of these phenomena, with subsequent construction and design of new devices or systems that can be used not only by industry (e.g., catalysis) but also in more personalized applications, such as batteries for smartphones, pacemakers, solar panels, and so on.

To facilitate understanding of these phenomena, **Basic concepts** part in this book present first a brief thermodynamic background allowing the reader to understand the electrified interfaces and electrolyte solution under equilibrium and steady-state conditions. Then it continues with a basic introduction to the structure of interfaces, followed by the charge transfer processes occurring at metal/electrolyte interface from the point of view of equilibrium and nonequilibrium phenomena, defining the so-called reversible and irreversible electrodes. The next part begins with an introduction to the selected electrochemical methods applied in the analytical and material chemistry that benefit from the essential knowledge gained in the previous chapters.

The next part of this book guides through the **Electrochemistry in material science – selected topics**. It aims at describing in more detail how the electrochemistry can be used in corrosion science, catalysis, deposition of new material on a conducting support, allowing to get insight into the mechanism of deposition and its kinetics. This chapter will lead to the nanostructured materials of different dimensions, organization, and topologies. Then, a very broad area, extremely important nowadays for our population, namely – the energy storage and conversion, will be set forth in order to show how the electrochemistry can be applied to understand the working of batteries and fuel cells, to stimulate and push forward their development and design, thus improving their reliability and durability. This part will conclude with the applications of electrochemistry in biology and medicine, thereby improving the quality of life of patients and providing information on the energetics of living organisms.

Finally, the last part, **Photoelectrochemistry in material science – selected topics**, will guide the reader through the selected topics of photoelectrochemistry, where light acts as a power source for electrical energy generation, photocatalysis, and photoelectrocatalysis, including the nanoscale processes at the semiconductor nanoparticles.

The goal of this book is to show all graduate and PhD students that electrochemistry not only has many applications for understanding of various phenomena in nowadays life but also has many applications in practical devices and can also stimulate new science-enabled technologies, nourishing leaps from bench-top to large-scale industries, providing also means for protecting our environment. Our book is based on lectures given at the University of Warsaw, Faculty of Chemistry. It addresses advanced

https://doi.org/10.1515/9783111160986-202

students and PhD students of chemistry, physics, engineering, and related subjects and also those scientists who want to get a solid background knowledge of this area.

We hope that this background will be useful to the interested reader, encouraging him/her to take the next several steps into the attractive area of applied electrochemistry. Our aim is to guide him/her through the selected topics of contemporary life in which electrochemistry progressively improves our knowledge and quality of life.

Acknowledgments: **Krystyna Jackowska** dedicates her part of this book to Polish and valued Russian friends (Dr. R.V. Ivanova and Prof. M.N. Rodnikova from Moscow) for their encouragement and support. Special thanks are due to Prof. Magdalena Skompska and Prof. Marek Szklarczyk from Warsaw University, and to Prof. G.A.Tsirlina and Prof. O.A. Petrii from Moscow State University for some reviews and references. She is also very grateful to her family for patience.

Paweł Krysiński highly acknowledges the understanding and forbearance of his family, with special thanks due to his wife, Hanna, for her extreme patience, kindness, support, and help.

The authors thank Dr. Magdalena Osial for her skills and time devoted for preparing all the graphics in this book. Without her work and willingness to share some of her experimental data, this book will be largely delayed, to say nothing about its form.

The authors are also greatly indebted to the editorial staff of De Gruyter.

Preface to the 2nd edition

In the still growing awareness of the necessity of minimizing the deteriorating impact on the environment by all fundamental activities of mankind, electrochemistry can provide tools for environmental control, pollution avoidance, and destruction of pollutants as well as clean energy conversion and storage. Moreover, the use of electrons in electrochemical processes is considered a "***green***" model in its preventive mode (electrosynthesis, hydrogen production, power storage, metal recovery/recycling) and the control/remedial mode (wastewater treatment, air pollution control, destruction of organic pollutants). In all these processes, the electron as a reagent frequently leads to cleaner technology by simple avoidance of the use of hazardous chemicals and allows for subsequent recovery, recycling, or at least pollution control.

Therefore, in this second edition of the Applied Electrochemistry textbook, we added a brief review of current and emerging approaches utilizing electrochemistry for a cleaner environment, the so-called "green" electrochemistry. This review is contained in **Part IV**, entitled **Electrochemistry in Environmental Science – selected topics.**

This part entails the utilization of electrochemistry to monitor the environment for certain most commonly polluting gases **(Chapter 13. Sensing the environment)**. This chapter classifies and shows the applications of electrochemical sensors in monitoring the safety of life through the development of rapid, selective, and miniaturized detection techniques that can trigger further remediation processes. As described in **Chapter 14** entitled **Green fuel – hydrogen production**. Environmental pollution, and global warming, have triggered vast research for new, promising sources of clean, "green" energy, such as hydrogen. One of the simplest methods for hydrogen production is water electrolysis. Therefore, this chapter encompasses methods and technologies currently used in water electrolysis, stressing out its thermodynamics, efficiency, and energy losses. A brief description of a new method of water electrolysis that is based on bioelectrochemical systems, such as microbial electrolysis, is also included. Furthermore, this chapter presents an outline of solar energy utilization in water electrolysis, including the semiconducting nanostructured systems and again, thermodynamics and efficiency for such devices.

Then, we will describe the electrochemical processes and methods in the elimination of the most common pollutants (**Chapter 15. Electrochemical and photocatalytic methods in pollutant removal**). An outline is given to the electrochemistry in the removal and recovery of heavy metals, followed by the removal of organic pollutants via the electrochemical treatment. This chapter finally will summarize the application of solar energy for a cleaner environment with a description of the processes utilizing this renewable and sustainable source of natural energy in the removal and degradation of some inorganic and organic pollutants.

To better understand all processes involved in the application of electrochemistry for a greener environment, some modifications were also necessary to previous chap-

https://doi.org/10.1515/9783111160986-203

ters, particularly to the **Basic concepts** part. And so, the **Basic Concepts** part, as before, consists of **Chapter 1.1 Structure of interfaces**, which introduces a brief thermodynamic background for understanding the electrified interfaces and electrolyte solution under equilibrium and steady-state conditions. As in the first edition it continues with a basic introduction to the structure of interfaces, followed by the charge transfer processes occurring at metal/electrolyte interface from the point of view of equilibrium and non-equilibrium phenomena, defining the so-called "reversible" and "irreversible" electrodes. However, the **Basic concepts** part in the current edition was expanded to include **Chapter 1.2 Structure of the bulk of electrolytes. Conductivity,** followed by **Chapter 1.3 Nonaqueous electrolytes,** describing the structure of the bulk of aqueous electrolytes, conductivity, as well as nonaqueous electrolytes, including ionic liquids, molten salts, and solid electrolytes. Finally, **Part I** was amended with a brief background to **Membranes and membrane potentials (Chapter 1.4)**, both under equilibrium conditions and non-equilibrium conditions, necessary for a better understanding of the design of batteries and other storage devices.

We hope that this 2nd edition, completed with new information contained in the additional chapters will be found interesting and useful, by giving the background knowledge and directions in the applications of electrochemistry for cleaner, "greener" and sustainable environment and quality of life.

Contents

Part II: **Electrochemistry in material science - selected topics**

Part I: **Basic concepts**

1 Basic concepts

1.1 Structure of interfaces

To begin with, it is necessary to establish a solid foundation for all topics presented in the following chapters. This foundation can be derived from the laws of thermodynamics that provide tools not only for qualitative and quantitative description of systems and processes but also capabilities to predict their further development. There are four state functions in thermodynamics, namely, the internal energy, U, enthalpy H, entropy, S, Gibbs' free energy, G (also called the *thermodynamic potential*), and the Helmholtz' free energy, F. Together with their parameters of state – V, p, T, n_i, these functions describe precisely the state of a given system or process. For the purpose of this book, let us choose Gibbs' free energy G (the thermodynamic potential) for subsequent chapters of this book. As mentioned earlier, this function is the state function, meaning that its change depends only on the initial and final state of the system. From the mathematical point of view such extremely small change can be written as the total differential versus the state parameters of Gibbs' free energy: $G = G(p, t, n_i)$. Thus,

$$dG = \left(\frac{\partial G}{\partial p}\right)_{T, n_i} dp + \left(\frac{\partial G}{\partial T}\right)_{p, n_i} dT + \Sigma\left(\frac{\partial G}{\partial n_i}\right)_{p, T} dn_i \tag{1.1}$$

Subscripts next to parentheses show that the remained parameters of state are constant. Thus, the meaning of this total differential is that we can sum partial differential of G versus p, keeping T and n_i constant, controlling the change dp, and so on, and then sum all partial differentials to get the overall change of G, dG. As long as we do not assign the physicochemical meanings of all three partial differentials in the above equation, it remains purely mathematical. However, taking into account the laws of thermodynamics, we can identify the meanings of these partial differentials, rewriting the above equation as follows:

$$dG = Vdp - SdT + \Sigma\mu_i dn_i, \tag{1.2}$$

where V is volume, S – entropy, and μ_i is the chemical potential of species "i" in the system under consideration.

The detailed arguments behind this transformation is beyond the scope of this book. Interested reader is directed to the textbooks on thermodynamics.

Now, let us identify the system that is used in electrochemistry – the electrode. Typically, an ideal conductor (semiconductor electrodes will be discussed in Section 10.2) and its properties do not depend on the bulk (volume), but surface behavior, where all of the excess charges originating, for example, from the external polarization, are localized. An experimentally measurable parameter that can be initially used for an interface with zero electric charge is the so-called surface or interface tension. The latter is more ap-

https://doi.org/10.1515/9783111160986-001

propriate, because there are no free surfaces, but surfaces separating two phases in contact. For our purposes, we will use the term "surface tension" throughout the text. Quantitatively, a work δw, required to increase a surface area by δA, is given by the following equation:

$$\delta w = \gamma dA, \tag{1.3}$$

Proportionality coefficient γ [J/m^2 = N/m] is called the *surface tension* (interfacial tension)

Thermodynamics of such an interface can be described with the help of the change of selected state function like Gibbs' free energy, which will depend only on the initial and final state. For an open, multicomponent system of a surface area A, we can write:

$$G = G(p,\ T,\ n_i,\ A), \tag{1.4}$$

and its total differential:

$$dG = \left(\frac{\partial G}{\partial p}\right)_{T,n_i,A} dp + \left(\frac{\partial G}{\partial T}\right)_{p,n_i,A} dT + \Sigma\left(\frac{\partial G}{\partial ni}\right)_{p,T,A} dn_i + \left(\frac{\partial G}{\partial A}\right)_{p,T,n_i} dA \tag{1.5}$$

or

$$dG = Vdp - SdT + \Sigma\mu_i dn_{i+}\gamma\, dA \tag{1.6}$$

Therefore, at constant p, T, n_i, we can obtain the thermodynamic definition of surface tension:

$$dG = \gamma\, dA \tag{1.7}$$

Until now we have considered the total value of Gibbs' free energy change (thermodynamic potential change) of a system with the surface. Now let us turn to the description of the surface (interface) itself. In doing this we will use the so-called Gibbs model, in which the ideal surface has no volume and the two phases separated by this surface are at equilibrium. This is illustrated in Fig. 1.1.

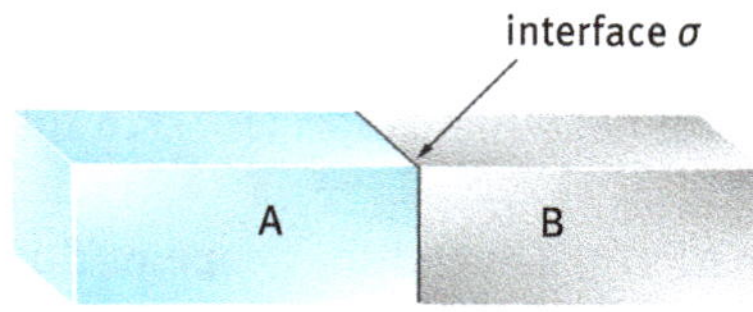

Fig. 1.1: The Gibbs model of two contacting phases A and B at equilibrium, represented by two slabs of different colors and separated by an ideal surface σ of $V^\sigma = 0$. See text for details.

All these regions are at equilibrium, and so chemical potentials of any component "i" that can distribute itself between these three regions are also at equilibrium:

$$\mu_i^A = \mu_i^B = \mu_i^\sigma = \mu_i \tag{1.8}$$

This model formally assumes that the total internal energy of such a system can be described by a sum of selected constituents: phase A, phase B, and their contact surface σ:

$$U_{\text{tot}} = U_A + U_B + U^\sigma \tag{1.9}$$

The total volume is due solely to a sum of volumes A and B, $V^\sigma = 0$, so

$$V_{\text{tot}} = V_A + V_B \tag{1.10}$$

Moreover, any component n_i of the system partitions itself between phase A, phase B, and surface σ. All these regions are at equilibrium, and the equilibrium condition is fulfilled also by the equilibrium of chemical potentials of species "i" in the system:

$$\mu_i^A = \mu_i^B = \mu_i^\sigma = \mu_i. \tag{1.11}$$

With the help of some thermodynamic relations that are beyond the scope of this book, we can finally write the expression for the Gibbs' free energy change of the surface σ:

$$dG^\sigma = -S^\sigma dT - Ad\gamma + \Sigma\mu_i\, dn_i{}^\sigma \tag{1.12}$$

This equation describes the equilibrium change of Gibbs' free energy of the interface (surface) separating two phases, relating it to the change of parameters T, γ, and n_i. Setting T = const and dividing by A to obtain the specific values per unit surface, we finally get after some rearrangement:

$$d\gamma = -\Sigma\,\Gamma_i^\sigma\, d\mu_i \tag{1.13}$$

or, in a more convenient form for the case of adsorbing species on a solid surface:

$$\Gamma = -(a/RT)\delta\gamma/\delta a \tag{1.14}$$

where Γ stands for the surface concentration of adsorbing species [mol/m^2], and a is the activity of this species in the solution, often replaced by its bulk concentration, c [mol/dm^3]. What we just derived is the quantitative relationship between the amount of a substance accumulated at the surface with its activity (or concentration) in the bulk of the solution. This equation, known as *Gibbs' adsorption isotherm*, plays a crucial role in, for example, separation techniques and catalysis.

Now, having all necessary instruments provided by the thermodynamics of the uncharged systems, it is time to utilize the above knowledge for the systems, in which the surface can be charged in a controlled manner – the electrodes [1–6]. Most if not all of the areas of applied electrochemistry contain electrode/electrodes as an integral part of the systems, determining the systems' properties, applications, and performance. Thus, let us introduce here the new state parameter – charge, q. At constant temperature, by virtue of the approach described earlier in this chapter, we can write:

$$dG^{\sigma} = -Ad\gamma + \Sigma\mu_i\, dn_i^{\sigma} + \left(\frac{\partial G^{\sigma}}{\partial q}\right)\gamma, n_i \tag{1.15}$$

The last term describes Gibbs' free energy (work at const. T), required for bringing a charge q from infinity into the system.[1] As we should remember from physics classes, this is the definition of the electrical potential E. Therefore,

$$dG^{\sigma} = -Ad\gamma + \Sigma\mu_i dn_i^{\sigma} + Edq \tag{1.16}$$

Because simultaneously the integral form for Gibbs' surface free energy is

$$G^{\sigma} = \Sigma\mu_i n_i^{\sigma} + E \times q \tag{1.17}$$

and therefore

$$dG^{\sigma} = \Sigma\mu_i dn_i^{\sigma} + \Sigma n_i^{\sigma} d\mu_i + Edq + qdE, \tag{1.18}$$

we finally get the relationship between the changes of surface tension, concentration changes in the bulk of a solution, surface concentration of species "i" and the surface charge of a metal electrode, σ^{Me}:

$$d\gamma = -\sigma^{\mathrm{Me}} dE - \Sigma\Gamma_i d\mu_i \tag{1.19}$$

Before we go deeper into the models describing the structure of the interface between the metallic electrode and electrolyte, we should place these efforts in a broader picture. First, we should stress that the metal/solution interface should be considered as model interface, easy to control experimentally (e.g., by controlling the polarization of the electrode), and the theory derived from the investigations of such an interface should be useful also for other types of interfaces, including the nanostructured interfaces and biointerfaces. But why the interfaces should be charged, is this process a spontaneous one? Yes, it is spontaneous, except for the case of external polarization. The reasons for the appearance of a charge on the interfaces can be summarized below:

- external polarization (electrodes);
- adsorption and orientation of solvent dipoles;
- ionic adsorption from the solution;
- dissociation of surface functional groups

  Of course, all these effects can occur simultaneously.

Essentially, one can say that there are no uncharged interfaces. The consequences of this charge will be manifested in chapters that follow. Here, we will present the model of metal/solution interface that allows us to get insight into the structure of the interfacial

---

1 Since our charged phase (metal electrode) behaves like an ideal conductor, this charge will be accumulated on its surface.

region, but more important from the point of view of applications – its utilization in the construction and design of modern power sources and electrode processes.

1.1.1 Electrical double layer at interfaces: metal/electrolyte

What is the electrical double layer (e.d.l.)? It is a region of molecular dimensions at the interface between the two phases/substances, in which the electric field is created (e.g., by polarizing an electrode or by dissociation of surface groups). Of course, both phases in contact have to contain charged species (electrons, ions, or polar molecules).

In the e.d.l., charges of opposite sign attract themselves and have the tendency to ***accumulate*** at the interface. Finite sizes and solvation result in the separation of these particles. Charges of the same sign repel themselves leading to an unequal distribution of opposite charges with respect to the charges of the same sign within the interfacial region. Taken together, these interactions generate the electric field in the interfacial region, regardless of the nature of two phases in contact (with the proviso discussed above).

In order to describe quantitatively the profiles of potential and the distribution of charges in the interfacial region, we will take advantage of the relation introduced above for the metal/solution interface (eq. 1.19). For the sake of simplicity, let us assume that we do not change the electrolyte concentration nor its content so that $d\mu_i = 0$. Then, the above equation can be simplified to

$$d\gamma = -\sigma^{\mathrm{Me}} dE \tag{1.20}$$

Let us now focus on the model electrified interface of ideally polarizable electrode/aqueous electrolyte solution. By an ideally polarizable electrode, we will understand an electrode that within a given potential range can only accumulate charge (being polarized or charged) with no redox reaction (electron transfer across the interface). Changing its polarization by dE, we also change the interfacial tension by $d\gamma$, with proportionality coefficient being the charge σ^{Me}, introduced onto the electrode surface (ideal conductor) as a result of its polarization. The negative sign tells us immediately that the interfacial tension decreases with an introduction of charge onto the surface. This is intuitively understandable, because the charges of the same sign on the surface repel each other with Coulomb forces, thus decreasing the cohesion interactions between the surface atoms. Furthermore, one can predict that at zero surface charge, the surface tension will be the largest. Such interfacial behavior has its further consequences, for example, in the electrocatalysis (Chapter 4), corrosion (Chapter 3), or in the underpotential deposition (Chapter 6), and will be discussed in the appropriate chapters. Here, we will continue to develop a model describing the interfacial property responsible for charge storage and electrical power generation.

The above equation can be rearranged, so we can get the following:

$$d\gamma/dE = -\sigma^{Me} \tag{1.21}$$

So, if we know the experimental dependence of γ versus E, from the slope of this dependence $E = f(\gamma)$ we can obtain the value of surface charge σ^{Me} of a metallic surface for each value of the applied potential E. The best known and thoroughly experimentally elaborated dependence of γ versus E, also known as the *electrocapillary curve*, is that for the interface of mercury/aqueous electrolyte solution. However, other interfaces were also thoroughly investigated. These include air/solution interface or the two immiscible electrolyte solutions. All such interfaces were investigated not only from the point of view of basic research in order to achieve some insight into the structure of various interfacial systems, but also from the point of view of applications in the adsorption and separation techniques. The shape of the electrocapillary curve can be drawn as in Fig. 1.2(a).

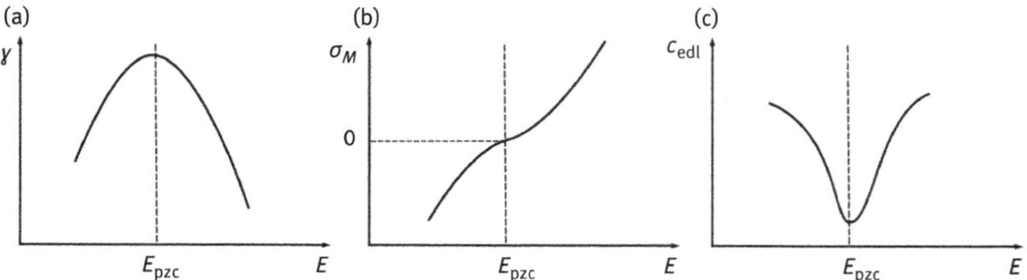

Fig. 1.2: Graphs representing: (a) the electrocapillary curve γ versus polarizing potential E, (b) surface charge σ, and (c) differential capacitance, $C_{e.d.l.}$. Please note the maximum on the electrocapillary curve corresponding to the potential of zero surface charge, E_{pzc} (inflection point on curve **b** and to the minimum on curve **c**).

Second derivative of the equation $d\gamma/dE = -\sigma^{Me}$ at constant chemical potential μ yields

$$d^2\gamma/dE^2 = -d\sigma^{Me}/dE = C_{e.d.l.} \tag{1.22}$$

where $C_{e.d.l.}$ is the differential capacitance of the electrical double layer formed in the interfacial region. The shape of the relations σ^{Me} versus E and $C_{e.d.l.}$ versus E is shown in Fig. 1.2(b) and (c), respectively. Now we see that the interface behaves as a system that stores the electric charge. The corresponding device is well known from electronics – a dielectric capacitor, in which the charge accumulated on its plates results in a potential gradient between these plates. Also the change of surface charge density by a factor $d\sigma^{Me}$ will result in a change of potential drop by dE (and vice versa), proportional to the capacity.

In the case of metal/solution interface, one of the condenser plate is easily identified – this is the electrode, whereas the description of a second "plate" that is formed by the ions of the opposite sign that are present in the solution depends on the model of the electrical double layer mentioned above. One condition, regardless of the accepted model, has to be fulfilled: all the interfacial region has to be electrically neutral, meaning that the charges on the metal surface (i.e., electrons) are compensated by the charges in the solution part of the e.d.l. (i.e., ions). This electroneutrality condition reads:

$$-\sigma^{\mathrm{Me}} = \sigma^{s} \text{ (superscript "s" stands for "solution")} \tag{1.23}$$

According to the simplest model proposed by Helmholtz (1879), in the solution the ions of the opposite charges with respect to the charge of an electrode (counterions) adsorb electrostatically via the Coulomb forces, directly on the electrode surface, neutralizing its charge. They form firm so-called Helmholtz' layer, with linear potential drop within its space, normal to the electrode surface. This is shown schematically in Fig. 1.3. Obviously, the electric field within such interfacial region is limited to the thickness this layer. Beyond it, inside the solution there is no electric field. Also, since we treat the metal as an ideal conductor, within bulk of the metal there is no electric field, too. This model corresponds directly to the flat capacitor.

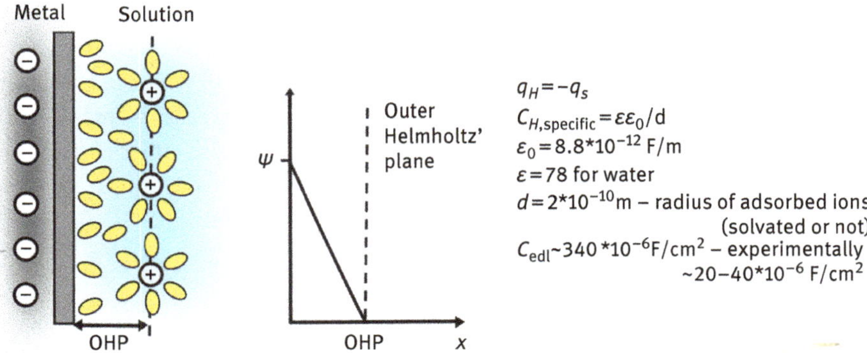

Fig. 1.3: The Helmholtz model of the electrical double layer with solvent (water) molecules and solvated ions at their closest approach to the electrode surface, forming the so-called outer Helmholtz plane (OHP).

This model predicts that the capacitance should be independent of the concentration of ions and electrode polarization (charge density). As it is found experimentally, it describes relatively well the electrical double layer behavior for large concentrations and electrode polarization.

Accounting for the thermal motions of ions in the electrolyte led independently L.G. Gouy and D.L. Chapman (1920) to the so-called diffuse electrical double layer. Due to the thermal motions, the ions cannot form the compact, firm layer, but are diffused, dispersed out of the interface forming layer of ions with exponentially changing concentration as a function of distance. These thermal motions counteract and overcome the organizing force of Coulomb interactions. Other assumptions underlying the Gouy–Chapman model are listed as follows:

- The ions are point charges (no dimensions).
- Solvent – dielectric continuum, characterized by electric permittivity ε (no molecular structure).
- Charge distribution in the volume dV in the radius dr from the interface is given by the Maxwell–Boltzmann distribution.
- The thermal motions are stronger than the electrostatic forces.
- The relationship between the space charge density ρ and the potential distribution in this region is given by the Poisson equation:

$$\frac{d^2\psi}{dx^2} = -\frac{\rho(x)}{\varepsilon\varepsilon_0} \tag{1.24}$$

$$\rho(x) = F\sum_i z_i c_i(x) \tag{1.25}$$

For an infinite, flat, ideally polarizable electrode, the potential profile along the normal to the electrode into the bulk of the electrolyte, $\psi(x)$, is given as follows:

$$\psi(x) = \psi_0 e^{-\kappa x} \tag{1.26}$$

where ψ_0 is the electrical potential at the surface and $\kappa^{-1} = \lambda_D$ is called the *Debye screening length* (Debye thickness), at which the potential ψ_0 decreases e-times (orange line in the inset in Fig. 1.4). Practically, beyond this distance, the effect of electrode polarization is negligible, screened by the counterions; however, within the e.d.l. region, the counterions are concentrated, while the coions diluted.

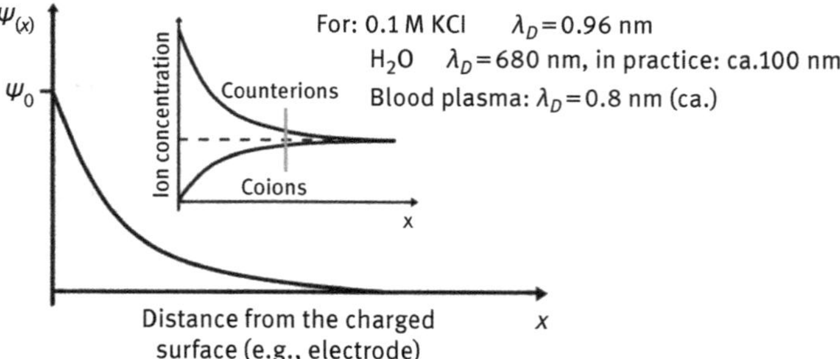

Fig. 1.4: Potential profile (distribution) along the normal from the electrode surface, according to the Gouy–Chapman model. Inset shows the coions and counterions distribution according to this model. The orange line marks the Debye thickness (length).

This distance strongly depends upon the ionic concentration, dielectric permittivity of the solvent and valency of ions according to the following:

$$\lambda_D = \left(\frac{e^2}{\varepsilon\varepsilon_0 k_B T}\sum\nolimits_i c_i^0 z^2\right)^{-0.5} \quad [\mathrm{m}] \tag{1.27}$$

where $\rho(x)$, volume charge density of ion "i" at a distance x [C m^{-3}]; $\psi(x)$, the electrical potential at distance x [V]; ψ_H, the electrical potential at Helmholtz layer [V]; ε, relative dielectric permittivity [no units]; ε_0, dielectric permittivity of vacuum [J^{-1} C^2 m^{-1}]; z, valency of ion; e, elementary charge [C]; k_B, Boltzmann constant [J K^{-1}]; T, temperature [K]; F, Faraday constant [C mol^{-1}].

The charge density at the surface is related to the potential at $x = 0$ by

$$\sigma^{\mathrm{Me}} = -\varepsilon\varepsilon_0\left(\frac{d\psi}{dx}\right)_{x=0} \tag{1.28}$$

$$\sigma(x) \approx \frac{2\kappa\varepsilon\varepsilon_0 k_B T}{ze}\sinh\left(\frac{\mathrm{ze}\psi(\mathrm{x})}{2\mathrm{k_B T}}\right) \tag{1.29}$$

where $\sigma(x)$ is the surface charge density at a distance x [C m^{-2}].

Finally, the recognition that joint models of Helmholtz and Gouy–Chapman explain better the observed experimental data, such as the behavior of the e.d.l. as capacitor, led O. Stern to the following (Fig. 1.5):

- e.d.l. contains two parts (*but this is not the reason why it is called "double," we will discuss this later*);
- Helmholtz layer – ca. of thickness of adsorbed ions or polar molecules, essentially bound to the surface, linear potential decay;
- Gouy–Chapman layer – from the Helmholtz layer to the bulk, diffuse, with exponential potential decay.

It is necessary to remind at this point that, regardless of the model of the e.d.l. potential profile, its value in the solution bulk (ψ_S) always equals zero. Now the interface has two regions in series that can accumulate charge: the Helmholtz layer with linear potential drop and the Gouy–Chapman layer with exponential potential decay. Thus, it can be represented as the two capacitors in series (Fig. 1.6).

As it is well known from physics, the inverse of total capacitance of the two capacitors in series equals to

$$\frac{1}{C_{\mathrm{e.d.l.}}} = \frac{1}{C_H} + \frac{1}{C_{G-Ch}} \tag{1.30}$$

Now, let us divide formally the overall potential decay within the e.d.l., $\psi_{\mathrm{Me}} - \psi_S$ into two parts: within the Helmholtz layer $\psi_{\mathrm{Me}} - \psi_H$, and within the Gouy–Chapman layer $(\psi_H - \psi_S)$; thus, $\psi_{\mathrm{Me}} - \psi_S = (\psi_{\mathrm{Me}} - \psi_H) - (\psi_H - \psi_S)$; keeping in mind that $\psi_S = 0$.

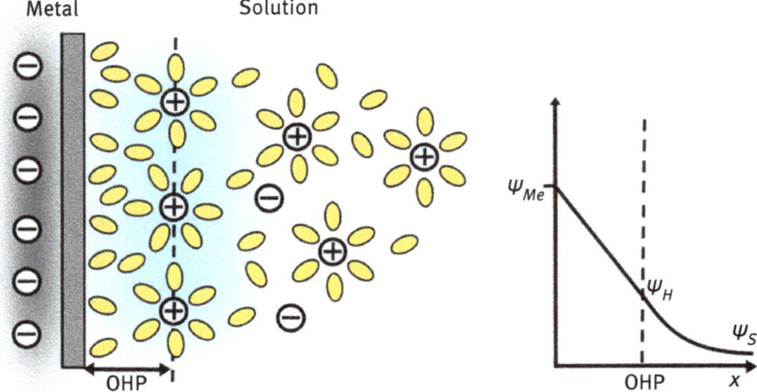

Fig. 1.5: The Stern model of e.d.l., being a joint model of Helmholtz and Gouy–Chapman models.

Fig. 1.6: Two capacitors in series representing two separate regions of charge aggregation and separation: the Helmholtz and Gouy–Chapman regions. This approach explains the experimental results in Fig. 1.7.

Then

$$\Delta\psi_{\text{e.d.l.}} = \Delta\psi_H + \Delta\psi_{G-Ch} \tag{1.31}$$

Dividing this last relation by a total charge on a surface and calling upon the definition of a capacitor, we can get the relation for two capacitors in series, as described earlier:

$$\frac{\partial\Delta\psi_{\text{e.d.l.}}}{\partial\sigma^{\text{Me}}} = \frac{\partial\Delta\psi_H}{\partial\sigma^{\text{Me}}} + \frac{\partial\Delta\psi_{G-Ch}}{\partial\sigma^{\text{Me}}} \tag{1.32}$$

or

$$\frac{1}{C_{\text{e.d.l.}}} = \frac{1}{C_H} + \frac{1}{C_{G-Ch}} \tag{1.33}$$

The first term describes the differential capacitance of Helmholtz layer (if present) with a linear potential decay, generally independent of the ionic concentration, whereas the second is the capacitance of Gouy–Chapman layer with an exponential potential decay, being strongly affected by the concentration of ions (Fig. 1.7), as shown in eqs. (1.27)—(1.29).

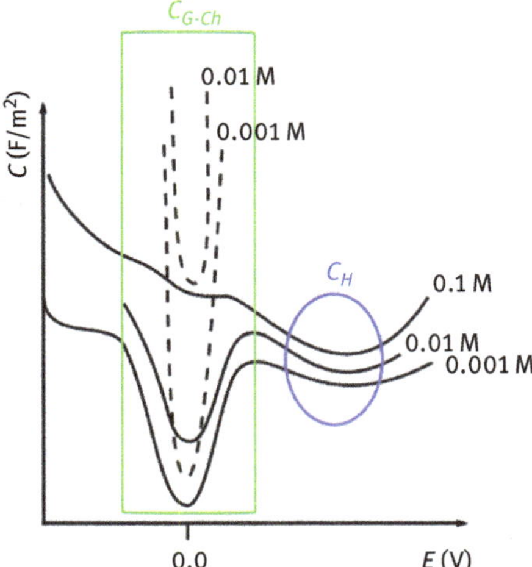

Fig. 1.7: Experimental data for differential capacitance on Hg electrode at various concentrations of aqueous 1:1 electrolyte (e.g., KCl, solid lines) and predictions of Gouy–Chapman theory (dotted lines). Green frame – the Gouy–Chapman region, blue oval – the Helmholtz region. (Adapted from: D.C. Graham, *Chem. Rev.* 41 1947 441–501).

The total differential capacitance, $C_{\text{e.d.l.}}$, is a very important quantity allowing to experimentally verify our model and subsequently utilize it in the designing of new charge storage devices, such as the e.d.l. supercapacitors. Therefore, let us evaluate its specific value per unit area, assuming some real quantities characteristic for aqueous electrolyte solutions. For this purpose we model the C_H as a plate capacitor formed by the metallic surface and hydrated ions adsorbed on it. Assuming the radius of such ions as ca. 2 Å, the capacitance of this layer would be $C_H = \varepsilon_H\varepsilon_0/r_{\text{ion}}$. The question under continuous dispute is: how to estimate the value of the dielectric permittivity of the Helmholtz layer, ε_H. For the purpose of this example, let us assume its value equal to 10. This in turn yields the value of 0.45 Fm^{-2} as the estimated capacitance C_H (45 μFcm^2), whereas the experimental values are typically 20–150 μFcm^2. The same time, the values of C_{G-Ch} depend on the concentration, varying from 7.2 μFcm^2 for 1 mM to ca. 200 μFcm^2 for 1 M monovalent aqueous electrolytes (the experimental values for the case of mercury electrode are 6.0 μFcm^2 to ca. 23 μFcm^2, respectively).

Regardless of these discrepancies, this brief insight discussed above explains why the theory of the electrical double layer has found its applications in designing the charge storage, energy conversion, and power devices for microelectronics. The e.d.l. applications are not limited to microelectronics, but also in analytical separation, rectification, and miniaturized field analytical devices, nanofluidics, leading finally to the so-called lab-on-a-chip, in which phenomena related to the properties of e.d.l. play crucial role in their operational capabilities and performance. Below we will exemplify the role of the e.d.l. for the field-effect control in nanofluidic transistors [7]. This example in the cited work shows the modulation of channel conductance that mimics the behavior of classical metal-oxide semiconductor field effect transistors (MOSFET). MOSFETS' typical applications are electronic switches, amplifiers, logic gates, and so on. In the discussed example, where the e.d.l. is put to work, under a certain source–drain bias, "gate" voltage induces concentration gradients switching directions following changes in gate-voltage polarity. Another example [8] shows the rectifying behavior upon the ionic conductance of nanochannels with charged walls. This time the negative charges originate from the functional groups present on the channel walls. At sufficiently small channel diameter, the electrical double layers formed on both walls start to overlap, repelling the coions (in the cited case – Cl^-) from the channel and allowing only the passage of counterions (in this case – K^+ ions) [7, 8] through the channel, making it highly selective.

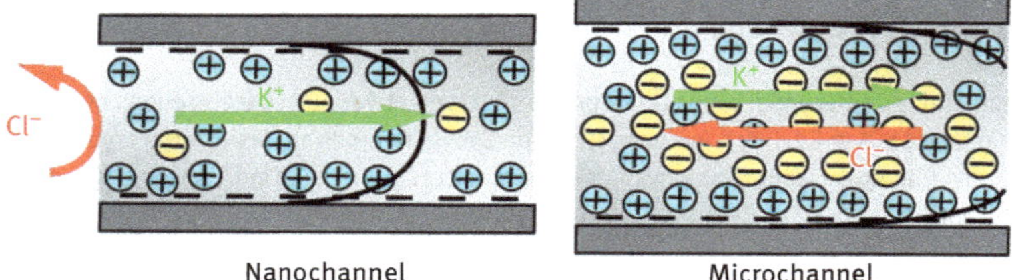

Fig. 1.8: Surface charge and e.d.l. effects in nanochannels and microchannels. Left panel: A scheme of two monolayers overlapping in a nanochannel of negatively charged walls with sufficiently narrow diameter, smaller than two Debye lengths of the e.d.l. In this case a rectifying behavior is observed as shown in the scheme with the red and green arrows for Cl^- and K^+ ions conductance, respectively. The reason of this behavior is the coions and counterions distribution in the e.d.l. electric field (see inset in Fig. 1.4). Right panel: Both ions freely moving through the microchannel that is either uncharged or much wider that the two Debye lengths.

The theory of e.d.l. is also useful in gaining insights into the surface properties of nanomaterials, such as colloidal suspensions, where surface charge can result either from the adsorption of ions or ionizable molecules stabilizing such suspension via the Coulomb repulsion forces, or the charge can originate from the structural surface functionalities (e.g., carboxylic groups), being inherent part of chemistry of nanoparticles. The first case, namely, the e.d.l. formation due to the charges adsorbing on a colloidal nanoparticle, is illustrated in Fig. 1.9.

1.1.2 Electrochemical potential – potentials at interfaces: internal, surface, external potential

The electrostatic potential inside a given phase, called the *inner potential*, φ, is defined by the work, w_e, performed by external forces, required to bring an electric, positive

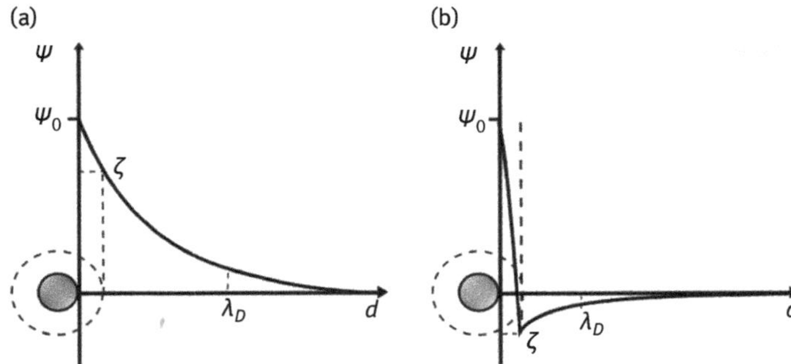

Fig. 1.9: e.d.l. structure according to Stern and potential distribution in the e.d.l., for (a) colloidal particle, positively charged and suspended in an electrolyte, (b) the same particle, positively charged but with strongly adsorbed negative ions (e.g., halogen ions); their adsorption changes the resultant surface charge (σ) and zeta potential (ς) of the nanoparticle, measured at the so-called plane of shear, appearing only when the nanoparticle moves in relation to the suspending medium.

test charge q_t from infinity and from a vacuum into the bulk of this phase, far from the phase boundary, divided by the value of this charge:

$$\varphi = \frac{w_e}{q_t} \tag{1.34}$$

Please note that the units of electrostatic potential are Volts [V], because J/C = V.

The assumption underlying this process is that the test charge is sufficiently small, so it does not affect the electrostatic field associated with this phase, and the work performed is only that necessary to overcome the electrostatic forces. All other forces, such as the chemical forces, remain unaffected during the process.

By analogy to the electrostatic potential, the chemical potential for particles "*i*" inside the same phase is defined as the work, w_{ch}, performed against all chemical forces during the transfer of n_i moles of particles "*i*" into such phase, divided by the number n_i of these particles. Again, the assumption is made that this virtual transfer does not affect the total concentration of these particles:

$$\mu_i = \frac{w_{ch}}{n_i} \tag{1.35}$$

Note that, in this case, the units of μ_i are Joules [J] and if n_i corresponds to the number of moles, then the units of a chemical potentials are J/moles.

The sum of chemical and electrostatic energies of charged particles "*i*" in a given phase (per mole, with charge z_iF) is called the *electrochemical potential of species "i"*:

$$\widetilde{\mu}_i = \mu_i + z_i F \varphi \tag{1.36}$$

here F is the Faraday constant and z_i is the valency of ion.

Figure 1.10(a) shows two metals Me_1 and Me_2 (treated as ideal conductors) brought into contact and at equilibrium, pictured by the arrow pointing in two opposite directions.

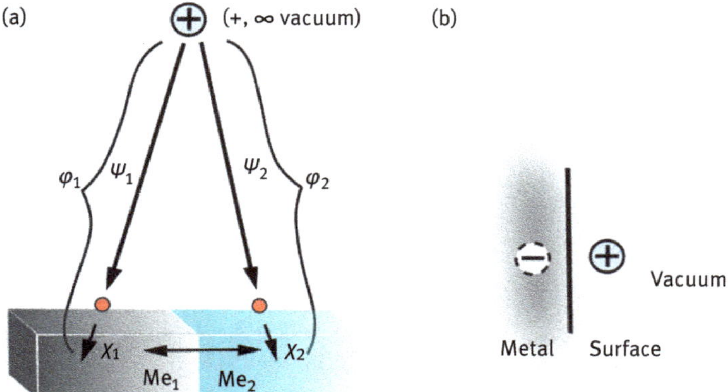

Fig. 1.10: (a) Two metal slabs at equilibrium, illustrating the definitions of inner, surface, and outer potentials. (b) This intends to explain the "image" forces appearing at close approach of a charge to the surface of an ideal conductor. See text for details.

The only charged components of these two metals that can partition between the two phases, achieving finally an equilibrium are the electrons (the nuclei do not mix under ambient conditions, preserving the identities of the two different metals). We now can use the definition of the electrochemical potential to quantitatively formulate this equilibrium as the state where the electrochemical potentials of electrons are the same in the two contacting metals:

$$\tilde{\mu}_e^{(1)} = \tilde{\mu}_e^{(2)} \tag{1.37}$$

We return to the consequences of eq. (1.37) later in the text. Now, let us focus on the inner (internal) potential φ. The transport of the test charge from infinity in vacuum can take place formally in two steps, shown in Fig. 1.10(a). First, the charge is brought close to the interface (red dot in Fig. 1.10(a)). The potential defined by this work is called the *external potential* ψ. (Please note that while discussing the electrical double layer we also used the symbol ψ. In this case, however, the reference point is not in vacuum but in the bulk electrolyte solution, far away from the interface, where $\psi_S = \psi(\infty) = 0$). Since the point of close proximity and the reference point in infinity are located in the same phase (in this case: the vacuum), the external potential ψ can be measured.

The statement "close to the interface" needs to be defined more precisely. When the charge is being moved from infinity toward the surface of charged conductor, the work is performed, primary due to the electrostatic field produced by the charge of the conductor. As the test charge approaches the surface, image forces begin to affect the test charge, distorting also the electrostatic field produced by the conductor. The term "image forces" reflects the induction of an image of the test charge, negative in sign, because the approach of positive test charge induces an accumulation of electrons at the surface of conductor due to Coulomb interactions, forming an image charge of opposite sign. Thus, the surface of ideal conductor acts as a "mirror," reflecting the approaching charge (Fig. 1.10(b)). Such induction creates additional dipolar potential drop across the surface affecting the original field of the conductor. Calculations of the cutoff distance of the image forces between the surface and the test charge give the value of such distance of ca. 1 μm.

In the next step the charge is brought through the interface and into the phase interior. This step defines the so-called surface potential χ. It is determined by surface charges and by dipoles aligned at the interface of the conductor. So finally we have:

$$\varphi = \psi + \chi \tag{1.38}$$

As can be immediately seen, the surface potential χ defined across the boundary (surface) can be related to the electrical double layer potential drop. One has to keep in mind, however, that the definitions of φ, ψ, and χ were given for an ideal conductor in contact with vacuum (ideal dielectric), whereas the e.d.l. models describe the interfaces between the two phases containing freely moving charges (ions, electrons, holes). In contrast to the external potential, the values of surface potential and inner potentials refer to points in different phases, and therefore cannot be measured.

Let us now return to eq. (1.37). Taking into account eq. (1.36), the equilibrium conditions for electrons in contacting metals Me_1 and Me_2 can be written as follows:

$$\mu_{e,1} - F\varphi_1 = \mu_{e,2} - F\varphi_2 \tag{1.39}$$

where $\mu_{e,1}$ and $\mu_{e,2}$ denote the chemical potentials of electrons in phase 1 and 2, respectively, a minus sign is due to the negative charge of electron ($z_e = -1$). After regrouping, we get

$$\Delta\varphi = \varphi_2 - \varphi_1 = (\mu_{e,2} - \mu_{e,1})/F \tag{1.40}$$

The potential difference, $\Delta\varphi$, between the two phases defined by this equation is called the *Galvani potential* (the term recommended by IUPAC) and cannot be measured directly, because it refers to the points in different phases; in the case of Fig. 1.10(a), these are the Me_1 and Me_2. Taking into account eq. (1.38), we can rewrite the eq. (1.40) in a form, showing its two components:

$$\Delta\varphi = \varphi_2 - \varphi_1 = (\psi_2 + \chi_2) - (\psi_1 + \chi_1) \tag{1.41}$$

or

$$\Delta\varphi = \Delta\psi + \Delta\chi \tag{1.42}$$

$\Delta\psi$ is called the *Volta potential*, whereas $\Delta\chi$, the surface potential difference, is frequently called the *surface potential*. So, how to obtain the value of $\Delta\varphi$ between the two metals Me_2 and Me_1? We already said that $\Delta\psi$ is measurable, but how to account for $\Delta\chi$ – the surface potential difference between the two inherently different metals (phases)? Let us play a trick and add one more set of interfaces, adding the same metal Me_1 on the other end of metal Me_2 as in Fig. 1.11. For clarity in the description that follows, let us mark it as $Me_{1'}$. This $Me_{1'}$ possesses the same surface properties as Me_1; therefore, $\chi_1 = \chi_1' = \chi$.

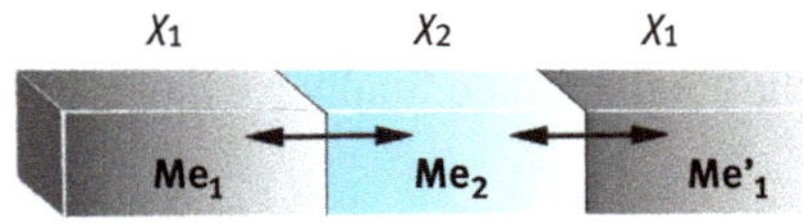

Fig. 1.11: Slab model representation of a system enabling measurements of the Galvani potentials.

So now, at equilibrium, we will have:

$$\mu_{e,1} - F\varphi_1 = \mu_{e,2} - F\varphi_2 \text{ and } \mu_{e,2} - F\varphi_2 = \mu_{e,1}' - F\varphi_1', \text{ but also } \mu_{e,1} - F\varphi_1 = \mu'_{e,1} - F\varphi_1'$$

Therefore, the value of $\varphi_2 - \varphi_1$ can be expressed in terms of $\varphi_1' - \varphi_1$:

$$\varphi_1' - \varphi_1 = \frac{\mu_{e,1}' - \mu_{e,1}}{F} \tag{1.43}$$

Let us now use eq. (1.41) and we obtain:

$$\varphi_1' - \varphi_1 = \psi_1' + \chi_1' - (\psi_1 + \chi_1) \tag{1.44}$$

When Me_1 and $Me_{1'}$ are the same metals, we can easily assume that their surface properties are the same. Therefore, $\chi_1' = \chi_1$ and under such conditions:

$$\varphi_1' - \varphi_1 = \psi'_1 - \psi_1 \tag{1.45}$$

the Galvani $(\varphi_1' - \varphi_1)$ potential is equal to Volta potential $(\psi_1' - \psi_1)$. Moreover, being localized in the same phase it is now is measurable.

1.1.2.1 Metal–solution interface at equilibrium: Nernst equilibrium potential

We are now ready to apply the above information to the system commonly found in electrochemistry: the interface between a metal (electronic conductor) and electrolyte solution (ionic conductor). For the sake of clarity, we introduce here the simplest possible system that is the metal at equilibrium with aqueous solution of its ions, such as $Cu|Cu^{2+}$ or $Zn|Zn^{2+}$. The counterions in the solution (anions) can be of any soluble salts of metal under consideration (e.g., SO_4^{2-}, NO_3^-). Let us rewrite eq. (1.37) describing the equilibrium condition for electrons in two contacting metals, in order to describe the equilibrium condition of Me^{z+} in the solution with the same metal cation inside the metal phase:

$$M^{z+}\,(\text{solution}) \leftrightarrow M^{z+}\,(\text{metal}) \tag{1.46}$$

$$\tilde{\mu}(M^{z+})^S = \tilde{\mu}(M^{z+})^M \tag{1.47}$$

or

$$\mu(M^{z+})^M + z_+F\varphi^M = \mu(M^{z+})^S + z_+F\varphi^S \tag{1.48}$$

After regrouping, we have

$$\varphi^M - \varphi^S = \Delta\varphi^{M/S} = [\mu(M^{z+})^S - \mu(M^{z+})^M]/z_+F \tag{1.49}$$

Since $\mu(M^{z+})^S = \mu^0(M^{z+})^S + RT\ln a_{M^{z+}}$, and the chemical potential of metal cation in its metal phase defines its standard chemical potential: $\mu(M^{z+})^M = \mu^0(M^{z+})^M$, we can finally write:

$$\Delta\varphi^{M/S} = [\mu^0(M^{z+})^M - \mu^0(M^{z+})^S]/z_+F + (RT/z_+F)\ln a_{M^{z+}} \tag{1.50}$$

The internal potential difference $\Delta\varphi^{M/S}$ between metal (M) and solution (S) is called the *Galvani potential*, as was discussed above, whereas the difference between the standard chemical potentials (square brackets in eq. (1.50)) is the change in standard molar Gibbs free energy, Δ_rG^0, of redox reaction eq. (1.46). This equation, in a form,

$$E_{eq} = E^0 + \frac{RT}{nF}\ln a_{M^{z+}} \tag{1.51}$$

is known as the *Nernst equation*, or electrode equilibrium potential, E_{eq}. Similar considerations lead to a more general equation, describing the Nernst equilibrium potential for redox couple of activities a_{ox} and a_{red} for oxidized and reduced forms, respectively, exchanging only electrons with metal electrode. The latter does not participate chemically in this process and is sometimes called the *redox electrode*. In these cases, the Nernst equilibrium potential will read:

$$E_{eq} = E^0 + \frac{RT}{nF}\ln\frac{a_{ox}}{a_{red}} \tag{1.52}$$

Equation (1.52) shows that the Nernst equilibrium potential is equivalent to Galvani potential (internal potential difference) of metal/solution interface at equilibrium. It depends on the activities of ions participating in the electrode processes. The activity of ion "i" in the solution, a_i, is related to its concentration, c_i, via the activity coefficient, γ_i: $a_i = \gamma_i c_i$, and so, eq. (1.52) can be written as follows:

$$E_{eq} = E^f + \frac{RT}{nF} \ln \frac{c_{ox}}{c_{red}} \tag{1.53}$$

where E^f is the so-called formal potential accounting for the activity coefficients (the effect of interactions of solution components on the standard electrode potential E^0), and as such it does not have thermodynamic meaning, contrary to E^0, being equal to:

$$E^0 = -\frac{\Delta_r G^0}{zF} \tag{1.54}$$

As we discussed earlier, the absolute values of Galvani potential or Nernst potential are not accessible experimentally. However, they can be assessed in relation to the reference electrodes. This will be addressed in Chapter 8.

1.1.2.2 Electron work function

A wealth of information on surface properties of conducting or semiconducting materials in modern applied electrochemistry and electronics is needed for pursuing the technological development of various devices and processes. One of such information is the electron work function, or simply the work function, Φ. It is defined as the minimum amount of work that has to be performed by the external forces in extracting one electron from a metal, across its surface and into vacuum, where it becomes a free electron with no kinetic energy. It is worth to realize that the work functions of all metals are positive; otherwise the electrons would escape from the material spontaneously. Because of the effect of surface properties, the work function depends on the position of the point to which the electron is transferred. We addressed this problem defining the internal, external, and surface potential, and as previously, the point in the vacuum, "just outside" the metal, is regarded as the final point of electron transfer. For such conditions, the work function Φ is equal to the difference in electrochemical potentials of the electron at this point in vacuum, $\tilde{\mu}^0$, and any point in the metal $\tilde{\mu}^M$. Since chemical potential of electron in the vacuum is equal zero: $\mu_e^0 = 0$, therefore the electrochemical potential of an electron in vacuum $\tilde{\mu}_e^0 = -e\psi$. Minus sign in this relation is due to the negative valency of electron. Therefore,

$$\Phi = \tilde{\mu}_e^0 - \tilde{\mu}_e^M = -e\psi - \left(\mu_e^M - e\varphi\right) = -\mu_e^M + e\chi^M \tag{1.55}$$

The same relation can be obtained directly from the definition of the electrochemical potential of the electron in a metal:

$$\tilde{\mu}_e^M = \mu_e^M - e\,\varphi^M = \mu_e^M - e\,(\psi^M + \chi^M) = (\mu_e^M - e\chi^M) - e\,\psi^M = \alpha^M - e\,\psi^M \tag{1.56}$$

According to the above equation,

$$\Phi = -\alpha = -\mu_e^M + e\cdot\chi^M \tag{1.57}$$

and

$$\alpha = -\Phi \tag{1.58}$$

As we stated above, Φ, depends on the material and the state of its surface (including crystallographic structure). The following table exemplifies the values of work functions for several metals. Since this is the energy required to withdraw the electron, its units are also the energy units in eV.

Pt	**5.12 – 5.93 eV (or 6.1 for Pt(111))**
Au	5.1–5.47 eV
Ag	4.27–474 eV
Pd	5.12–5.93 (Pd(111)) eV
W	4.32–5.63 eV
Ni	5.04–5.35 eV

The measurements of ***work functions*** of conducting and semiconducting materials find their applications in various areas of applied science, for example, electrocatalysis (cf. Chapter 4), materials characterization, semiconductor science, polymer science, electrical characterizations, nanostructured materials, soft samples, and even biosciences; various techniques have been developed to obtain such information. These include the photoemission spectroscopy, measurements of current induced by electron beam, scanning electron microscopy, and Kelvin probe/vibrating plate capacitor techniques. Interested reader is directed to the review work, describing the latter electrochemical technique [9].

1.1.3 Charge transfer processes across the metal/electrolyte interface

1.1.3.1 Basic concepts of nonequilibrium thermodynamics

The previous paragraph provided some basic thermodynamic definitions and concepts for the systems at equilibrium. Even in the case of the e.d.l. formation, one can easily assume that from the point of view of other processes of interest in the applied electrochemistry, the structure of the interface is attained so fast that can be considered as being at "equilibrium." This will be shown in the text below. However, most of the electrochemical processes that are currently employed or observed in the energy storage and transformations, electrodeposition, corrosion, or photoelectrochemistry are related to the transport of charge, mass, and energy (e.g., heat), and, as long as this transport continues, we cannot describe it using the equilibrium formalism [10]. Or, perhaps we can, modifying appropriately such formalism?

Let us summarize briefly the systems at equilibrium:

- It is characterized by *uniform* values of intensive state parameters and "potentials" throughout the system, such as p, T, μ, $\tilde{\mu}$; in other words, all intensive state parameters (X) are constant and their gradients are zero: $\frac{dX}{dt} = 0, \nabla X = 0$.
- Densities of conjugate extensive quantities, for example, V, N_i, S, can be inhomogeneous at equilibrium (here: N_i and S are the number of molecules of species "i" and the entropy of the system, respectively).
- S (entropy) is maximized at constant internal energy U of the system.
- For the closed system $d_iS = dS - \frac{Q_{el}}{T}$, where d_iS is called the *entropy production*, and Q_{el} is the heat exchange between the system and its surroundings during the equilibrium process. It is immediately evident that at equilibrium where the second law of thermodynamics is fulfilled, $d_iS = 0$.

Generally, thermodynamics describes what cannot happen (the first and the second law of thermodynamics). But how to apply its concepts for systems that are not at equilibrium?

Irreversible thermodynamics applies the principles of thermodynamics for nonequilibrium systems and gives the relaxation rules for a system to achieve the *equilibrium state* or *stationary state*. Let us introduce two terms that are basic for the formalism that follows: the ***flux*** J_i of energy, current, mass, and so on and the ***thermodynamic force*** X_i. Its impact results in the flux J. For the case of mass flux, $J_m = \frac{dm}{dt} \cdot \frac{1}{A}$. The fluxes are vectors, whereas the thermodynamic forces can be either vectors or scalars. If there is only one thermodynamic force responsible for a given flux, it is called the *conjugated force*. Before we put all the above at work, let us formulate below several postulates that will be useful for further discussion.

Postulate 1

Generally, the beauty of thermodynamics lays in the fact that it is possible to apply the state functions defined for reversible systems and to use principles of continuum limit to define meaningful, useful, ***local*** values of various thermodynamic quantities, for example, chemical potential μ_i. Therefore, useful theories can be derived by assuming a functional relation between the ***rate*** of a process and the ***local*** deviation from equilibrium (the "driving force").

Postulate 2

For any irreversible process, the rate of entropy production is ***everywhere*** ≥ 0

$$\frac{d_iS}{dt} \geq 0 \tag{1.59}$$

Postulate 3

For small perturbations we can assume a linear coupling between fluxes J_i and thermodynamic forces X_i:

$$J_i = L_{ii}X_i \tag{1.60}$$

Here, "*i*" or subscripts denote the different types of fluxes, for example, heat Q, charge q, current I, and so on, and L_{ii} are the so-called Onsager phenomenological coefficients. Now, if there are multiple fluxes in the system and each one is somehow dependent on the other, then for a selected flux "*i*" we can write:

$$J_i = \sum_{i,j} L_{ii}X_iX_j \tag{1.61}$$

Developing the above for many fluxes,

$$J_i = L_{ii}X_i + L_{ij}X_j + \ldots + L_{in}X_n \tag{1.62}$$

$$J_j = L_{ji}X_i + L_{jj}X_j + \ldots + L_{jn}X_n \tag{1.63}$$

....

$$J_n = L_{ni}X_i + L_{nj}X_j + \ldots + L_{nn}X_n \tag{1.64}$$

or

$$J_i = L_{ii}X_i + \sum_{\substack{k=1 \\ k\neq i}}^{n} L_{jk}X_k \tag{1.65}$$

For a better understanding of this approach, let us consider only two fluxes: the flux of charge (current) and the flux of heat. It is a common knowledge that when a current flows through a conductor (wire), this conductor warms up in a way related to the magnitude of current: the larger the current, the warmer the wire. In the above formalism, we can describe these two interrelated fluxes as

$$J_q = L_{qq}X_q + L_{qQ}X_Q \tag{1.66}$$

$$J_Q = L_{Qq}X_q + L_{QQ}X_Q \tag{1.67}$$

Here L_{qq} and L_{QQ} are called the *direct coefficients*, because they conjugate the thermodynamic force directly to the flux it causes, whereas L_{qQ} and L_{Qq} are the *coupling coefficients* because they indirectly couple all other possible fluxes with the selected single flux. In our case the flux of current is coupled to the flux of heat and vice versa: if one heats up the metallic conductor, its resistivity increases, and so the flux of current decreases. The latter observation leads us directly to Onsager symmetry postulate.

Postulate 4

Onsager symmetry postulate: $L_{ij} = L_{ji}$ (microscopic reversibility). This postulate is equivalent to

$$\frac{\partial J_q}{\partial X_Q} = \frac{\partial J_Q}{\partial X_q} \tag{1.68}$$

Table below summarizes some examples of fluxes and their conjugated thermodynamic forces.

Process	Flux J	Conjugated force	Empirical law
Energy (heat) transport	$J_Q = \frac{1}{A}\frac{dQ}{dt}$	$\nabla\left(\frac{1}{T}\right) = -\frac{\nabla T}{T^2}$	$J_Q = -K\nabla T$ Fourier
Charge transport	$J_q = \frac{1}{A}\frac{dq}{dt}$	$-\nabla\varphi$	$J_q = -\rho\nabla\varphi$ Ohm
Mass transport	$J_i = \frac{1}{A}\frac{dn_i}{dt}$	$-\nabla\mu_i$(at constant T)	$J_i = -M_i c_i \nabla\mu_i$ Fick

Let us now put all the above together for some "edible" applications in electrochemistry. First, we focus only on mass flux of charged species "i," identifying it as independent of other possible fluxes and assigning to it the thermodynamic driving force X_i equal to the electrochemical potential gradient of this species. Moreover, the empirical Onsager proportionality coefficient L_i equals to the product of concentration of species "i", c_i, and its mobility, u_i, defined as the ability of charged particles to move through a medium in response to an electric field. Thus, we can write

$$\begin{aligned} &J_i = L_i \cdot X_i \\ &X_i = -\mathrm{grad}\tilde{\mu}_i; \qquad L_i = u_i \cdot c_i \\ &J_i = -u_i \cdot c_i \cdot \mathrm{grad}\tilde{\mu}_i \end{aligned} \tag{1.69}$$

Minus sign in the description of the thermodynamic driving force is due to the inherent mathematical definition of a gradient which is positive from "low" values to "high" values, whereas flux J occurs from "high" to "low."

Assuming only one direction of flux along the "x" axis and substituting $\tilde{\mu}_i$ with its definition (eq. 1.36), we will get (at uniform, constant temperature throughout the system):

$$J_i = -u_i c_i \left[RT\frac{d\ln c_i}{dx} + z_i F\frac{d\varphi}{dx}\right] \tag{1.70}$$

where all symbols have the same meaning as before. This equation, called the *Nernst–Planck equation for flux*, can be rewritten as follows:

$$J_i = -u_i RT\frac{dc_i}{dx} + u_i c_i z_i F\frac{d\varphi}{dx} \tag{1.71}$$

This is one of the fundamental equations that can be applied to various areas of electrochemistry where the movement of charges (or mass) is involved. To show the usefulness of the Nernst–Planck equation for solving various problems, we now put it under the test, setting forth different limiting conditions describing different processes in electrochemistry.

1. Let us assume that there is no electric potential gradient in our system, that is, $\frac{d\phi}{dx} = 0$. This condition describes the situation, where the charged species (e.g., ion) is far away from the electrified interface (e.g., electrode). This is also a situation where the ions are beyond the thickness of the e.d.l. discussed in the previous section. Then, if there is still transport of the charged species, under the isothermal conditions, this transport (flux) will be described by

   $$J_i = -u_i RT\frac{dc_i}{dx} = -D_i\frac{dc_i}{dx} \tag{1.72}$$

   where D_i is the diffusion coefficient of species "i." If we consider 1 mole of charged species, then $u_i RT = z_i FD$

   So, from the Nernst–Planck equation, we obtained the well-known Fick's law for mass transport due to the concentration gradient. Such systems work under diffusion control.

   But what we could get if there is no concentration gradient, but these are the electrons and/or ions that flow due to the potential gradient?
2. Let us require that only electrical potential gradient is of interest. Then $\frac{dc_i}{dx} = 0$, and

   $$J_i = -u_i c_i z_i F\frac{d\varphi}{dx} \tag{1.73}$$

   Taking into account Faraday's law of equivalence of mass and charge transport:

   $$j_{el} = \sum_i z_i F J_i \tag{1.74}$$

   and merging these two equations, we will have

   $$j_{el} = -\sum_i U_i c_i z_i^2 F^2\frac{d\varphi}{dx} = -\sum_i U_i c_i z_i F\frac{d\varphi}{dx} \tag{1.75}$$

   The right-hand side of this equation is exactly the first Ohm's law if we set the resistance $R = -\kappa^{-1}$. For the case of ion transport, however, we should take into account the stoichiometry coefficients, v_i, of electrolyte dissociation, and so we have the straightforward definition of electrolyte conductivity:

$$\kappa = \sum_i |z_i| c_i F U_i, \text{where } U_i \text{ is called the electrolytic mobility.} \tag{1.76}$$

We can multiply the examples of applications of the Nernst–Planck equation, but the two above examples will show us its usefulness in applied electrochemistry. Let us elaborate a little more on the first example: the first Fick's law and diffusion-controlled processes. This approach finds its applications in the electroanalytical chemistry, because the solutions of differential equations of Fick's first and second laws (the latter we will discuss in, e.g., Chapter 2), under appropriately chosen boundary conditions, lead to the direct functional relations between the measured current and concentration of the electroactive species in the solution, food or environment. Therefore, this provides precise control over the composition or possible contamination of the investigated samples or pollutants. In many cases, it is also possible to determine the identity of an investigated electroactive substances. The monitoring capabilities of electrochemistry is utilized frequently in pharmacy and medicine (e.g., body fluid analyses).

The second example is more obvious: Ohm's law is one of the fundamentals of electronics. On the other hand, the case of charge transport in the electrolyte describing its conductivity is one of the essential factors in constructing the charge storage and converting devices. A proper choice of charges moving between the electrodes in batteries and cells, providing their high mobilities and high conductivities, is one of the technological problems continuously requiring improvement as the demand for such devices grows.

Having all this basic information introduced in the sections above, we now describe processes that can be involved in the electron transfer across the electrode/solution interface – the redox process. They are visualized in the graph below, with the *proviso* that there is no adsorption/desorption of electroactive species, nor the consecutive reactions take place. In the description below, *Ox* represents the *oxidized* form of an electroactive species (e.g., $Fe(CN)_6^{3-}$) while *Red* represents the *reduced* form of an electroactive species (e.g., $Fe(CN)_6^{4-}$).

In Scheme 1.1:

- **Step 1** describes the diffusion of the electroactive species (*Ox* for the reduction and *Red* for the case of oxidation) to/from the Helmholtz plane (OHP) at the electrode (compare Section 1.1), this transport is characterized by the mass transfer coefficient k_D [cm/s] and will be discussed below.
- **Step 2** approximates a reorientation of ionic “atmosphere” that accompanies a charged electroactive species and formed of nonelectroactive counterions always present in the solution (e.g., K^+ ions). Typically, this process takes ca. 10^{-8} second (tens of nanoseconds).
- **Step 3** reorganization of solvent dipoles, ca. 10^{-11} second (tens of picoseconds).
- **Step 4** accounts for the distance changes between the “central” ion and its “ligands” (e.g., $Fe^{3+/4+}$ and CN^- in our example below), time frame ca. 10^{-14} second.
- **Step 5** is the electron transfer itself, ca. 10^{-16} second.

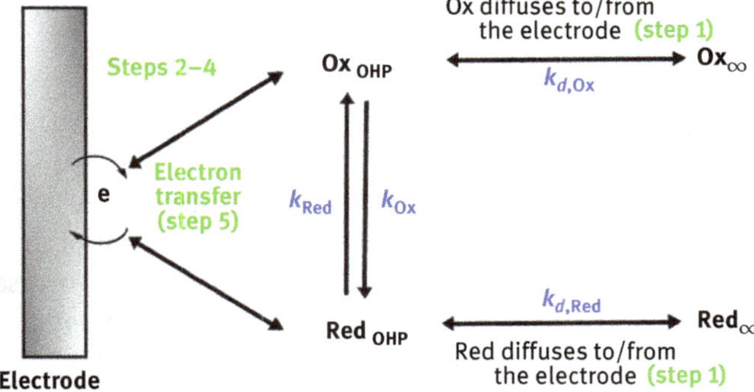

Scheme 1.1: A scheme showing processes that can be involved in the electron transfer across the electrode/solution interface – the redox processes. No adsorption/desorption of electroactive species take place, and no the consecutive reactions are involved. More details are given in the text.

For further description of the electrode processes, we will always follow the convention (Stockholm convention) to write the redox reaction toward reduction; therefore,

$$\mathrm{Ox} + \mathrm{ne} \rightarrow \mathrm{Red}$$

$$\mathrm{Cu}^{2+} + 2\mathrm{e} \rightarrow \mathrm{Cu}^{0}$$

$$\mathrm{Zn}^{2+} + 2\mathrm{e} \rightarrow \mathrm{Zn}^{0}$$

or

$$\mathrm{Fe(CN)_6}^{3-} + 1\mathrm{e} \rightarrow \mathrm{Fe(CN)_6}^{4-}$$

where ne stands for a number of electrons involved in the unit redox process (half-reaction).

1.1.3.2 "Reversible" electrode processes

By "reversible" electrode processes we understand the electrode processes that are limited by the rate of diffusion of a redox species to the electrode (**Step 1** in the graph above) and not by the rate of electron transfer between the redox species and the electrode (or other processes **Steps 2–5**). From this point of view, the redox process itself is at the dynamic equilibrium, meaning that the rate of oxidation is equal to the rate of reduction. There are two obvious conditions of reversibility: reversibility of mass and reversibility of energy. The first one states that the mass converted in the "forward" process has to be equal to the mass converted back during the "backward" process. The second condition means that the energies necessary for "forward" and "backward" processes are equal. At constant pressure p and temperature T it means that the Gibbs' free energy change of the redox reaction, $\Delta G_{p,T} = 0$.

Some quantitative considerations: Let us call upon the Nernst–Planck equation for flux for the case of diffusion processes (reduction of Ox):

$$J_{Ox} = -u_{Ox}RT\frac{dc_{Ox}}{dx} = -D\frac{dc_{Ox}}{dx} \tag{1.77}$$

If we now set that the volume containing the electroactive species is sufficiently large, we can assume that the concentration gradient is constant and independent of time (stationary state), which gives us immediately that the J_{Ox} is constant, too (of course within the limited time frame). Then, introducing Faraday's law, the equation for current, with c_{Ox} dependent only on the distance from the electrode, $c_{Ox}(x)$, will read:

$$i = -nFDA\left[\frac{dc_{Ox}(x)}{dx}\right] \tag{1.78}$$

Further assumption of a linear concentration profile in a constant diffusion region of thickness δ (see Fig. 1.12) can be written as follows:

$$\frac{dc_{Ox}(x)}{dx} = \frac{c_{Ox,x=0} - c^0_{Ox}}{\delta} \tag{1.79}$$

where $c_{Ox,\,x\,=\,0}$ is the concentration of Ox at the electrode, c^0_{Ox} is the concentration of Ox in the bulk of the solution, and δ is the thickness of the diffusion layer, as mentioned earlier.

Thus the equation for current is now

$$i = -nFDA\left[\frac{c^0_{Ox} - c_{Ox,x=0}}{\delta}\right] \tag{1.80}$$

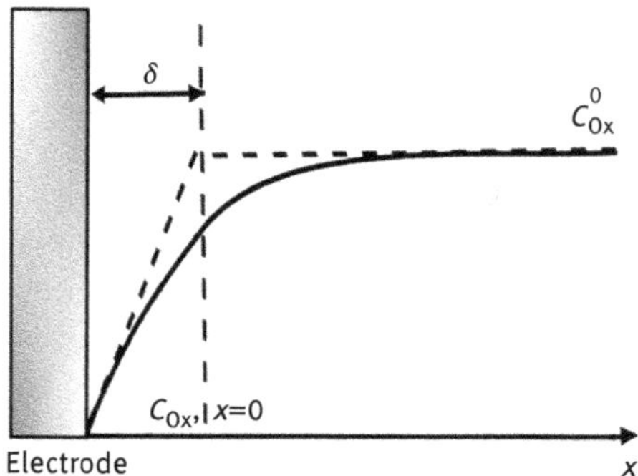

Fig. 1.12: Graph representing a "frozen in time" concentration profile versus the distance x from the electrode surface ($x = 0$). As a consequence, a constant diffusion region of thickness δ is developed. Bulk concentration of Ox is denoted here as c^0_{Ox}, whereas its concentration at the electrode surface is $c_{Ox,x=0}$. This means that the potential applied to the electrode is sufficiently large to reduce immediately all arriving Ox form.

It should be obvious for the reader that there exist two limiting conditions for the magnitude of the concentration gradient $\frac{dc_{\text{Ox}}(x)}{dx}$:

- $\frac{dc_{\text{Ox}}(x)}{dx} = 0$, meaning that no resultant flux appears to/from the electrode. This situation reflects the equilibrium conditions; no reduction (oxidation) reaction takes place at the electrode surface or, more correctly, the rate of reduction equals the rate of oxidation. This means that the concentration of Ox in the bulk ($x = \infty$) equals to its concentration at the surface of the electrode ($x = 0$), $c^0_{\text{Ox}} = c_{\text{Ox},x=0}$.
- $c_{\text{Ox},x=0} = 0$, meaning that the redox reaction is so fast that it literally consumes all of Ox form diffusing to the electrode from the bulk, converting it into *Red* form. This corresponds to the maximum value of the concentration gradient, and according to the eq. (1.80), this results in the maximum value of ionic flux J_{Ox} and (Faraday law) yields constant, limiting value of current, i_l:

$$i_l = nFAD\left(\frac{c^0_{\text{Ox}}}{\delta}\right) \tag{1.81}$$

The equilibrium conditions of redox processes at the metallic electrodes are quantitatively described by the well-known Nernst equation, relating the equilibrium potential, E_{eq}, of such electrodes with the concentration ratio of forms *Ox* and *Red* present in the solution:

$$E_{\text{eq}} = E^f + \frac{RT}{nF}\ln\frac{c_{\text{Ox}}}{c_{\text{Red}}} \tag{1.82}$$

where E^f is the so-called formal potential of the electrode if the ratio $c_{\text{Ox}}/c_{\text{Red}} = 1$ (see also Section 1.1.3); other symbols have their usual meanings.

But what drives the formation of a concentration gradient, finally reaching its maximum value for the case of immediate reduction (oxidation) of the electroactive species at the electrode surface, so that $c_{\text{Ox},x=0} = 0$ ($c_{\text{Red},x=0} = 0$)? The reason is the external polarization of the electrode, for example, from an external voltage source, such as potentiostat. To drive the reduction, the electrode should be polarized negatively with respect to its equilibrium, Nernst potential. Under such conditions, the electrode acts as a source of electrons for the reduction process to proceed. The more negative the electrode potential, the more *Ox* form is being consumed at the surface. Quantitatively, this is described by an overpotential, η. The overpotential is the potential difference between the potential at which the redox event proceeds experimentally, E, and the thermodynamically determined equilibrium potential (Nernst equation), E_{eq}:

$$\eta = E - E_{\text{eq}} \tag{1.83}$$

Since the diffusion to the electrode is the slowest process, the electrode polarization develops the concentration gradient, the overpotential is called the *concentration overpotential,* η_c. Until now we were discussing the situation of disturbing the electrode from its equilibrium by polarizing it from an external source, changing the concentration of electroactive species in the electrode close vicinity. In this case, we can control the value of η_c by controlling the value of E. However, it is necessary to say, at this moment, that the concentration overpotential can also be created in the situation when our electrode is connected via an external resistance to the other electrode of different potential. Thus, we build an electrochemical cell being discharged by the resistance. Current i flows through the circuit causing the concentrations of the electroactive species at the electrode surfaces to also change: on the anode the oxidation process will increase the concentration of Ox^I decreasing the concentration of Red^I, whereas, on the cathode, the reduction will build the concentration of Red^{II}, decreasing the concentration of Ox^{II}. Thus the concentration overpotential will be created whether we externally polarize the electrode (electrolysis), or the electrode will spontaneously depolarize itself performing electrochemical work in the cell.

Let us now return to the diffusion-limited electrolysis on the electrode. The problem we are now facing is: how does the concentration profile developed with the applied overpotential correspond to the measured electronic current flow. In other words, can we use the current measured during the electrode polarization as a “fingerprint” of the type of redox process occurring at this electrode?

To answer this question, we relate the electrode potential E to the concentration of Ox at the electrode surface $c_{\mathrm{Ox},x=0}$. We then can get

$$E = E^f + \frac{RT}{nF}\ln\frac{c_{\mathrm{Ox},x=0}}{c_{\mathrm{Red}}} \tag{1.84}$$

Now, after subtraction from the above relation the equilibrium potential E_{eq} defined by the Nernst equation, we get the relation of the concentration overpotential with the natural logarithm of the ratio of the concentration at the surface of the electrode to the concentration in the bulk. Because $c_{\mathrm{Ox},x=0} < c^0$ when the reduction proceeds, the value of overpotential for the reduction will be negative:

$$\eta = \frac{R}{nF}\ln\frac{c_{\mathrm{Ox},x=0}}{c^0} < 0 \tag{1.85}$$

Introducing to this equation the expressions for current i and limiting current i_l we will finally obtain the desired relation between the concentration overpotential and current measured experimentally:

$$\eta_c = \frac{RT}{nF}\ln\frac{i_l - i}{i_l} \tag{1.86}$$

or

$$i = i_l\left[1 - \exp\left(\frac{nF\eta_c}{RT}\right)\right] \tag{1.87}$$

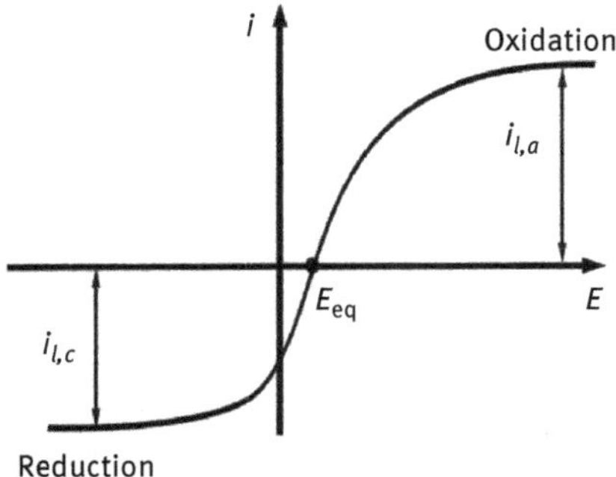

Fig. 1.13: Current–potential profile resulting from eq. (1.87), applied for both the reduction and oxidation processes. Characteristic points, enabling immediate identification of E_{eq} and i_l for the investigated process, are shown.

Of course, the above considerations can be carried out for the anodic processes, replacing Ox with Red as the electroactive species that diffuse to the electrode surface.

There are two characteristic values that can be immediately obtained from the above dependence of current versus potential, which are characteristic of the investigated redox process:

- the equilibrium potential, E_{eq} of the investigated redox couple, for the case of zero current measured in an external circuit ($\eta = 0$), and
- the limiting current, i_l for large overpotentials, both cathodic and anodic. The last one is related directly to the diffusion coefficient D and the thickness of the diffusion layer δ by eq. (1.81). The ratio $D/\delta = k_D$ is the so-called mass-transfer coefficient.

Scrupulous reader will easily see that the above considerations are valid only for the steady-state conditions, where J_{Ox} (or J_{Red} for oxidation reaction) are constant because in the above description of the diffusion-limited, linear, semi-infinite processes we did not account for the time of the electrolysis process. In fact, the graph representing the "frozen-in-time" concentration profile should be supplemented with other graphs, showing the evolution in time of the diffusion layer at constant applied potential, and/or graph showing the potential-dependent concentration change at the electrode surface (Fig. 1.14).

As it is shown in this graph, as the time of electrolysis increases at constant electrode potential, the thickness δ of the diffusion layer also increases, expanding from the electrode surface toward the bulk of the electrolyte, so $\delta_1 < \delta_2 < \delta_3$. With this, the concentration gradient, being inversely proportional to δ, decreases (compare eq. 1.81), slowing down the diffusion of the electroactive species to the electrode surface. This, in turn, decreases the current flow through the external circuit.

All this can be accounted for if we include in the formula for the current flow during polarization of the electrode not only the overpotential domain but also the time domain. We will discuss it in the next parts, presenting transient methods used in applied electrochemistry.

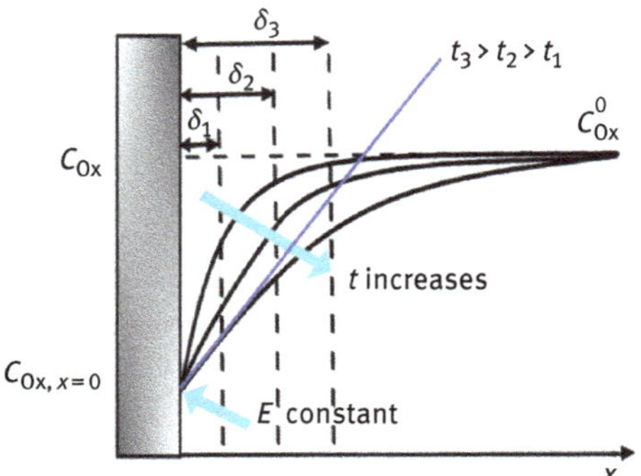

Fig. 1.14: Time evolution of concentration profiles of the diffusion layer at the electrode surface at constant applied potential.

1.1.3.3 "Irreversible" electrode processes: basic concepts of electrochemical kinetics

Let us now assume that **Step 5** – the electron transfer itself – in the scheme (Scheme 1.1) of processes involved in the redox reaction is the sluggish one. The main reason validating such an assumption is the kinetic limitation of the heterogeneous electron transfer rate. However, one should take into account **Steps 2–4** in the detailed description. Such considerations, including changes of distances or geometry in the solvation and/or coordination shells of the two chemical species, led to the development of Marcus, Hush, Levich, Dogonadze (MHLD) theory (R.A. Marcus, Nobel Prize in chemistry 1992).

In other words, there exists an activation energy barrier limiting kinetically the transfer of electrons across the metal/solution interface. If so, we can draw upon the kinetic equations that should account for the electron transfer processes under such conditions.

First, keeping in mind the accepted convention of redox processes: Ox + *ne* → Red, and bearing in mind that the overall "concentration" of electrons in their source (e.g., metallic connectors) during such reaction is independent of time, we can consider the redox process as described by the first-order kinetics:

$$v = -\frac{dc_{Ox}}{dt} = \frac{dc_{Red}}{dt} = kc \tag{1.88}$$

As we stated before, Faraday's law relates the flux of charges in time (the rate of the reaction, v), with current flow: $i = nFAv$, or, introducing current density $j = i/A$.

$$v_{Red} = k_1 c_{Ox} \quad \text{and} \quad v_{Ox} = k_2 c_{Red} \tag{1.89}$$

and therefore

$$j_C = nFk_1 c_{Ox} \quad \text{and} \quad j_A = nFk_2 c_{Red} \tag{1.90}$$

where k_1 and k_2 are the reduction and oxidation rate constants, respectively. Other symbols were defined before. At equilibrium Ox + *ne* ↔ Red, the potential of such electrode reaches the Nernst' value and the rate of reduction, v_{Red}, equals the rate of oxidation, v_{Ox}, and so the cathodic and anodic currents are equal: $j_C = j_A = j_O$. The last value, j_O, is called the *exchange current density* and characterizes the redox process for a given system (electrode, electrolyte solution, and redox species). Considering the electrochemical processes occurring on the electrode, both the reduction and oxidation processes should be considered at any time and the overpotential value, even though one predominates the other (except at equilibrium). During the reduction, the cathodic current is larger than the anodic one, and vice versa, during the oxidation, the anodic current is larger. What is measured during the experiment in the external circuitry is the resultant, net current. According to the IUPAC convention, the net current is

$$j = j_A - j_C \tag{1.91}$$

Since the concentrations of Ox (or Red) forms in the bulk of electrolyte do not change in time in the applied potential range, the presence of net current corresponds to the changes of the reaction rate constants k_1 and k_2. Let us now apply the Eyring–Polanyi theory (transition-state theory) for the electrode process to account for the dependence of rate constants of the energetics of the electron transfer rate. We also introduce the concept of standard rate constant k^0 and standard Gibbs free energy of activation, $\Delta G^{0\#}$:

$$k^0 = \frac{k_B T}{h} exp\left(-\frac{\Delta G^{0\#}}{RT}\right) \tag{1.92}$$

Here, k_B is the Boltzmann constant, h denotes the Planck constant, and other symbols have their usual meanings.

If the net observed current is a sum of cathodic and anodic ones, in light of eq. (1.90), it will depend on the two rate constants: k_1 for reduction and k_2 for oxidation:

$$k_1{}^0 = \frac{k_B T}{h} exp\left(-\frac{\Delta G_1^{0\#}}{RT}\right) \tag{1.93}$$

$$k_2{}^0 = \frac{k_B T}{h} exp\left(-\frac{\Delta G_2^{0\#}}{RT}\right) \tag{1.94}$$

where $\Delta G_1{}^{0\#}$ and $\Delta G_2{}^{0\#}$ are the standard Gibbs free energies of activation for reduction and oxidation reactions, respectively.

According to Fig. 1.15, $\Delta G_1{}^{0\#} > \Delta G_2{}^{0\#}$, so the reduction rate constant, $k_1{}^0$, is smaller than the oxidation rate constant, $k_2{}^0$, and the reduction current is smaller than the oxidation one. The electric field $\Delta G_E = nFE$ decreases the value of $\Delta G_2{}^{0\#}$ to $\Delta G_{2,E}{}^{0\#}$ by a fraction $(1-\alpha)nFE$ accelerating even further the oxidation reaction, slowing down the reduction by an increase of $\Delta G_1{}^{0\#}$ by a fraction αnFE. The coefficient α, called the *charge transfer coefficient* or *barrier symmetry coefficient*, falls within the range

$0 \le \alpha \le 1$ reflects the influence of electrode polarization on the Gibbs free activation energy of the reaction. Taking this into account, we can rewrite eqs. (1.93) and (1.94) in form:

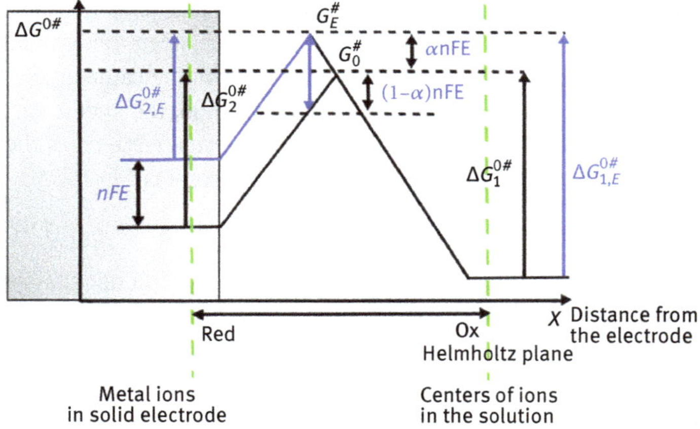

Fig.1.15: Schematic drawing of the standard Gibbs free energy of activation in the absence of electric field *E* (black) and in the presence of electric potential, *E* (red), polarizing the electrode anodically. Detailed description is given in the text.

$$k_1 = \frac{k_B T}{h} \exp\left(-\frac{\Delta G_1^{0\#} + \alpha nFE}{RT}\right) \tag{1.95}$$

and

$$k_2 = \frac{k_B T}{h} \exp\left(-\frac{\Delta G_2^{0\#} - (1-\alpha) nFE}{RT}\right) \tag{1.96}$$

These two equations can be rewritten to separate the part with no electric field ("chemical"), described by k_1^0 and k_2^0, respectively, from the part containing the effect of polarization of the metal/solution interface ("electrical"):

$$k_1 = k_1^0 \exp\left(-\frac{\alpha nFE}{RT}\right) \tag{1.97}$$

and

$$k_2 = k_2^0 \exp\left(\frac{(1-\alpha) nFE}{RT}\right) \tag{1.98}$$

Now, introducing these equations into eq. (1.90), we will get the equations for cathodic and anodic currents that flow through the electrode at potential *E*. At equilibrium, the potential of an electrode is equal to Nernst equilibrium potential, $E = E_{eq}$. Moreover, the rates of oxidation and reduction reactions are equal (eq. 1.89), and so

$$v_{Red} = v_{Ox} = k_1 = k_2 \tag{1.99}$$

Thus, under equilibrium conditions we now have

$$k_1{}^0 \exp\left(-\frac{\alpha nFE_{eq}}{RT}\right) = k_2{}^0 \exp\left(\frac{(1-\alpha)nFE_{eq}}{RT}\right) = k_s \tag{1.100}$$

Rewriting the above two equations in terms of current densities (eq. 1.90) and keeping in mind that the resultant current flow is a net flow of anodic and cathodic ones (eq. 1.91), we will have

$$j = nFc_{Red}k_2{}^0 \exp\left(\frac{(1-\alpha)nFE}{RT}\right) - zFc_{Ox}k_1{}^0 \exp\left(-\frac{\alpha nFE}{RT}\right) \tag{1.101}$$

Now, calling upon the definition of the overpotential (this time called the *kinetic overpotential*) $\eta = E - E_{eq}$, or $E = E_{eq} + \eta$, we can rewrite the above equation in a form:

$$\begin{aligned} j = nFc_{Red}k_2^0 \exp\left(\frac{(1-\alpha)nFE_{eq}}{RT}\right)\exp\left(\frac{(1-\alpha)nF\eta}{RT}\right) \\ - nFc_{Ox}k_1^0 \exp\left(-\frac{\alpha nFE_{eq}}{RT}\right)\exp\left(-\frac{\alpha nF\eta}{RT}\right) \end{aligned} \tag{1.102}$$

Finally, after some rearrangement we have the so-called Butler–Volmer equation:

$$j = j_0\left[\exp\left(\frac{(1-\alpha)nF\eta}{RT}\right) - \exp\left(-\frac{\alpha nF\eta}{RT}\right)\right] \tag{1.103}$$

where

$$j_0 = nFk_s c_{Red}{}^{\alpha} c_{Ox}{}^{1-\alpha} \tag{1.104}$$

is the so-called the exchange current density, describing the rate of oxidation and reduction currents at equilibrium, and k_s is the standard rate constant described above (eq. 1.100).

Equation (1.103) is valid when the electrode reaction is controlled by the kinetics of electron transfer at the electrode (not the diffusional mass transfer described earlier), taking advantage of Arrhenius and Eyring transition state theory (eqs. 1.93 and 1.94). The utility of the Butler–Volmer theory in electrochemistry is wide and we can show it later in the text for the case of, for example, corrosion processes. The overpotential η appearing in this equation is called the *activation overpotential* and is different from the concentration overpotential in its origin, as discussed above for the case of diffusion-limited currents (Section 1.3.2). The most important parameter describing the rate of electron exchange across the electrode interface is the exchange current i_0 (or exchange current density j_0). It is characteristic for a given electrode material, redox system present, and electrolyte solution; the higher its value the faster the system reaches equilibrium state and equilibrium potential, approaching the limiting conditions described by the diffusion of electroactive species (eq. 1.87). If the exchange current is small, we can assume the irreversibility of the electrode process and the equilibrium state of the electrode cannot be achieved. The exchange current is the equilibrium current (electron self-exchange between the Ox and Red states or electrode material and Red or Ox) and does not require any additional energy due to the polarization of the electrode). The effect of electrode polarization is contained in the exponential part of the Butler–Volmer equation. Therefore, let us consider that we are measuring the current density at high activation overpotentials, either for a cathodic, reduction process (η_C) or anodic, oxidation process (η_A). For sufficiently high overpotentials eq. (1.103) will take the form:

$$\text{Cathodic process:} \qquad j_C = j_0 \exp\left[\frac{-\alpha nF\eta_c}{RT}\right] \tag{1.105}$$

$$\text{Anodic process:} \qquad j_A = j_0 \exp\left[\frac{(1-\alpha)nF\eta_A}{RT}\right] \tag{1.106}$$

Frequently, the above equations are shown in a linearized form of $\ln j_C$ or $\ln j_A$:

$$\ln j_C = \ln j_0 - \frac{\alpha nF\eta_c}{RT} \tag{1.107}$$

$$\ln j_A = \ln j_0 + \frac{(1-\alpha)nF\eta_A}{RT} \tag{1.108}$$

This notation often called the *Tafel relations* allows one for easy evaluation of the exchange current j_0, as shown in Fig. 1.16.

For small values of activation overpotential (η < 25 mV), eq. (1.103) can also be linearized, because for small x, the exponential $e^x \approx 1 + x$. Thus, we can get the linear dependence of current density and η:

$$j = j_0 \frac{nF\eta}{RT} \quad \text{or} \quad \eta = \frac{RT}{zF}\frac{j}{j_0} \tag{1.109}$$

Because the units of fragment $\left(\frac{RT}{zF}\frac{1}{j_0}\right)$ are Ohms, the right-hand side of this equation can be rewritten in the form:

$$\eta = R_e j \tag{1.110}$$

where R_e is called the *resistance of the electrode reaction* or *polarization resistance* (for corrosion process discussed later).

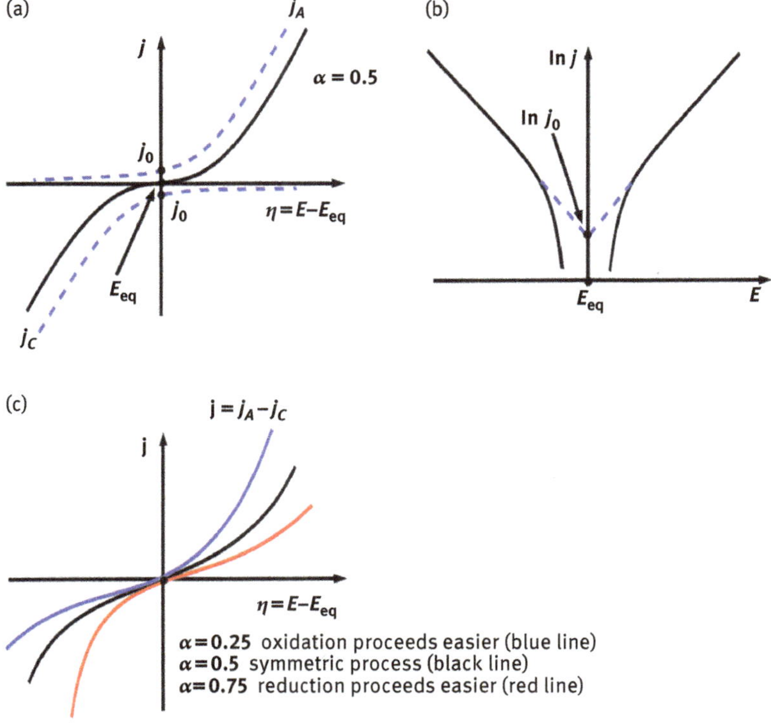

Fig. 1.16: (a) Graphical representation of Butler–Volmer equation (eq. 1.103), showing also the anodic (j_A) and cathodic (j_C) contributions to the measured current, as well as the nonmeasurable exchange current j_0 for the case of charge transfer coefficient $\alpha = 0.5$. (b) Linearization of eq. (1.103), showing the evaluation of the exchange current j_0. The slopes are $-\frac{\alpha zF}{RT}$ and $\frac{(1-\alpha)zF}{RT}$ for the cathodic and anodic currents, respectively. From these slopes we can evaluate the value of charge transfer coefficient, α. Typically, its value falls within the range of $0.25 < \alpha < 0.75$, and its effect on the redox current is exemplified in c).

Summarizing the above mechanisms of redox reaction at the electrodes that are responsible for current flow in the external circuit, the latter can be represented graphically (Fig. 1.17), with separate regions assigned to different limiting conditions: kinetic and diffusional.

1.1.3.4 Briefly on Marcus, Hush, Levich, Dogonadze (MHLD) theory of electrode processes

One may notice that in the above description of electrode processes, **Steps 2–4** shown in Scheme 1.1, were completely neglected. Although in Eyring–Polanyi theory (transition-state theory) for the electrode process the electrode and the Ox or Red forms become strongly coupled forming the so-called activated complex, the electrode can be only weakly coupled with the substrates. In such a case the electroactive forms (Ox or

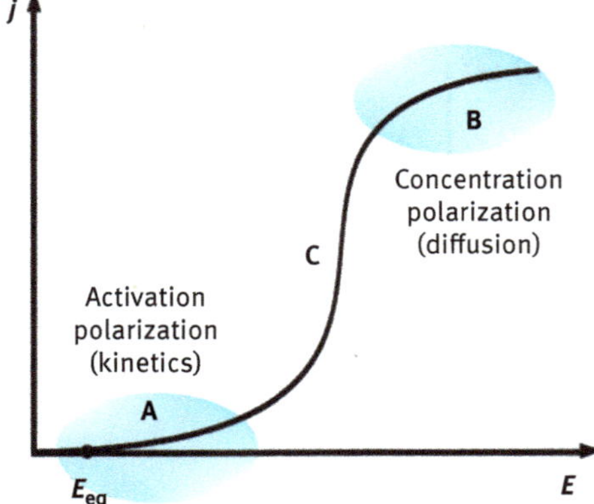

Fig. 1.17: Anodic (oxidation) current summarizing the mechanisms of redox reaction with separate regions assigned to different limiting conditions: kinetic and diffusional.

Red) retain their individual chemistries. It is the reorganization of the surroundings (induced thermally), the solvent (outer sphere), and the solvent together with ligands (inner sphere) that create favorable situation before the electron transfer. This concept takes advantage of Marcus electron transfer theory (published 1956, Nobel Prize in Chemistry 1992), which employs a classical, purely electrostatic model. It was further developed, extended, and refined significantly by Hush, Levich, and Dogonadze to address a variety of different types of charge transfer reactions. We address here only brief insight into this theory.

The effects of such structural changes were accounted for by the introduction of the ***reorganization energy***, λ, corresponding to the energy required to change the substrate (Ox) surroundings into the structure characteristic for the product (Red) at equilibrium. It is equivalent to the energy required for the products to achieve the configuration of substrates (or vice versa). If we assume that the energy curves can be approximated by quadratic functions (parabolas) as in Fig. 1.18, then the Gibbs free energy of activation, $\Delta G^{\#}$, is equal to

$$\Delta G^{\#} = \frac{1}{4\lambda}(\lambda + \Delta G_r)^2 \tag{1.111}$$

Moreover, an assumption is made that these parabolas are identical in shape for substrates and products. Thus, the rate constant of electron transfer during the unit redox process, k_{ct}, can be written in a form equivalent to eq. (1.92), but including the reorganization term λ:

$$k_{ct} = \frac{kT}{h}\exp\left(-\frac{\Delta G^{\#}}{RT}\right) = \frac{kT}{h}exp\left(-\frac{(\lambda + \Delta G_r)^2}{4\lambda RT}\right) \tag{1.112}$$

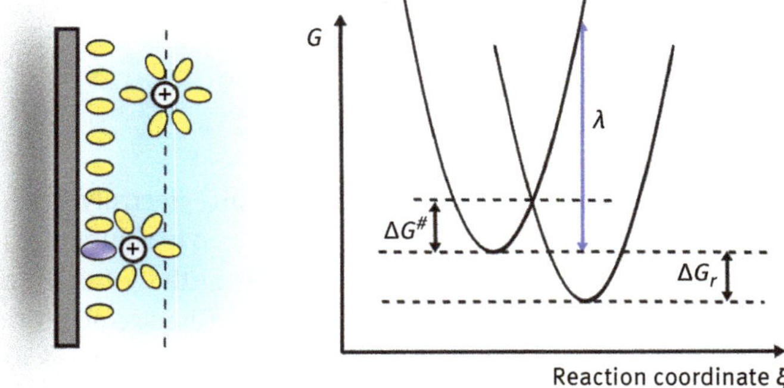

Fig. 1.18: Left panel: Outer sphere electron transfer (upper part): reaction between the electrode and the electroactive species does not require changes in the original sphere and the e.d.l. Inner sphere (lower part): Reaction with the formation of an activated complex with structural changes of the original coordination sphere and e.d.l. structure. Note that even if there is no strong interaction with the electrode, an outer sphere reaction can depend on the electrode material because of the double layer effects and the effect of the metal on the structure of the Helmholtz layer. Right panel: Description of symbols used in eq. (1.111).

MHLD theory found its implications not only in electron transfer between inorganic, organometallic, and organic molecules, but also in bioelectrochemistry of proteins, enzymes, and nucleic acids.

Bibliography

[1] Bocris J O'M, Reddy AKN. Modern electrochemistry. V2 New York, USA, Plenum Press, 1970.

[2] Bagotsky V.S. Fundamentals of electrochemistry. Hoboken, USA, J Wiley & Sons Inc., 2006.

[3] Gileadi E. Physical electrochemistry. Fundamentals, techniques and applications. Weinheim, FRG, Willey-VCH Verlag GmbH & Co, 2011.

[4] Koryta J, Dvorák J, Kavan L. Principles of electrochemistry, John Wiley & Sons Inc, ed., Chichester, 1993.

[5] Bard AJ, Inzelt G, Scholtz F, eds., Electrochemical dictionary, 2nd Edition, Springer-Verlag, 2008, 2012.

[6] Brett CMA, Oliveira Brett AM., Electrochemistry: Principles, methods, and applications, Oxford Science Publ., 1993.

[7] Karnik R, Fan R, Yue M, Li DY, Yang PD, Majumdar A. Electrostatic control of ions and molecules in nanofluidic transistors. Nano Letters 2005, 5, 943–948.

[8] Wang C, Wang S, Chen G, Kong W, Ping W, Dai J, Pastel G, Xie H, He S, Das S, Hu L. Flexible, bio-compatible nanofluidic ion conductor. Cem. Mater. 2018, 30, 7707–7713.

[9] Melitz W, Shen J, Kummel AC, Lee S. Kelvin probe force microscopy and its applications, in Surface Science Reports, 2011, 66, 1–27.

[10] Kjelstrup S, Bedeaux D, Johannessen E, Gross J. Non-equilibrium thermodynamics for engineers, 2nd edition. World Scientific,. Singapore, 2017.

1.2 Structure of the bulk of electrolytes: conductivity

1.2.1 Interactions in aqueous electrolytes

1.2.1.1 Ion–solvent interaction; solvation

For the purpose of this text, we will discuss primarily the aqueous electrolyte solutions, but the potential reader should keep in mind that the electrolyte solutions can use also other solvents of sufficiently high relative dielectric permittivity (dielectric constant), ε, allowing for the dissociation of solute molecules into ions. And so the electrolyte solutions are the systems containing a solvent and a substance that is dissociated partially (weak electrolytes) or totally (strong electrolytes) into ions. These solutions conduct the electric current and the charge carriers are the dissociated ions moving in the external electric field. Of course, the electroneutrality condition has to be fulfilled. The choice of whether the electrolyte can be classified as "strong" or "weak" depends on the so-called dissociation coefficient (constant), α, being equal to 1 for all strong electrolytes, such as ionic salts, hydrochloric acid, alkaline metals' hydroxides, etc. In the case of weak electrolytes, the dissociation constant is below 1 and depends on the concentration of such electrolytes. The lower the concentration, the higher the dissociation constant, approximating 1 for infinitesimally small concentrations. It is also worth mentioning, that the dissociation constant depends strongly on the properties of a solvent affecting the interactions between the ions and solvent molecules. In the absence of an external electric field, one should consider at least two types of interactions: ion–solvent molecules interactions and ion–ion interactions. In the classical theories of electrolyte solutions, ions are considered as particles moving in a dielectric continuum of a solvent.

Therefore, the main deviation from an ideal gas behavior was due to the Coulomb electrostatic interactions, determining the nonideal behavior of electrolytes. In general, one should consider also the electrostatic and/or Van der Waals interactions between the ions and solvent molecules, contributing largely to the behavior of electrolyte solutions. These interactions are generally described as solvation (hydration in the case of aqueous solutions) of ions and lead to the formation of a solvation "sphere" around an ion. The importance of such interactions becomes evident when comparing the hydration energies of monovalent ions and neutral molecules: in the former case, the enthalpy of hydration is in the range of several hundred kJ/mol, whereas for simple nonpolar molecules, such as methane, it is in the range of tens of kJ/mol. Therefore, it is evident that such interactions should lead to the structuring of electrolyte solutions. The simplest description of ion–solvent interactions was proposed by Max Born (ca. 1920). It allows us to estimate the electrostatic component of Gibbs free energy of solvation of an ion, by treating it as a sphere of an "effective" radius r, of $z_i e$ unit charge in a continuous dielectric medium of a relative dielectric constant ε. This estimation is based on the following three-step rationale:

a) discharging the ion in a vacuum requires a work $W_1 = \int_{z_ie}^{0} \psi_1 dq$, where ψ_1 is the electrical potential at the surface of an ion: $\psi_1 = \frac{q}{4\pi\varepsilon_0 r}$. After integration, we get:

$$W1 = -\frac{(z_ie)^2}{8\pi\varepsilon_0 r} \tag{1.113}$$

b) discharged ion is now introduced to the solvent of a relative dielectric constant ε_r without any work loss or gain,
c) re-charging of an ion in this new environment (solvent), to the electric potential at the surface $\psi_2 = \frac{q}{4\pi\varepsilon\varepsilon_0 r}$. This process requires a work $W_2 = \int_0^{z_ie} \psi_2 dq = \frac{(z_ie)^2}{8\pi\varepsilon\varepsilon_0 r}$.

Under constant pressure and temperature conditions, the work required for both processes carried out for 1 M of ions is equal to the change of the Gibbs free energy:

$$\Delta G = -\frac{N_A z^2 e^2}{8\pi\varepsilon_0 r_0}\left(1 - \frac{1}{\varepsilon_r}\right) \tag{1.114}$$

Even though this interpretation is oversimplified, it is frequently used when comparisons of different solvation behavior of solvents are considered, as can be the case in the evaluation of interfacial electrochemical processes and selection of solvents during the construction of fuel cells. On the other hand, for a given solvent, the "effective" ionic radius and ionic charge are factors leading sometimes to the so-called structure breaking of a solvent. This effect increases with ionic radius, for example, $Li^+ < Na^+ < Rb^+ < Cs^+$ or, $Cl^- < NO_3^- < I^- < ClO_3^-$ and with an increasing charge, for example, $Li^+ < Be^{2+} < Ca^{2+}$. The solvation of ions has to be considered when discussing the electrolyte's electric conductivity/ionic resistance in batteries and fuel cells because it plays a more important role in determining electrolyte properties than the ion–ion interactions described below.

1.2.1.2 Ion–ion interactions

By eq. (1.35) we have already defined the chemical potential of species "i," μ_i. Now, we can use it for our considerations of ion–ion interactions. For an ideal solution, the chemical potential of its component, μ_i^{id}, can be described as:

$$\mu_i^{id} = \mu_i^0 + RT\ln c_i, \tag{1.115}$$

where μ_i^0 is the standard chemical potential of species "i" and c_i is its concentration. However, due to the ion–ion interaction in the real solutions the "real" concentration of species "i" is different and is reflected by the activity coefficient γ_i. Thus,

$$\mu_i^{re} = \mu_i^0 + RT\ln c_i\gamma_i, \tag{1.116}$$

where μ_i^{re} is the chemical potential of "i" in a real solution. Therefore, the difference between μ_i^{re} and μ_i^{id} can be considered as molar Gibbs free energy loss for the electrostatic interactions between ions, resulting in the first approximation, in the deviations of electrolytes from ideal solutions:

$$\mathrm{RTln}\gamma_i = \mu_i^{re} - \mu_i^{id} \tag{1.117}$$

Activity coefficients of single ions cannot be measured experimentally because an electrolyte solution must contain both positively charged ions and negatively charged ions. Instead, a mean activity coefficient for the strong, totally dissociated electrolyte A_nB_m is defined as:

$$\gamma_\pm = \left(\gamma_-{}^n \gamma_+{}^m\right)^{1/(n+m)} \tag{1.118}$$

And so, the multiple-charge generalization from electrostatics gives the expression for the potential energy of the entire solution

$$RTln\gamma_\pm = \mu^{re} - \mu^{id} \tag{1.119}$$

Some thermodynamic considerations, easily found in other textbooks, can lead to the equation, named the Debye–Hückel limiting equation (Peter Debye, Erich Hückel, 1923). The underlying theory assumes a simplified model of electrolyte solution, nevertheless giving quite good predictions of mean activity coefficients for ions in dilute solutions. Under similar assumptions as in the case of the Gouy–Chapman model described earlier in this chapter, it is a linearized Poisson–Boltzmann model (eqs. (1.24) and (1.25)) and provides a good starting point for modern approaches to the description of real electrolyte solutions. However, what differs the Debye–Hückel model from the Gouy–Chapman electrical double layer model is, that instead of a flat, uniform metal | solution interface, now we have a "central" point charge (be it a cation or anion), surrounded by a spherical cloud of counterions' contribution, balancing to zero the net charge of a central ion and its ionic cloud. The central ion and its ionic cloud are electrically neutral, and the radius of such a cloud is described by the very same equation as the Debye length of the e.d.l., *vide supra*, eq. (1.27).

The evaluation of the electrostatic energy of interactions between the central ion and its ionic cloud (Coulomb interactions) is the key point of the D–H theory. As a consequence, it allows for the evaluation of such interactions with the limiting equation, valid for strong electrolytes under infinite dilution, containing monovalent ions (e.g., KCl):

$$RTlog\gamma_\pm = -A|z_+z_-|\sqrt{I} \tag{1.120}$$

where A is a temperature-dependent constant characteristic for the solvent at a given temperature T, z_+ and z_- are the valencies of ions, and I is the ionic strength of the solution:

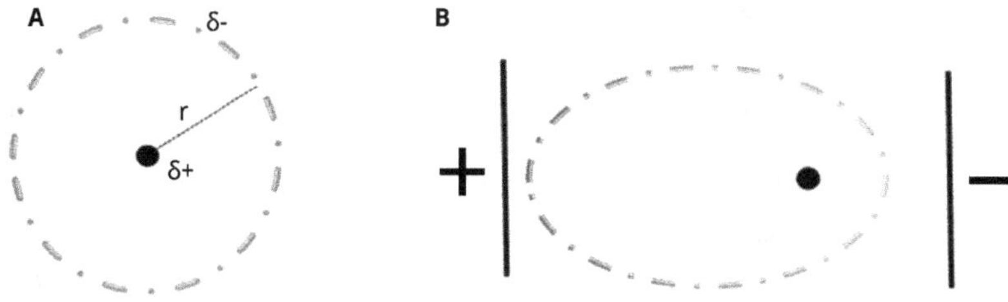

Fig. 1.19: An ionic atmosphere at rest (**A**). A spherical cloud of charge $\delta+$ (or $\delta-$) surrounds point charge ions in solution. Its radius is identical to the Dybye length of the e.d.l., eq. (1.27). (**B**). Asymmetrical cloud in an external electric field; positively charged central ion and net-negatively charged ionic cloud.

$$I = \frac{1}{2}\sum_i c_i z_i^2 \tag{1.121}$$

For a more concentrated solution, the minimum distance between the ions, ***a***, is introduced (the distance of closest approach of ions), an empirical parameter, changing the form of the limiting Debye–Hückel equation:

$$log\ \gamma_\pm = -\frac{A|z_+z_-|\sqrt{I}}{1+aB\sqrt{I}}, \tag{1.122}$$

here B is also a constant dependent on the solvent parameters and T. Finally, for even higher ionic strength, one more empirical parameter C is added, accounting for the effect of ionic strength on the dielectric constant of a solvent.

$$log\ \gamma_\pm = -\frac{A|z_+z_-|\sqrt{I}}{1+aB\sqrt{I}} + CI \tag{1.123}$$

All these modifications are illustrated in Fig. 1.20.

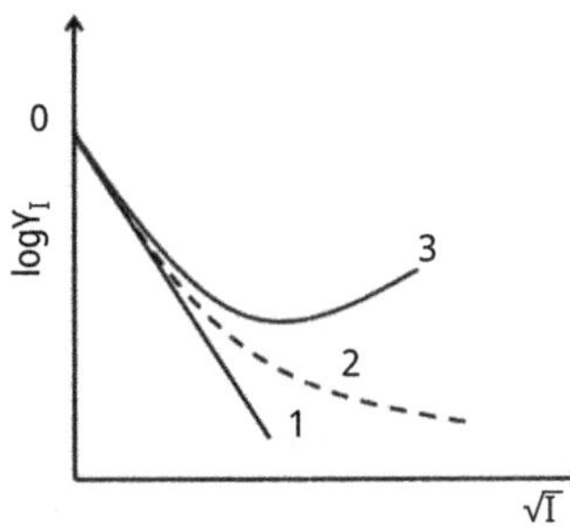

Fig. 1.20: Graphical representations of subsequent modifications of the D–H theory: curve **1** – the limiting equation (straight line), **2** – including the distance of closest approach, **3** – influence of the ionic strength on the dielectric behavior of a solvent.

Therefore, the activity coefficients are themselves functions of concentration as the amount of inter-ionic interaction increases as the concentration of the electrolyte increases.

1.2.2 Conductivity of electrolytes

It was pointed out in the previous section, that apart from the mass transport during the electrode reactions (reversible and irreversible), the Nernst–Planck equation (eq. 1.70) describes also a substantial part of ion mass/charge transport through the bulk of the electrolyte solution under the influence of the external electric field, called migration, $J_{m,i}$. In addition to this, this equation should be expanded to include also the convective transport of ions, defined as the transport of substances with a moving medium (e.g., in a liquid flow due to the thermal gradient – convection). Under typical conditions, the total flux of species "i" is then the algebraic (vector) sum of densities of all flux types:

$$J_i = J_{d,i} + J_{m,i} + J_{cv,i} \tag{1.124}$$

Here: $J_{d,i}$ describes the diffusion transport (1st Fick's law, eq. (1.72)) and $J_{m,i}$ is the migration flux, eq. (1.73). The convection flux density, $J_{cv,i}$, description is beyond the scope of this textbook. Since we have described in detail the limiting case of diffusion current control of the electrode processes (Section 1.1.3.2), let us now focus on the migration flux of ions. Taking into account eq. (1.73) (*vide supra*), we can get the straightforward definition of a specific conductivity of electrolyte solution, $\kappa = F \sum_i |z_i| c_i U_i$, (compare eq. (1.76)), where $F z_i$ corresponds to the charge transferred by 1 mol of "i" ions with valency z_i. Other symbols were described previously. What can be immediately seen from this relation is that all effects occurring in the electrolyte that decrease the ionic mobility will also decrease the conductivity of the electrolyte. Keeping the ionic conductivity as high as possible is of crucial importance in designing power sources and storage systems, fuel cells in particular (Chapter 8, cf. eq. (8.11a)). One may also think that since the specific conductivity scales linearly with concentration, it might be sufficient to increase the conducting behavior of a specific electrolyte by simply increasing its concentration. Let us verify this assumption by introducing of molar conductivity of an electrolyte, Λ:

$$\Lambda = \frac{\kappa}{c} \tag{1.125}$$

$$\Lambda = F \sum_i |z_i| U_i \tag{1.126}$$

By doing this, we eliminate the effect of electrolyte concentration, so the molar conductivity should be constant, and concentration-independent. This type of behavior can be observed only at almost infinite dilutions of electrolytes, where at constant temperature T, the electrolyte ions can move independently of one another in the external electric field. This principle is often referred to as Kohlrausch's law of independent migration. Under such conditions, each ionic species' contribution to the conductivity of the electrolyte solution depends only on the nature of that particular ion. This approximation is valid also for weak electrolytes because, in the infinite dilution, even these electrolytes are totally dissociated into the respective ions. Thus, the Kohlrausch law can be written as:

$$\Lambda^0 = \sum_i \nu_i \lambda_i^0 \tag{1.127}$$

where Λ^0 is the limiting molar conductivity of the electrolyte, ν_i is the stoichiometric coefficient of an ion "*i*," and λ_i^0 is its individual limiting conductivity.

Comparing the above two equations (eqs. (1.125) and (1.126)), it is evident that the molar conductivity of an individual ion "*i*" can be written as:

$$\lambda_i = z_i FU_i \tag{1.128}$$

where U_i denotes the mobility of ion "*i*" in the solution at a given concentration (U_i^0 will be its mobility at infinite dilution). The mobility of ion at a given external electric field strength, E [V/m], is given by:

$$U_i = \frac{\mathrm{v_i}}{E}, \tag{1.129}$$

where $\mathrm{v_i}$ is the velocity of an ion [m/s].

Therefore, apart from the concentration of an electrolyte, also the individual ionic mobilities (conductivities) at a given concentration do contribute to the overall charge-transfer properties of an electrolyte. The above discussion (and eqs. (1.76), (1.125–1.127)) gives a background to understanding the conducting behavior of aqueous electrolytes illustrated by the graphs in Fig. 1.21A, B.

There are several information that can be drawn from this graph:

1. The specific conductivities of weak electrolytes are much smaller compared to the strong ones. This is a direct effect of smaller concentration of "free" ions concerning the analytical concentration of such electrolyte, due to the incomplete dissociation of, for example, acetic acid molecules into the respective ionic species.
2. For very dilute electrolytes, regardless of whether it is strong or weak, the dependence of κ vs. c approaches linearity, as predicted by eq. (1.76), because the number of ions increases with the concentration.
3. For weak electrolytes, one can observe that the specific conductivity reaches a shallow plateau and then decreases, because of a decrease in its dissociation constant (for a more detailed description, see Ostwald dilution law).
4. Similar behavior can be observed for strong electrolytes, however, the maximum value at these curves is much higher and attained much later. This observation can be explained by a smaller "effective" concentration of ions forming associates of different stoichiometries and charges, for example, the ionic pair has a net charge

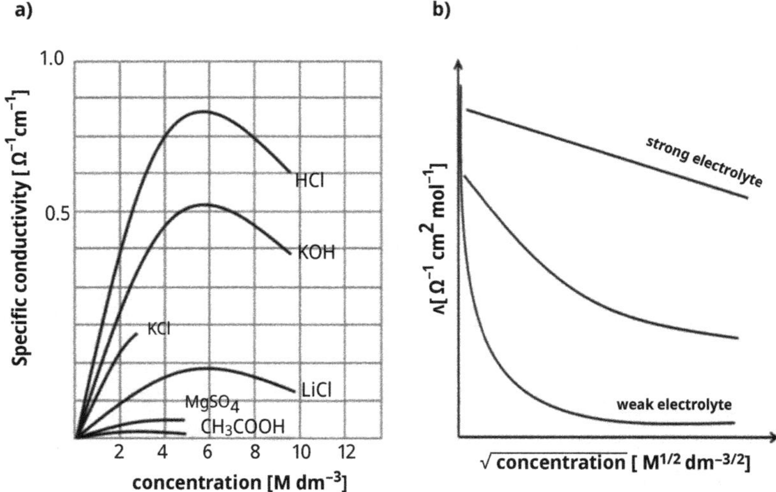

Fig. 1.21: A presents the dependence of specific conductivity, κ (eq. (1.76)), of concentrations of aqueous solutions of strong (HCl, KOH, KCl, and LiCl, dissociation constant = 1) and weak ($MgSO_4$, CH_3COOH, dissociation constant < 1). **B** – molar conductivities of strong, medium, and weak electrolytes vs. square root of their concentrations (see text for details).

zero and therefore does not contribute to the overall conductivity, whereas the ion-triplet has a net charge ± 1, but its dimensions are much larger than that of contributing ions and so it moves slower in the external electric field. The ion-pair formation due to the electrostatic interactions was quantitatively formulated by Niels Bjerrum (1926).

However, a drastic decrease in the specific conductivity of strong electrolytes after reaching a maximum cannot be explained by this relatively simple kind of interaction between ions of opposite charge. As the electrolyte concentration increases, the aggregates formed can be held together also by chemical forces, thus attaining special features, such as shorter interionic distances, partial desolvation of ions contained in such structures, or even forming a new solvent sheath. These structures can be seen distinctly in spectroscopies (IR or Raman). All these give rise to the two effects that are responsible for a drastic decrease of conductivity in concentrated strong electrolytes: the relaxation effect and the electrophoretic effect, both impeding strongly the ionic mobilities. The first one, in its simplest form, results from the presence of an ionic atmosphere around a "central" ion. Under the influence of an external electric field, the central ion moves to the electrode of the opposite sign, whereas its ionic cloud moves toward the other electrode. Therefore, during its movement, the central ions lose and rebuild their ionic cloud in specified time, called the relaxation time, which depends on the type of ions, solvent, external electric field strength, and temperature.

This also leads to a temporal deformation of the ionic cloud (see Fig. 1.19B) and such structure generates a local electric field of opposite direction with respect to the external electric field in which the ions move, conducting current. This effect hinders the movement of ions, decreasing their mobilities and decreasing electrolyte conductivity. The other effect – the electrophoretic effect, is related to the movement of solvated ions: cations and anions in opposite directions against each other in the external electric field. The friction between these moving solvated ions diminishes the ionic mobility, scaling with the square root of ionic strength and inversely proportional to the solvent parameter – the so-called dynamic viscosity (internal friction coefficient), η, of a solvent. The higher the value of this parameter, the "thicker" the solvent imposing higher resistance to the moving ions, and of course – higher resistance of the electrolyte solution. The consequence of both effects on the specific conductivity of electrolyte solutions is their dependence on concentration, solvent viscosity, dielectric constant, and temperature. These parameters have to be taken seriously into account, for example, in the construction of new power sources or power storage: The lower the specific conductivity of the electrolyte, the higher the internal resistance in a such device, affecting its performance, costs, and usability (Chapter 8).

For better visualization of the ionic effective mobilities in solutions, Fig. 1.21B shows the dependence of molar conductivity of strong to weak electrolytes in an aqueous solution as a function of the square root of the ionic strength. The form of this function predicting linear behavior for strong univalent (1:1) electrolytes was derived empirically by Kohlrausch:

$$\Lambda = \Lambda^0 - K\sqrt{I} \tag{1.130}$$

where K is the empirical constant, sometimes named the Kohlrausch coefficient. This equation allows for the evaluation of limiting molar conductivities of strong electrolytes by extrapolation of Λ as a function of $\sqrt{c}$ to $c \to 0$. This approach is valid only for strong electrolytes, as illustrated by Fig. 21B. A more detailed description, allowing for the assignment of a physical meaning of parameter a, led to the Debye–Hückel–Onsager (D–H–O) theory. This theory considers also the parameters characterizing not only the electrolyte, such as the valencies of ions and their individual limiting conductivities, λ_i^0 , but also those of the solvent, such as dielectric conductivity ε, and its dynamic viscosity, η. The latter parameter is necessary to describe the motion of ions in the solvent. We stress again that the D–H–O theory is limited only to strong electrolytes and cannot be used for evaluation of limiting conductivities of weak electrolytes. One has to consider the degree of dissociation of an electrolyte, α, and its dissociation constant, $K_{a.}$ The interested reader is encouraged to get a deeper understanding of the interrelations between the Debye–Hückel theory (ions are immobile) and the Debye–Hückel–Onsager theory (ions are moving in the external electric field) for both types of electrolytes, by reading some basic electrochemistry textbooks referred to at the end of this chapter [1].

1.3 Nonaqueous electrolytes

Until now, we were discussing primarily the aqueous electrolyte solutions that are obtained by the dissolution of salts in water. In such systems, the energy necessary for the dissociation of salt into ions was provided primarily by the solvent–salt lattice interactions. However, the lattice dissociation into ions can be achieved also by providing heat: the salt can be melted down without any additional component acting as a solvent. Such a system is called molten salt for a temperature well above 100 °C or ionic liquid (IL) if the melting temperature is around or even below ca. 100 °C. Anyway, as time goes by, definitions of both liquids tend to overlap with each other. Currently, however, the term molten salts is usually applied to melts of inorganic compounds, composed of cations and anions, being in a solid state at room (and slightly above) temperature. With an increase in temperature, the short-range and long-range ordering of solids is overcome by thermal energy, preserving only the long-range ordering in a molten, liquid state. In this state the ions possess certain mobilities, keeping approximately constant density of liquid due to the cohesion forces between the constituent particles. However, before going further, we have to stress that the theories describing the structure of the interface between the electrode and aqueous electrolyte, as well as those referring to the conductivity behavior of the ion–ion and solvent–ion interactions cannot be applied directly to ionic liquids and molten salts. First of all, there is no solvent involved in dissociation and solvation. Secondly, the "infinite dilution" limit and dimensionless charge approximation (point charge limit) that we were using in Section 1.1.1 and in describing the structure of aqueous electrolyte, do not hold in the case of ILs and molten salts. The sizes of ions are not to be ignored (the so-called excluded volume), particularly for bulky constituents of ionic liquids, nor the high viscosity of such electrolytes, affecting their conducting properties. The interested reader is referred to the literature at the end of this chapter, also for the structures of most common ILs [2, 5].

1.3.1 Ionic liquids

Exploration of environmentally friendly green solvents as alternatives to aqueous, high-temperature molten salts or volatile organic compounds in electrochemical energy storage, and electrolytic processes has been persistently pursued. A class of electrolytes, called ionic liquids (ILs) is widely studied from the point of view of replacing aqueous electrolytes for such purposes. Ionic liquids are typically defined as compounds totally composed of ions in their pure forms, meaning that they are overall neutral assemblies of charged particles. They are characterized by melting points below or around the boiling point of water under ambient pressure. Their hybrid organic-ionic behavior and intermolecular interactions give rise to a set of complex phenomena that are beyond the scope of this book. Therefore, here we will briefly outline only those features that are directly related to the electrochemical aspects of ILs.

Ionic liquids (ILs) possess extraordinary properties such as nonflammability, and relatively high ionic conductivity at a level of ca. 10–30 mS/cm, low toxicity, the wide electrochemical window of ca. 4 V, and electrochemical and thermal stability; all these features make them ideal candidates for electrolytes in electrochemical devices like batteries, double-layer capacitors, fuel cells, photovoltaics, actuators, and other electrochemical devices [2–4].

In addition, ILs have been widely used in electrodeposition, electrosynthesis, electrocatalysis, and electrochemical capacitors, making them ideal solvents for a wide range of electrochemical applications. Since ionic liquids contain only charge carriers, means that when they are used as solvents, they also act as supporting electrolytes, minimizing costs and wastes by reducing the use of hazardous and polluting organic solvents for a greener environment. The use of ILs in so many different applications is determined by their intrinsic properties. There are many feature reviews dealing with different aspects of ILs, in all these areas of applied electrochemistry and the reader is directed to these excellent publications [2–5]. Of particular interest are ILs existing in the liquid phase around room temperature. A typical room temperature IL with a bulky organic cation (e.g., alkylammonium, N-N-dialkylimidazolium, alkylimidazolium, N-alkylpyridinium, and alkyl-phosphonium) is weakly coordinated to an organic or inorganic anion, such as BF_4^-, Cl^-, I^-, PF_6^- $AlCl_4^-$, etc. The ionic conductivity of ILs (at least ten times smaller than that of conventional aqueous electrolytes) is highly dependent on temperature (as is the case for all solvents) because as viscosity increases, the charge carriers must overcome a greater frictional force, decreasing their mobilities, and, as a result, conductivities. These relations can be represented in Arrhenius-type plots, but instead of a straight line, ILs show a certain curvature that can be explained by the temperature-dependent changes in the structural properties of bulky ions. Nonetheless, the choice of appropriate cation and anion of the ILs can alleviate to some extent such dependency, improving their transport properties, adjusting also the potential window of IL's stability and its capability to dissolve salts with metallic cations for, for example, metal-ion batteries using ionic liquids as solvents. It is also worth mentioning that the specific conductivity, κ, of conventional strong electrolyte solutions is proportional to the number of charge carriers. However, the assumption of ILs consisting entirely of ions as charge carriers does not hold true. Rather, as in the case of the strong aqueous KCl electrolyte specific conductivity graph shown in Fig. 1.21A, where after reaching maximum value a decrease in conductivity is observed with increasing concentration, one should expect, that IL ions of opposite sign may form relatively stable aggregates of nullified resultant charge. If so, these aggregates will not participate in the transfer of charges. Such ion-ion interactions may vary for different ionic liquids and their mixtures. Therefore, it is difficult to define the number of charge carriers by simply knowing the IL chemical structure. Literature data estimation leads to the conclusion that for many commonly used RTILs (room-temeprature ionic liquids), the degree of ion pair dissociation at room temperature varies between 50% and 70%.

The features of RTILs that make them special as electrolytes can be summarized below:

- RTILs are electrolytes without solvents that stay liquid in a broad temperature interval around the room temperature.
- The ions of RTIL are typically large organic ions (at least one of them), containing a substantial amount of hydrophobic constituent functionalities. These features reduce the freezing temperature of the ILs, because the Coulomb interactions are weaker in the range of their action, whereas their complicated shape hinders reaching ordered crystal structures.
- If RTILs are composed of less reactive ions, they are more inert electrochemically compared to many standard electrolyte solutions; therefore they can sustain higher voltages (larger potential window).
- Various RTILs can be mixed with each other to obtain desired electrochemical behavior, such as lower viscosity and thus increased ion mobility.

1.3.2 Molten salts

Generally speaking, molten salts consist of cations and anions with various valencies, being regarded as extremely concentrated solutions or solutes without solvents (such as high-temperature ILs) at elevated temperatures and include alkali halides and salts such as nitrates, sulfates, and carbonates. Currently, this is often expanded to comprise silicates, borates, and organic salts. Of course, one can easily find many more potential candidate cations and anions for molten salt; the only limitation in such a creative approach is their melting temperature and conductivity. If binary and ternary systems are considered, this number is, in fact, quite large. Molten salts dissociate completely into ions and therefore they are considered strong electrolytes. The high mobility of the ions at elevated temperatures results in high ionic conductivity. Therefore, molten salts are good conductors exhibiting 10–50 times higher conductivity than the aqueous electrolyte or ionic liquids. This in turn results in a remarkable decrease in the internal resistance of the electrochemical power sources and storage devices described later in this book (Chapter 8). Additionally, molten salt electrolytes are characterized by wide electrochemical windows, low raw material costs, and nonflammability. The features used to characterize molten salts are mostly the same as those of ionic liquids mentioned above, except for the ions that are mainly inorganic with no additional chemical functionalities, and therefore a thermal energy is required to overcome the lattice association forces solidifying the electrolyte structure. However, the need to keep the high temperature of a device utilizing the molten salt as an electrolyte, requires special construction of such a device, additional energy lost for heating, as well as some safety issues and thermal management related to the regime of high-temperature operation. This may lead to increased weight and cost. Nevertheless, the elevated temperatures required for electrochemical devices utilizing molten salts can be advantageous for operation in harsh environments, such as metallurgy, the oil and gas industry, internal combustion engines, hybrid transportation vehicles, or even space exploration.

1.3.3 Solid electrolytes

Since there has been a continuously growing demand for more efficient, reliable, and environmentally friendly materials for use as electrolytes, particularly for solid-state energy storage systems and environmental sensors described later in this book, solid electrolytes (SEs) are considered one of the key components for these systems. Their advantages over the electrolytes described above are thermal and chemical stability, lower toxicity, wide electrochemical window, and good mechanical properties. However, the main issue is the ionic conductivity, especially at ambient temperatures. Solid electrolytes are generally classified into two main groups: inorganic electrolytes (IE), and polymer electrolytes (PE). Further, the inorganic electrolytes incorporate oxides, and sulfides (both amorphous and crystalline), whereas the polymer electrolytes include, for example, polymer-salt complexes, gel polymer electrolytes, and composites. Comparing their conductivities, the inorganic electrolytes provide much higher ionic conductivities (up to 10^{-2} S/cm for sulfide-based electrolytes), comparable to those of organic liquid electrolytes, whereas polymer electrolytes show conductivities still not sufficient for their subsequent wide applications. Therefore, we will focus below on the conductivity behavior of inorganic electrolytes, outlining the migration of ions in a solid electrolyte at different scale lengths: from the atomic to a macroscopic scale.

Atomic scale. At the atomic scale, mobile ions, in the case of oxides and sulfides – cations (e.g., Li^+, Na^+, or Mg^{2+}) diffuse in solids along the lowest energy migration pathways and can be envisioned as ion hoping between stable sites and/or intermediate metastable sites of the crystalline framework of anions (e.g., O^{2-}, S^{2-}). In a crystalline case, cationic vacancies and interstitial sites are also considered mobile-charged species. Three main migration mechanisms can be distinguished: i) ion migration into a neighboring vacant site (vacancy migration in the opposite direction, ii) interstitial direct migration between not fully occupied sites, and iii) concerted mechanism in which the migrating ion from the interstitial site displaces its neighbor in the lattice into the next metastable cation site. This is shown graphically in Fig. 1.22 below

The conductivity, σ, of an ion in a solid electrolyte can be defined as:

$$\sigma = q \times n \times u \tag{1.131}$$

Here, q is the charge, n is a concentration of mobile charge carriers in a solid, and u is their mobility defined as before. We also should notice, that typically only a fraction

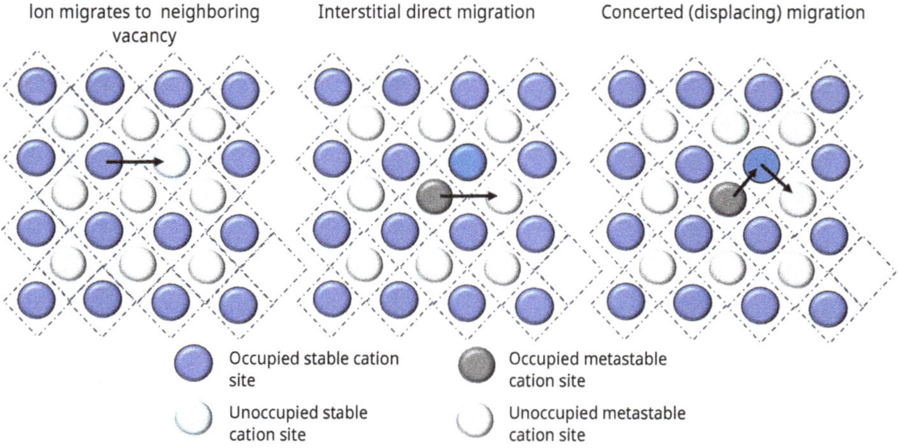

Fig. 1.22: Ion migration pathways in a crystal lattice of a solid electrolyte.

of total charges is mobile, thus a precise determination of n is at least ambiguous. As can be easily understood, ion conduction is a thermally activated process, usually described by a modified Arrhenius equation:

$$\sigma = \sigma_0 \times T^m e^{\frac{-E_a}{k_B T}}, \tag{1.132}$$

where σ_0 is the pre-exponential factor, m is typically equal to -1, k_B is the Boltzmann constant, T is temperature and E_a is a characteristic activation energy for ion conduction. In the simplest case, E_a includes the energy of forming the mobile defects and energy for their migration (highest energy barrier along the conduction path). Of course, the charge transfer, regardless of the type of electrolyte (liquid or solid) is linked to the mass transfer. In the case of solid electrolytes, it is also useful to define the diffusivity of charges, D, according to the relationship similar to eq. (1.72):

$$D = \frac{u}{q} k_B T, \tag{1.133}$$

and therefore, after substituting into the equation for conductivity:

$$\sigma = \frac{Dq^2}{k_B T} \times n \tag{1.134}$$

These basic equations can be utilized to understand and describe simple, monocrystalline materials. However, solid electrolytes currently used in various electrochemical devices, show an amorphous character, at least partially. Therefore, various concepts and theories have been developed for the description of such materials. Some of the problems that have to be accounted for are outlined below.

<u>*Macroscopic scale.*</u> At a macroscopic scale, the information on the ionic conductivity of solid electrolytes comes mainly from the impedance spectroscopy measurements (see Section 2.2). These measurements are sensitive to structural features, however, they provide only the total ionic conductivity. Resolving the contributions of each possible component to the obtained results requires some empirical hypotheses, careful modeling of each possible charge transfer mechanism, and fitting procedures. Despite continuous efforts, a theory for conduction in amorphous solids is not yet satisfactorily established. This is mainly because compositional and/or structural inhomogeneities, including impurities, can dominate the macroscopic ionic conductivity. The obvious examples of such inhomogeneities in polycrystalline materials are grain boundaries between identical crystals or contact surfaces between crystals of different orientations or misaligned grains. Another feature that strongly affects ionic conductivity in solid electrolytes is poor physical contact between neighboring solid particles, leading to porosity and inhomogeneous current densities, detrimentally affecting the conductivity of the sample. To get better insight into the fundamentals of solid-state electrolytes, the interested reader is referred to [6, 7].

Bibliography

[1] Bagotsky VS. Fundamentals of electrochemistry. Wiley & Sons Inc., 2006.
[2] Fedorov MV, Kornyshev AA. Ionic liquids at electrified interfaces. Chem. Rev. 2014, 114(2), 978–2036.
[3] Lee CP, Ho KC. Handbook of ionic liquids, properties, applications and hazards. New York, Nova Sci Publishers, Inc., 2012.
[4] Matsumoto H. Electrochemical aspects of ionic liquids. Ohno H, ed. Hoboken, NJ, John Wiley & Sons, Inc., 2005.
[5], Dong K, Liu X, Dong H , Zhang X, Zhang S. Multiscale studies on ionic liquids. Chem. Rev. 2017, 117, 6636–6695.
[6] Famprikis T, Canepa P, Dawson JA, Saiful Islam M, Masquelier C. Fundamentals of inorganic solid-state electrolytes for batteries. Nat. Mater. 2019, 18, 1278–1291. doi: 10.1038/s41563-019-0431-3.
[7] Karabelli D, Birke KP, Weeber M. Performance and cost overview of selected solid-state electrolytes: Race between polymer electrolytes and inorganic sulfide electrolytes. Batteries 2021, 7, 18. doi: 10.3390/batteries7010018.

1.4 Membranes and membrane potentials

1.4.1 Equilibrium potentials

The formation of a potential drop across metal | electrolyte solution interphase – the electrical double layer, can be considered a straightforward example of equilibrium potential drop. This is because the movement of electrons in the metal phase and ions in the electrolyte is so fast that any perturbation introduced to the system, such as polarization change, will readjust the structure of the e.d.l. much faster than the diffusion-controlled electrochemical process (see Scheme 1.1 and its description in the text). Even if the electrode process is irreversible, the substrate and product species, by virtue of their concentrations, do not affect significantly the electrical double-layer structure – so it remains essentially at equilibrium.

Apart from this situation, we can imagine and build systems in which, even though there is no external polarization or redox reactions, the potential drop across an interface will be formed. The key word here is the "interface." Thus, let us take into consideration a system, in which there are two aqueous electrolytes with equal concentrations of KCl (or other salt, acid, or hydroxide) that are separated by a semipermeable membrane. The ions of this salt can move freely through the membrane, so in this particular case, such a membrane is called a diaphragm. On the contrary, a semipermeable membrane means that it allows only certain ions to pass through it, but other ions cannot pass through such a membrane. The membrane can be porous and its semipermeability (permselectivity) may stem from the structural features, such as the functional groups embedded inside its matrix, the pore size, or the charge of the membrane surface and pores. Well-known examples are the dialytic membranes used in clinics and laboratories. So, let us put them to work, as in Fig. 1.23.

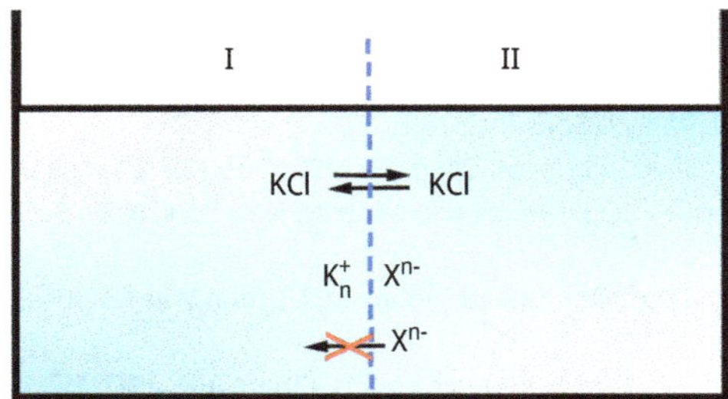

Fig. 1.23: The semipermeable membrane separating two aqueous solutions of KCl of initially equal concentrations in compartments **I** and **II**. The initial equilibrium condition is established. Next, the potassium salt of a polyvalent anion, $K^+{}_nX^{n-}$, is added to compartment **II**. See text for details.

We have already discussed above the equilibrium conditions under constant p, and T. This was done, for example, in the case of electrons in the contacting metal 1 and metal 2, eq. (1.37). This condition is valid also for any ionic species "i" capable of penetrating the semipermeable membrane. For these, we can write for our KCl salt:

$$\tilde{\mu}_i^I = \tilde{\mu}_i^{II} \tag{1.135}$$

Using the definition of electrochemical potential, we will have for both potassium and chloride ions on both sides (I and II) of the semipermeable membrane:

$$\mu_{K^+}{}^I + z_{K^+}F\varphi^I = \mu_{K^+}{}^{II} + z_{K^+}F\varphi^{II} \tag{1.136}$$

and for anions:

$$\mu_{Cl^-}{}^{I} + z_{Cl^-} F\varphi^{I} = \mu_{Cl^-}{}^{II} + z_{Cl^-} F\varphi^{II} \tag{1.137}$$

Introducing next the definition of chemical potentials for both ionic species and keeping in mind that $z_{K^+} = 1$, whereas $z_{Cl^-} = -1$, we will have after some regrouping:

$$\varphi^{II} - \varphi^{I} = \frac{RT}{F} ln \frac{c^{I}_{K^+}}{c^{II}_{K^+}} = \frac{RT}{F} ln \frac{c^{II}_{Cl^-}}{c^{II}_{Cl^-}}, \tag{1.138}$$

or, more generally

$$\varphi^{II} - \varphi^{I} = -\frac{RT}{z_i F} ln \frac{c^{II}_{i}}{c^{I}_{i}} \tag{1.139}$$

Please note, that because there are two aqueous solutions on both sides, therefore, the standard chemical potentials of each ionic species, $\mu_i{}^0$, are the same on both sides and therefore they cancel out. However, if we consider the interface between the two immiscible solutions of the same salt that can be dissolved in both solvents (e.g., water | nitromethane system), we cannot state that the standard potential of species "*i*" is the same in both solvents. Therefore, the $\mu_i{}^{0,I} \neq \mu_i{}^{0,II,}$ and the above equation should include also a term taking into account such a situation. The difference in standard chemical potentials defines the partition coefficient, K_p of the extraction process frequently used in purification procedures:

$$\varphi^{II} - \varphi^{I} = -\frac{\mu^{0,II} - \mu^{0,I}}{z_i F} - \frac{RT}{z_i F} ln \frac{c^{II}_{i}}{c^{I}_{i}} \tag{1.140}$$

$$K_p = -\frac{\mu^{0,II} - \mu^{0,I}}{z_i F} \tag{1.141}$$

Anyway, in our system of aqueous solutions separated by a semipermeable membrane, the above equilibrium equation is called the Donnan equation. Since the KCl aqueous concentrations are equal on both sides of a membrane, the resultant potential difference is zero.

$$C^{I}_{K^+} = C^{I}_{Cl^-}$$

$$C^{II}_{K^+} = C^{II}_{Cl^-}$$

and

$$\varphi^{II} - \varphi^{I} = 0$$

This initial situation will change if we introduce to the right-hand side compartment a given concentration of potassium salt of multivalent, large anion, such as polymeric anion or protein, $K^+_nX^{n-}$, whose anion X^{n-} is large enough to be blocked by the membrane. For this ion, the membrane is impermeable, whereas K^+ and Cl^- still can move through the membrane. The electroneutrality condition will require the influx of K^+ ions with simultaneous efflux of Cl^- ions from compartment II, to achieve finally:

$$C^{II}_{K^+} = C^{II}_{Cl^-} + X^{n-,II}$$

From this, one can easily get that $C^{II}_{K^+} > C^{I}_{K^+}$, and $C^{II}_{Cl^-} < C^{I}_{Cl^-}$.

Therefore, the Donnan equilibrium potential established across a semipermeable membrane in such a system will be negative, using the notation: $\varphi^{II} - \varphi^{I} < 0$.

Of course, mirroring the system from Fig. 1.23, that is, adding the salt $K^+_nX^{n-}$ to the left compartment (instead of the right one) will result in a positive value of the Donnan potential. Please note that in the discussion above, the structure of a semipermeable membrane is unimportant. Its role is to mechanically separate polyanion X^{n-} from its penetration through the membrane.

Let us go now a step further. What happens right after introducing $K^+_nX^{n-}$, but before the equilibrium potential is established? This situation is described below.

1.4.2 Nonequilibrium potentials

For such a situation, we should use the full form of the Nernst–Planck equation, solving it with the help of Faraday's law under the electroneutrality conditions, for charged species of valence $z_i \neq 0$, able to move across the membrane

$$J_i = -u_i RT \frac{dc_i}{dx} - u_i c_i z_i F \frac{d\phi}{dx} \tag{1.142}$$

$$j_i = z_i F J_i = -z_i u_i \, RTF \frac{dc_i}{dx} - u_i c_i z_i^2 F^2 \frac{d\varphi}{dx} \tag{1.143}$$

Electroneutrality requires that the ionic current passing across the membrane at any time equals zero, therefore:

$$\sum_i J_i = \sum_i j_i = 0 \tag{1.144}$$

$$-RTF \sum_i z_i u_i c_i \frac{dlnc_i}{dx} - F^2 \sum_i z_i^2 u_i c_i \frac{d\varphi}{dx} = 0 \tag{1.145}$$

After regrouping:

$$RTF \sum_i z_i u_i c_i \frac{dlnc_i}{dx} = -F^2 \sum_i z_i^2 u_i c_i \frac{d\varphi}{dx} \tag{1.146}$$

If we now assume that the potential drop across the membrane is the dependent variable (one can equally well solve it with concentration profile as the dependent variable), we will get:

$$\frac{d\varphi}{dx} = -\frac{RT}{F}\left(\sum_i z_i u_i c_i \frac{dlnc_i}{dx}\right) \Big/ \sum_i z_i^2 u_i c_i \tag{1.147}$$

Let us now define the so-called ionic transference number (ion transfer number), t_i, as:

$$t_i = \frac{|z_i|U_i c_i}{\sum_j |z_j|U_j c_j}, \tag{1.148}$$

which describes how much charge is carried out by a given ion, "*i*," compared to the sum of all other ions "*j*" present in the solution, or, in other words, what is the fraction of the total electric current carried in an electrolyte by a given ionic species "*i*." Here $U_i = |z_i| F u_i$ is called electrolytic mobility.

After integration of the above differential equation from side I to side II, and assuming constant values of $\boldsymbol{c_i}$ and $\boldsymbol{t_i}$, being independent of distance from the membrane, for the univalent electrolyte we can obtain:

$$\varphi^{II} - \varphi^{I} = \Delta\varphi_L = -\frac{RT}{F}\left[\left(\frac{t_+}{z_+}\right) + \left(\frac{t_-}{z_-}\right)\right] ln\frac{c^{II}}{c^{I}} \tag{1.149}$$

or, keeping in mind the negative sign of z_-, in the case of monovalent ions:

$$\Delta\varphi_L = -\frac{RT}{F}[t_+ - t_-] ln\frac{c^{II}}{c^{I}} \tag{1.150}$$

The above equation, defines the so-called liquid junction potential, $\Delta\varphi_L$, and tells us that the value and sign of the potential developed across the semipermeable membrane depend on the difference between the ionic transference numbers (or their mobilities) in given electrolytes. If $c^{II} > c^{I}$ and cations are moving faster across the membrane compared to anions ($t_+ > t_-$), the value of $\Delta\varphi_L$ will be negative. Moreover, the properties of such a membrane are irrelevant according to this approach. However, most of the membrane separators used in the construction of electrochemical batteries, fuel cells, and electrochemical sensing devices require separators whose performance strongly depends on their structure and are specially tailored for their applications.

Let us have a closer look at the interface between the aqueous solutions II and I, differing in their composition. It seems logical, that a transition layer will develop within the separating membrane, within which the concentrations of each component "*i*" can exhibit a smooth, linear change from their values, $c_{i,o}$ and $c_{i,i}$ (see Fig. 1.24):

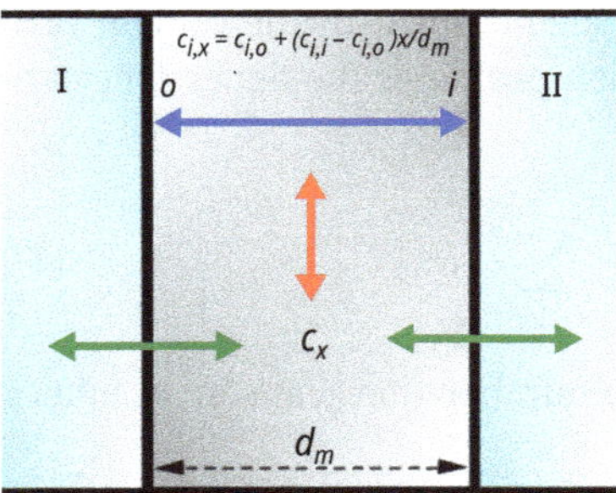

Fig. 1.24: A scheme of a permselective membrane of thickness d_m, separating compartments **I** and **II**. Linear concentration gradient is assumed between sides "*o*" and "*i*" of this membrane (blue arrow), with two Donnan equilibrium potentials developed across both interfaces (green arrows). An ion exchange reaction between bound and free ions within the membrane is also shown (red arrow). Other mechanisms can also be involved in the performance of such a system. See the text for a detailed description of this scheme.

$$c_{i,x} = c_{i,o} + (c_{i,i} - c_{i,o})x/d_m. \tag{1.151}$$

Here, $c_{i,x}$ is the concentration of "*i*" at point *x* inside a membrane, d_{m} is the thickness of a membrane, whereas $c_{i,i}$ and $c_{i,o}$ are the concentrations of "*i*" at the membrane surface "*i*" and "*o*." This summarizes the so-called Henderson approach, and introduced to eq. (1.150) for univalent electrolytes on both sides of such a membrane, yields the Lewis–Sargent equation:

$$\Delta\varphi_L = +\frac{RT}{F}ln\frac{u_+ + u_{-,II}}{u_+ + u_{-,I}} = \frac{RT}{F}ln\frac{\Lambda_{II}}{\Lambda_I} \tag{1.152}$$

However, this approach does not account for the role of membrane structure in its selectivity. Here, we may want to recall a rectifying behavior of a nanoporous membrane containing charged nanochannels. Because of the presence of the electrical double layer due to the charges on the channel walls, if such membrane separates two electrolytes, at sufficiently small channel diameters, the co-ions will be repelled from the channels, allowing only the flow of counterions through the channel, thus providing very high selectivity of ionic conductance for such separator (see Section 1.1.1, Fig. 1.8)

Let us now go a step further and consider a membrane of a given thickness with its chemical structure capable of exchanging ions between the membrane constituents and only selected ions from the aqueous electrolyte solutions. These are the so-called ion exchange membranes. The unique properties of ion exchangers can be explained by their structure – a framework carrying functional groups bearing positive or negative charge, compensated by ions of opposite charge (counter-ions). The counter-ions can move freely, and therefore they can easily be replaced by other ions with the same sign and at least similar affinity to the functionalities of the structural framework. Depending on the interactions and affinities between the functional groups and counter-ions, important characteristics of an ion exchange membrane can be designed – namely – its selectivity, defined as the ability of an ion exchange membrane to distinguish between different species of counter-ions.

The best-known and used example of such a system is a glass, pH-sensitive electrode. It is beyond the scope of this book to describe in detail the functioning of glass electrodes as pH-sensing (proton exchange) devices. The interested reader is referred to the literature examples at the end of this section [1, 2]. Here, we will only outline processes involved in the functioning of a tiny, glassy, noncrystalline solid membrane, leading to the potential development responsive to the concentration of hydrogen ions in the solution. This solid, based on a silicon oxide framework, is doped with a relatively high concentration of alkali and/or alkaline metal oxides. The metal ions possess significant mobilities within the framework, whereas the negatively charged functional groups are bound within the silicon oxide framework, and therefore, their mobility equals zero. In contact with water, the glass hydrates with metal ions quantitatively being exchanged for hydrogen ions. Therefore, the following system and processes can be exemplified in the case of a typical "sodium" glass membrane, and Na^+, and H^+ ions present in an aqueous solution:

Phase I ***o* membrane *i*** **Phase II**

Na^+ ($c_{Na^+,I}$), H^+ ($c_{H^+,I}$) | Na^+(c_{Na}), H^+ (c_H) | Na^+($c_{Na^+,II}$), H^+ ($c_{H^+,II}$)

Both types of ions can diffuse into the membrane with different mobilities u_{Na} and u_H. The membrane has a defined number of negatively charged, structural sites, where Na^+ and H^+ are the counterions and can be bound to these sites. The total potential developed across such a system is a sum of contributing potential differences:

$$\Delta\varphi_M = \varphi^{II} - \varphi^{I} = (\varphi^{II} - \varphi_i) + (\varphi_i - \varphi_o) + (\varphi_o - \varphi^{I}) \tag{1.153}$$

Moreover, the equilibria between ions present in the solutions and those bound in the membrane can be described by two equilibria: the Donnan equilibria between the ions in the solutions and those free, unbound within the membrane, and the ion exchange equilibrium between bound and free ions within the membrane.

The Donnan equilibria:

$$\varphi^{II} - \varphi_i = -\left(\frac{RT}{F}\right)\ln\left(\frac{c_{Na^+,II}}{c_{Na^+,i}}\right) = -\left(\frac{RT}{F}\right)\ln\left(\frac{c_{H^+,II}}{c_{H^+,i}}\right) \tag{1.154}$$

$$\varphi_o - \varphi^{II} = \left(\frac{RT}{F}\right)\ln\left(\frac{c_{H^+,I}}{c_{H^+,o}}\right) = \left(\frac{RT}{F}\right)\ln\left(\frac{c_{Na^+,I}}{c_{Na^+,o}}\right) \tag{1.155}$$

where $c_{Na^+,i}$, $c_{H^+,i}$, $c_{H^+,o}$, and $c_{Na^+,o}$ are the concentrations of free ions in the membrane, respectively, whereas $c_{Na^+,II}$, $c_{H^+,II}$, $c_{H^+,I}$, and $c_{Na^+,I}$ are the concentrations of these ions in solutions II and I (phase II and I), respectively.

The ion exchange equilibria of the type Na^+ (free) ↔ Na^+ (bound) and H^+ (free) ↔ H^+ (bound) are:

$$K_{Na^+} = C_{Na^+,o}/c_{Na^+,o} = C_{Na^+,i}/c_{Na^+,i}, \tag{1.156}$$

$$K_{H^+} = C_{H^+,o}/c_{H^+,o} = C_{H^+,i}/c_{H^+,i}, \tag{1.157}$$

where $C_{Na^+,o}$, $C_{Na^+,i}$, $C_{H^+,o}$, and $C_{H^+,i}$ denote the concentrations of bound ions, assuming their ideal behavior. Moreover, the sum of C_{Na^+} and $C_{H^+}$is constant as the constant is the number of ion exchange sites in the membrane (negatively charged, structural sites).

The ratio K_{Na^+}/K_{H^+} is obviously the equilibrium constant of the ion exchange reaction:

$$Na^+\,(\text{free}) + H^+(\text{bound}) \leftrightarrow Na^+(\text{bound}) + H^+(\text{free})$$

Now we have to describe the potential difference inside the membrane, between sides "*i*" and side "*o*" $(\varphi_i - \varphi_o)$. This is the diffusion potential and we will use again the Nernst–Planck equations:

$$J_{Na^+} = -u_{Na^+}RT\left(\frac{dC_{Na^+}}{dx}\right) - u_{Na^+}C_{Na^+}F\left(\frac{d\varphi}{dx}\right) \tag{1.158}$$

$$J_{H^+} = -u_{H^+}RT\left(\frac{dC_{H^+}}{dx}\right) - u_{H^+}C_{H^+}F\left(\frac{d\varphi}{dx}\right) \tag{1.159}$$

Since we assumed that there is no net current passing through the membrane and that the mobilities of structural anions (functional negative groups of the glass matrix) are equal to zero, $u_- = 0$, we can write that $J_{H^+} + J_{Na^+} = 0$. Therefore:

$$\frac{RTd(u_{Na^+}C_{Na^+} + u_{H^+}C_{H^+})}{dx} + (u_{Na^+}C_{Na^+} + u_{H^+}C_{H^+})F\left(\frac{d\varphi}{dx}\right) = 0 \tag{1.160}$$

Integrating the above equation from $x = o$ to $x = i$, we will finally get the diffusion potential inside the membrane:

$$\varphi_i - \varphi_o = -\frac{RT}{F}\ln\frac{u_{Na^+}C_{Na^+,i} + u_{K^+}C_{H^+,i}}{u_{Na^+}C_{Na^+,o} + u_{K^+}C_{H^+,o}} \tag{1.161}$$

Now we are ready to substitute all terms in eq. (1.153) for $\Delta\varphi_M$ with eqs. (1.154), (1.155), (1.161), and ion exchange equilibria (eqs. (1.156) and (1.157)). This will yield the following relation for the membrane potential, $\Delta\varphi_M$:

$$\Delta\varphi_M = -\frac{RT}{F} ln \frac{u_{Na^+}K_{Na^+}c_{Na^+,II} + u_{H^+}K_{H^+}c_{H^+,II}}{u_{Na^+}K_{Na^+}c_{Na^+,I} + u_{H^+}K_{H^+}c_{H^+,I}} \tag{1.162}$$

By dividing the nominator and denominator of this equation by $u_{Na^+}K_{Na^+}$, and setting $\frac{u_{H^+}K_{H^+}}{u_{Na^+}K_{Na^+}} = K_{H^+/Na^+}$, we get the so-called Nikolsky–Eisenman equation:

$$\Delta\varphi_M = -\frac{RT}{F} ln \frac{c_{Na^+,II} + K_{H^+/Na^+}c_{H^+,II}}{c_{Na^+,I} + K_{H^+/Na^+}c_{H^+,I}} \tag{1.163}$$

K_{H^+/Na^+} is called the selectivity constant of an ion-selective membrane for ions H^+ versus ions Na^+ as in our case, or in general, for ions of one type with respect to ions of different species. In the case of large K_{H^+/Na^+} and $\frac{c_{H^+,I}}{c_{Na^+,I}}$ sufficiently high, we can get Nerstian slope of $\Delta\varphi_M$ as a function of $logc_{H^+,I}$:

$$\Delta\varphi_M = const. + \frac{2.303RT}{F} logc_{H^+,I} \qquad (K_{H^+/Na^+} \gg \frac{c_{Na^+,I}}{c_{H^+,I}}) \tag{1.164}$$

The membrane is then specifically selective to H^+ ions. Thus, using the example of glass and pH electrodes, we were able to explain the selectivity behavior of ion exchange membranes. Of course, depending on the design and purpose of such membranes, the mechanisms involved in their performance will differ. The interested reader is directed to the literature already mentioned above [1, 2].

Here, we should mention other types of solid-state membranes widely used in industrial electrochemical applications, energy conversion, and storage technologies, and environmental monitoring systems. Examples of such membranes include metal oxides, metal sulfides, and mixed systems. Among metal oxides with high ionic conductivity, a prerequisite for their applications as all-solid-state electrolytes described in the previous chapters, zirconia-based membrane separators (ZrO_2) are widely utilized. For better stability and electrochemical performance zirconium oxide is frequently stabilized with yttrium (YSZ, yttria-stabilized zirconia) or scandium oxides (SSZ, scandia-stabilized zirconia). In these solid crystalline structures, the mobile charges are oxygen dianions in the crystalline network. At sufficiently high temperatures, the mobile oxygen ions are responsible for charge transfer within the solid-state membrane and the overall system separating two compartments of, for example, partial oxygen pressures pO_2^I and pO_2^{II}, can be outlined in a way similar to that already shown above:

Compartment I *o* **membrane** *i* **Compartment II**
O_2 (pO_2^I, gas) | YSZ(or SSZ)mobile O^{2-} ions | O_2(pO_2^{II}, gas)

In this case, the developed membrane potential will be given by:

$$\Delta\varphi_M = t\frac{RT}{2nF}ln\frac{pO_2^{II}}{pO_2^{I}}$$

where ***t*** is an average ionic transference number (eq. (1.148)), ***n*** is the valence of gas being responsible for potential development (in this case oxygen with **4e** redox reaction), and other parameters have their usual meanings.

Let us now summarize the picture that emerges from the above section, from the point of view of utilizations of such ion-exchange/ion conducting membranes in applied electrochemistry, including these applications that are involved in sustainability and a cleaner environment ("green" electrochemistry).

1. First, the ion exchange processes can be driven by concentration gradients, an electric field applied across the membrane (electrodialysis), or both. In such situations, the ions are not only exchanged but also cross the membrane.
2. Depending on their applications, the semipermeable membrane can be inorganic or organic ion-conducting solids (e.g., in fuel cells, electrolyzers), organic-inorganic composites, metal-organic frameworks, hydrogen-permeable metal membranes, but sometimes also a thin layer of supported solution separating two electrolytes (e.g., in ion-selective sensor electrodes).
3. When the membrane separates two compartments, it can work continuously. This is different from processes used in analytical applications and separations, where the ion exchange column or bed operates discontinuously with the need for regeneration.
4. In the case of membranous systems, some synthetic applications are possible, such as the recovery of compounds from wastewater or desalination of salt water.
5. When such a membrane is placed directly between the anode and cathode in an electrolyzer or a battery it acts as an electrolyte, sometimes the only electrolyte between electrodes.
6. Power can be generated in fuel cells by supplying suitable substrates/reactants to the electrodes separated by such a membrane.
7. Power can be stored when the two electrodes separated by such membrane are charged (rechargeable batteries, capacitors, supercapacitors, and hybrid cells).
8. These membranes can be used for various types of sensors for real-time control of the environment, industrial processes, and pollution detection/prevention (Part IV, Chapter 13).

Among the above applications, fuel cells and industrial electrolysis for hydrogen and oxygen "green" generation are the most known and demanding for the ion exchange membranes from the point of view of low electrical resistance, mechanical and chemical resistance to pressure, temperature, and other operating conditions. The design of each membrane's characteristics strongly differs per its application, due to the working conditions of the device utilizing such a membrane. Lately, the so-called bipolar membranes (BPMs) are gaining more interest, particularly in energy conversion and storage technologies. These membranes are composed of cation- and anion-exchange layers laminated together with the abrupt transition of ion-exchange behavior in an interfacial layer. The BPM provides conditions to operate in different electrolytes on either side, whereas the interfacial layers provide space for nanostructured catalysts for energy conversion and storage processes. The examples of applications of membranous systems designed for various domains of electrochemistry, including "green" electrochemistry, will be given later in this book.

Bibliography

[1] Ion-selective electrodes: Overview; Eric Bakker. In: Encyclopedia of analytical science (third edition). 2019.

[2] Głąb S., Hulanicki A. Ion-selective electrodes: Glass electrodes. In: Reference module in chemistry, molecular Sci. Eng. 2013.

2 Selected electrochemical methods applied in analytical chemistry and material science

Applied electrochemistry utilizes numerous methods in its efforts to improve progressively our quality of life through the understanding of charge transfer phenomena. These efforts lead to the construction and design of new devices and systems that can be used in industry (e.g., electrical energy storage and supply, catalysis and corrosion), and also in more personalized applications, such as batteries for smartphones, laptops, and solar panels. Electroanalytical techniques provide unique information on mechanisms and systems that are currently involved in the necessary development of the above areas. Both the instrumentation and theoretical fundamentals have been advanced such that nonspecialists can easily use them in search of further development in technology, industry, and everyday life. Below we lay a brief background of selected electrochemical methods, commonly used in material science for:

- characterization of interfaces;
- characterization of electrode processes, kinetics, mass transfer;
- corrosion studies;
- coatings and paints;
- investigations of membrane transport phenomena;
- electrocatalysis, adsorption, and electrosorption studies;
- studies of cells, batteries, fuel cells, supercapacitors;
- conductive polymers;
- semiconductors;
- photoelectrochemistry and others.

2.1 Transient methods

Before we start discussing selected electrochemical methods in this chapter, we have to summarize briefly some basics of the design of electrochemical cell that is typically used for such studies, as well as the electrode response to the applied potential, E, or introduced charge, Q, in terms of the measured current, i. First of all, since we already discussed that at the electrolyte/solution interface the electrical double layer is formed, accumulating and separating ionic charges, it behaves as a condenser. This is because if one polarizes the condenser with potential, E, it will accumulate/separate charges, Q, proportionally to its capacitance, C:

$$Q = C \times E \tag{2.1}$$

https://doi.org/10.1515/9783111160986-002

Assuming the constant value of C and differentiating the above versus time, we can get:

$$\frac{dQ}{dt} = i_c = C \times \frac{dE}{dt} \tag{2.2}$$

Since the derivative of Q versus time is current, this equation tells us that any change in potential in time $\left(\frac{dE}{dt}\right)$ will result in charging current, i_c, passing through the electrochemical cell. This current overlaps and adds to the so-called Faradaic current, being the result of the charge transfer reaction at the electrode. Therefore, it is necessary to separate these two currents in order to gain insights into the mechanisms of charge storage (e.g., supercapacitors) and charge transfer (e.g., electrocatalysis and electrodeposition). For now, it is sufficient to keep this in mind, but we will return to this in more detail later in the text.

Thus, in the absence of a charge transfer reaction (ideally polarizable electrode), the current flow results in charging (or discharging) the capacitance of the electrical double layer and is also called the capacitive current.

2.1.1 Capacitive current

The electrochemistry in material science usually begins in the electrochemical cell. Its design is dictated by the organization and the number of electrodes used, as well as their purpose. Typically, there are three electrodes: working electrode (WE), reference (Ref) electrode, and the counter/auxiliary electrode (CE). The CE should not affect the investigated flow of current, therefore the area of this electrode should be much larger than that of the working electrode. The working electrode is polarized against the Ref of constant potential, independent of the process being investigated on the working electrode. The current flow (both charging and Faradaic) as a response to the WE polarization/charging is measured between the WE and CE electrodes. Because the capacitance of other components of an electrochemical cell is much larger, they do not contribute to the investigated current flow. Therefore, this type of situation can be represented by an electrical equivalent circuit, containing the capacitance of the e.d.l. of WE electrode, C_{dl}, in series with the ionic (ohmic) resistance of the solution, in which all three electrodes are immersed, R_s. Now we can introduce the relationships between the electrode potential and capacitive current, i_c, for ideally polarizable electrode in the three most frequently used analytical methods as follows:

(a) Potential step (chronoamperometry), assuming that at time = 0 there is no charge at the electrode surface:

$$i_c = \left(\frac{E}{R_s}\right)\exp(-t/R_sC_{dl}) \tag{2.3}$$

This equation tells us, that after applying a potential step of magnitude, E, an exponentially decaying current will flow with its time constant. This is observed as current "spike" within the first few tenths of a second after application of a given potential value.

(b) Current step, assuming that at t = 0, i = 0 and then the RC circuit is charged by a constant current, i_c, giving rise to the potential E:

$$E = i_c\left(R_s + \frac{t}{C_{dl}}\right) \tag{2.4}$$

This relation allows for the determination of C_{dl}. This is also the principle of the determination of chemisorbed hydrogen on platinum or other catalysts by its oxidation.

(c) Potential scan (linear scan), where the potential is scanned from its initial value, E_i, to a final value, E_f, with a rate $v = \frac{dE}{dt}$ over the time t (the final potential value is determined by the duration of scan, t):

$$i_c = vC_{dl} + \left[\left(\frac{E}{R_s}\right) - vC_{dl}\right]\exp(-t/R_sC_{dl}) \tag{2.5}$$

It is easily seen that the current response for the ideally polarizable electrode (no Faradaic current) contains two components: a steady-state, vC_{dl}, and a transient one. Proportionality of i_c to v enables determination of not only the capacitance of e.d.l. but also any charge accumulated at the surface, such as electrochemically active surface layer, adsorbed atoms. This proportionality is of importance also for the thin layer cells. It may happen that during cyclic scans and cyclic voltammetry (CV) experiments, particularly at high scan rates, the capacitive current may exceed the Faradaic response, being proportional to $v^{1/2}$ as we will see next.

The most important current response directly related to the redox processes on WE electrode is the Faradaic response. In this case, in the equivalent circuit representing our experimental setup, we should introduce the charge-transfer resistance, R_{ct}, connected in parallel to the WE capacitance, allowing for the current flow across the electrode/solution interface. Now, we are ready to arrange all equivalent elements of our experimental setup in a simple scheme shown in Fig. 2.1.

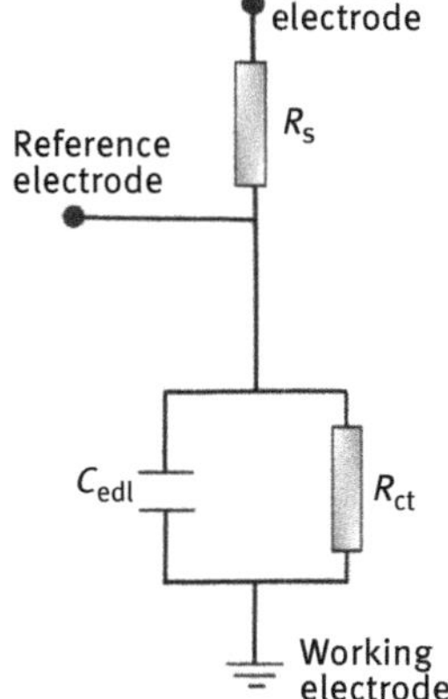

Fig. 2.1: A simple electronic scheme (equivalent circuit) of the electrochemical cell. R_s – electrolyte resistance, R_{ct} – charge transfer resistance related to the redox reactions (faradaic current) occurring at the working electrode (WE), C_{edl} – differential capacitance of the electrical double layer at the WE electrode.

2.1.2 Faradaic current

Most of the voltammetric techniques used in applied electrochemistry are based on a controlled change of the potential of the working electrode. Therefore, it is not possible to attain steady-state conditions, even for a longer period of time. In the case of transient techniques, such as chronoamperometry, linear scan, or cyclic voltammetry (CV), the steady-state conditions can never be reached. So, we cannot rely directly on the Nernst–Planck equation, nor on the first Fick's law, discussed in the previous part. However, we can use the second Fick's law that considers the flux of matter in time domain, t. For the case of a semi-infinite diffusion along the x-axis, normal to the electrode surface (x is also the distance from the electrode surface), the second Fick's law will read:

$$\frac{\partial c_{\mathrm{Ox}}(x,t)}{\partial t} = D_{\mathrm{Ox}} \frac{\partial^2 c_{\mathrm{Ox}}(x,t)}{\partial x^2} \tag{2.6}$$

This partial differential equation can be solved by setting boundary conditions that are appropriate for a selected electrochemical technique.

As mentioned earlier, the most frequently used transient techniques are based on controlled potential; therefore, we later focus on chronoamperometry and CV. However, the interested reader is encouraged to search in the suggested readings at the end of this paragraph.

(a) Potential step (chronoamperometry)

This is illustrated in the case of a reduction reaction in Fig. 2.2. This figure contains also the boundary conditions necessary to solve second Fick's law in order to get the current response over time.

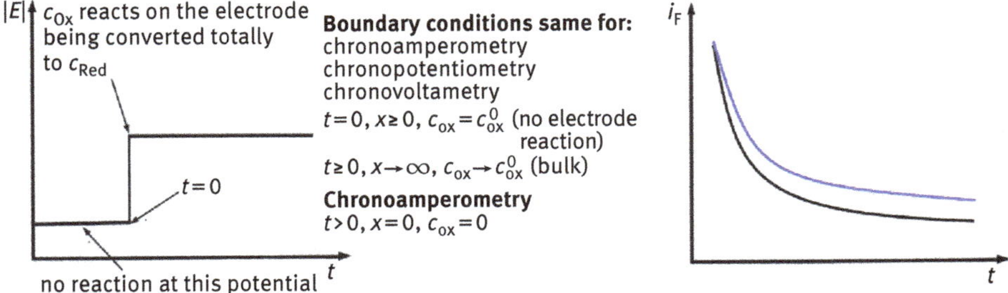

Fig. 2.2: Chronoamperometry. Left: potential step of magnitude sufficient to reduce (or oxidize) all substrate arriving to the electrode surface. Right: current response to this potential step described by the Cottrell equation. Blue curve – the effect of semispherical diffusion, eq. (2.9). See text.

Solving eq. (2.6) with these conditions will yield the so-called Cottrell equation:

$$i_F(t) = nFADc^0 \frac{1}{(\pi Dt)^{1/2}} \tag{2.7}$$

The term found in the denominator of this equation,

$$\delta_D = (\pi Dt)^{1/2} \tag{2.8}$$

describes the time evolution of the diffusion layer into the bulk of the solution; it increases with $t^{1/2}$.

The above equations are valid only for planar electrodes of "infinite" size from the point of view of the diffusion layer thickness, δ_D. For microelectrodes and spherical electrodes, we should change to radial coordinates and the current response will read:

$$i_F(t) = nFADc^0 \left[\frac{1}{(\pi Dt)^{\frac{1}{2}}} + \frac{1}{r_0}\right] \tag{2.9}$$

where r_0 is the radius of the spherical electrode or microelectrode. This equation will become the usual Cottrell equation for $r_0 \to \infty$. It is also important to note that for short t, the spherical contribution is minimal, whereas for longer times the first term in square brackets can be neglected and a steady-state current will flow.

(b) Cyclic voltammetry

CV is a type of potentiodynamic technique. In a CV experiment, the working electrode potential is changed linearly versus time and after the set potential is reached, the WE potential is reversed linearly to return to the initial potential. These cycles of potential scans may be repeated as many times as needed. The current at the working electrode is plotted versus the applied voltage (the WE potential) to give the cyclic voltammogram graph. CV can be used to study qualitative and quantitative information about electrochemical processes under various conditions, such as the presence of intermediates in oxidation–reduction reactions, or the reversibility of a reaction. It can be very useful in the developmental studies of power sources, for example, resolving the processes responsible for power storage and generation. Therefore, in the next paragraph, we will consider in more detail this technique.

Assume that only Ox is present initially; therefore, we will consider the cathodic polarization only (convention). The transport of Ox obeys the Fick's law. The reduction reaction will produce Red species, and then, upon the reverse scan (anodic polarization), the oxidation reaction will also obey the Fick's law:

$$\frac{\partial c_{\mathrm{Ox}}(x)}{\partial t} = D_{\mathrm{Ox}} \frac{\partial^2 c_{\mathrm{Ox}}(x)}{\partial x^2}; \qquad \frac{\partial c_{\mathrm{Red}}(x)}{\partial t} = D_{\mathrm{Red}} \frac{\partial^2 c_{\mathrm{Red}}(x)}{\partial x^2} \tag{2.10}$$

Faradaic current, i_F, is always "conjugated" with capacitive current, i_C, as discussed earlier, because of the charge accumulation at the electrode surface (e.d.l.):

$$i = i_C + i_F = C_{\mathrm{e.d.l.}} \times (dE/dt) + i_F = C_{\mathrm{e.d.l.}} \times v + i_F$$

In order to solve the above partial differential equations, we have to set the following boundary conditions:

1. At $t = 0$, at the electrode surface, $x = 0$, concentration of c_{Ox} is equal to the bulk concentration, while c_{Red} is equal to zero: $c_{\mathrm{Ox},\, x=0} = c^0_{\mathrm{Ox}}$, $c_{\mathrm{Red},\, x=0} = 0$
2. At $t > 0$ and far away from the electrode, c_{Ox} approaches the bulk concentration, whereas c_{Red} approaches zero: $x \to \infty$ $c_{\mathrm{Ox}} \to c^0_{\mathrm{Ox}}$, $c_R \to 0$
3. At $t > 0$, at the electrode surface ($x = 0$) both fluxes of Ox and Red compensate (what arrives, upon reaction must leave):

$$D_{\mathrm{Ox}} \left(\frac{\partial c_{\mathrm{Ox}}(x)}{\partial t} \right)_{x=0} + D_{\mathrm{Red}} \left(\frac{\partial c_{\mathrm{Red}}(x)}{\partial t} \right)_{x=0} = 0$$

4. At the time between the starting point ($t = 0$, E_i) and time of the reversal of the scan, λ, the potential sweep is linear with rate υ: $0 < t \le \lambda$, $E = E_i - \upsilon t$
5. At $t > \lambda$, the linear scan proceeds in the opposite direction with the same rate:

$$E = E_i - \upsilon\lambda + \upsilon(t - \lambda)$$

For reversible, diffusion-controlled processes, one more boundary condition should be introduced, derived directly from the Nernst equation:

$$\frac{c_{\mathrm{Ox},\, x=0}}{c_{\mathrm{Red},\, x=0}} = \exp\left[\frac{nF}{RT}\left(E - E^0\right)\right] \tag{2.11}$$

These boundary conditions can be used to derive the theoretical expression for the current–potential–time relationship, yielding the shape of a reversible voltammogram. It is beyond the scope of this book to present more theory; it can be found in the suggested readings. What is important, however, is the resultant shape of a diffusion-controlled cyclic voltammogram, presenting maximum and minimum peak currents for the reduction and oxidation reaction of a given redox species (Fig. 2.3).

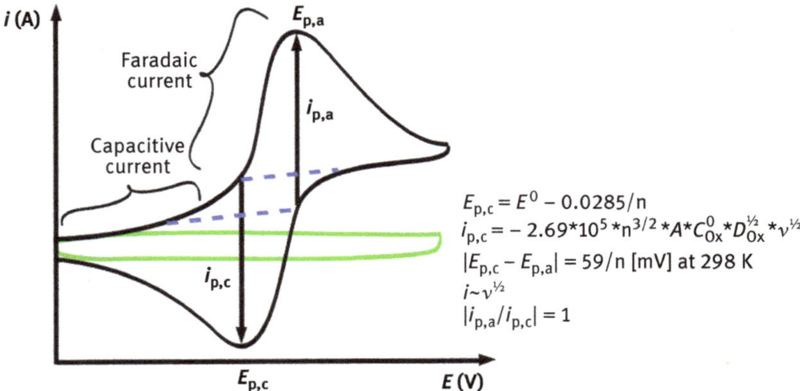

Fig. 2.3: Cyclic voltammogram showing the summary of characteristics for reversible electrode processes. The contributions of capacitive and Faradaic currents to the overall response are distinguished, as well as the so-called background current in the absence of a redox couple (green curve), due to the capacitive contribution of ideally polarizable WE. Reversibility criteria are also summarized.

Any change in CV shape, peak-to-peak separation, or peak currents ratio indicates that the observed electrode process changes from the diffusion-controlled, reversible regime into the irreversible, kinetically controlled one.

Below is the graph summarizing the above techniques and illustrating graphically the shape of the applied potential stimulus (first column), the measured current response (second column), and the concentration profile during the time of the applied potential stimulus. The first row shows the steady-state conditions (first Fick's law), independent of time, with constant diffusion layer thickness.

2.2 Electrochemical impedance spectroscopy

This section describes the basic principles of the electrochemical impedance spectroscopy (EIS), a very powerful electrochemical technique that found its applications in almost all areas of applied electrochemistry.

The classical electrochemical techniques measure charge, current, potential as a function of time, the latter can be expressed in terms of applied potential in potentiodynamic techniques, such as CV. Contrasting to these, EIS reports the measured impedance as a function of frequency (frequency domain) at a constant, selected po-

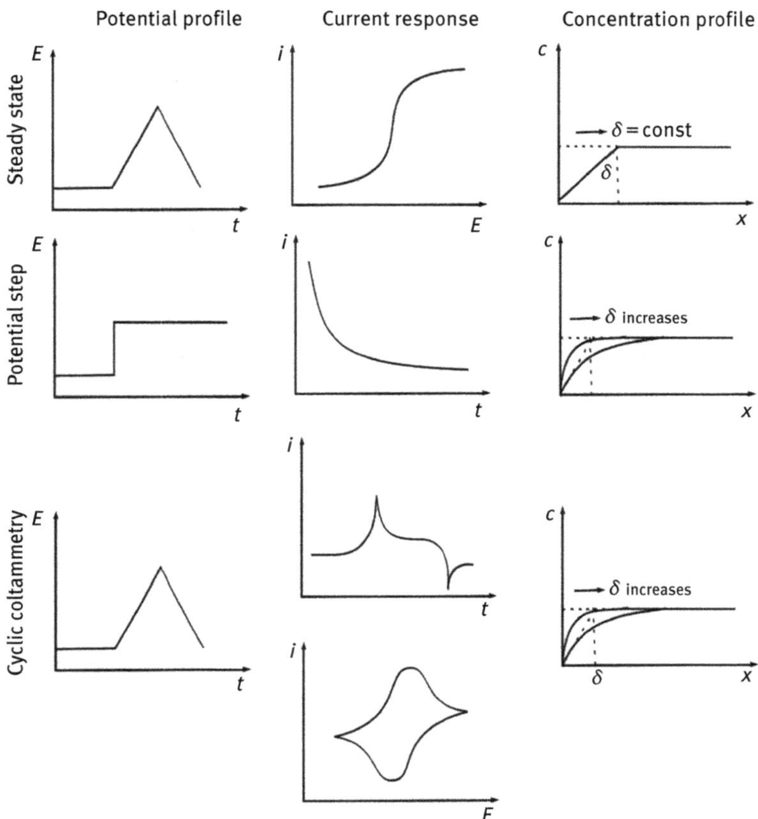

Fig. 2.4: Plots of potential profiles applied to the working electrode (left column), current–potential or/and current–time responses to the applied potential (middle column), and concentration profiles as a function of distance from the electrode surface (right column), for different types of electroanalytical techniques. Please note that in the case of cyclic voltammetry, the *X*-axis can be either time (*t*) or potential (*E*), because both are related by the potential scan rate.

tential. Understanding EIS may pose some difficulties, as it requires some knowledge of Laplace and Fourier transforms and complex numbers; nevertheless, the advantages of EIS are numerous. At each frequency and applied potential, EIS provides a large number of useful information that can be analyzed with the help of the electrical equivalent circuits, representing different electrochemical processes at the electrode interface and its vicinity. Impedance measurements allow us to determine the components of this circuit and their values. These electrical components are related to the physicochemical characteristics of the system. This will be clearly seen from the text and examples that follow.

Let us recall the well-known Ohm's law that relates the resistance of an electrical circuit element known as a resistor, to the current flow ***I*** under the influence of applied voltage *E*:

$$R = \frac{E}{I} \tag{2.12}$$

This relation, however, is valid only for the ideal resistor *R*. The properties of an ideal resistor are as follows:
- The Ohm's law is fulfilled at all voltages *E* and currents *I*
- The value of *R* is independent of frequency of the applied voltage *E*
- The alternating current (AC) and voltage signals are in phase with each other (see Fig. 2.5).

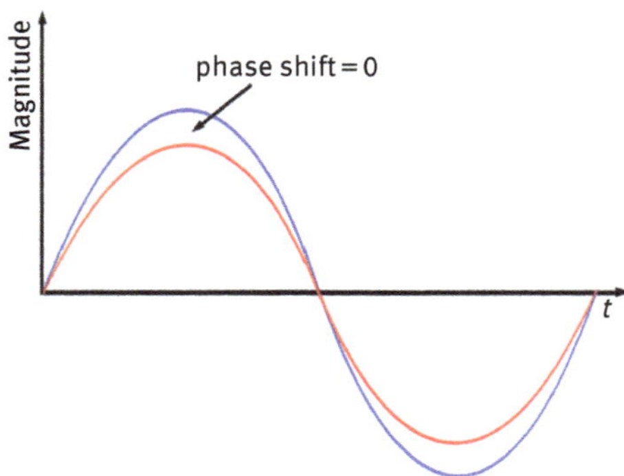

Fig. 2.5: Current and voltage signal profiles in the AC (alternating current) electrochemical impedance spectroscopy (EIS) in the case of pure electrical resistance, *R*, in the studied system. No phase shift between the two is observed.

However, real electrochemical cells and redox processes cannot be represented by the Ohm's law, but require a more general parameter – impedance, *Z*:

$$Z = \frac{E_{AC}}{I_{AC}} \tag{2.13}$$

As in the case of Ohm's law, the impedance *Z* is a measure of the relation between the applied voltage, *E*, and current response, I_{AC},, but unlike the resistance, *R*, it is not limited by the properties of an ideal resistor. And so, the electrochemical impedance is usually measured by applying an AC sinusoidal voltage of small amplitude and frequencies varying from 10^6 Hertz to 0.01 Hertz, to an electrochemical cell and measuring the current response of the investigated system. This current response can then be analyzed as a sum of sinusoidal functions (Fig. 2.6).

When an AC flows through a circuit, the relation between current and voltage across a circuit element is characterized not only by the ratio of their magnitudes, but also the difference in their phases. For example, in an ideal resistor, the moment when the voltage reaches its maximum, the current also reaches its maximum, as shown in the graph (Fig. 2.5) (current and voltage are oscillating in phase). But for a

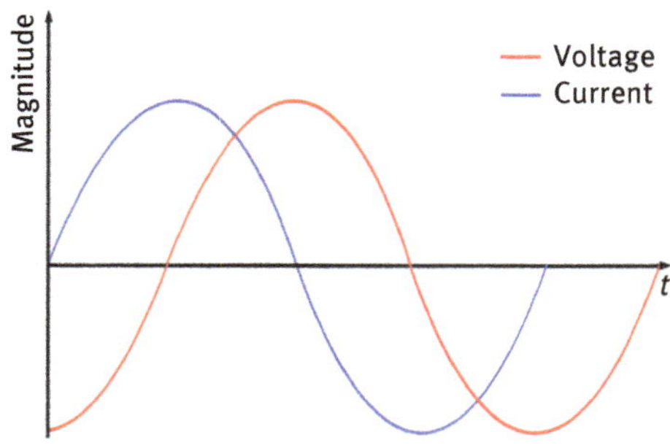

Fig. 2.6: Phase shift of current response for a sinusoidal potential applied to the electrode in the presence of a capacitive component to the studied system. The amplitude change is neglected.

capacitor, the maximum current flow occurs as the voltage passes through zero and vice versa (current and voltage are oscillating 90° out of phase, see Fig. 2.6). Complex numbers are used to keep track of both the phase and magnitude of current and voltage. The excitation signal, expressed as a function of time, can be written as follows:

$$E_t = E_0 \sin(\omega t) \tag{2.14}$$

Here, E_t is the potential at time t, E_0 is the amplitude of the potential (its value at a maximum), and ω is the radial frequency (in radians), related to the signal frequency f in Hertz, as follows:

$$\omega = 2\pi f \tag{2.15}$$

Assuming the linearity or pseudolinearity of current–voltage relation, valid only for a small excitation signals, the current response for the above excitation will also take sinusoidal form and should read:

$$I_t = I_0 \sin(\omega t + \varphi) \tag{2.16}$$

So, the current response I_t is shifted in phase by a phase angle φ and has a different amplitude than I_0. As mentioned earlier, the impedance Z is potential-to-current ratio (as in Ohm's law), so:

$$Z = \frac{E_t}{I_t} = \frac{E_0 \sin(\omega t)}{I_0(\sin\omega t + \varphi)} = Z_0 \frac{\sin(\omega t)}{\sin(\omega t + \varphi)} \tag{2.17}$$

Thus, as potential and current, the impedance is expressed in terms of magnitude Z_0 and phase shift φ. With the help of Euler's relationship:

$$\exp(j\varphi) = \cos\varphi + j\sin\varphi \tag{2.18}$$

we can replace the trigonometric forms of eqs. (2.14)–(2.17) in terms of imaginary relations:

$$E_t = E_0 \exp(j\omega t) \tag{2.19}$$

and the current response:

$$I_t = I_0 \exp(j\omega t - \varphi) \tag{2.20}$$

Finally, the impedance $Z(\omega)$ is a complex number:

$$Z(\omega) = \frac{E_t}{I_t} = Z_0 \exp(j\varphi) = Z_0(\cos\varphi + j\sin\varphi) \tag{2.21}$$

Equation (2.21) tells us that as a complex number, the total impedance of a system, $Z(\omega)$, is a sum of real, Z_{re}, and imaginary, Z_{im}, part:

$$Z(\omega) = Z_{re} + Z_{im} \tag{2.22}$$

or

$$Z(\omega) = Z' + jZ'', \text{where } j = \sqrt{-1} \tag{2.23}$$

2.2.1 Data presentation

Typically, the experimental data obtained during EIS experiments can be displayed in complex plane (imaginary part Z'' vs. real part Z'), called the Nyquist plot, but they can also be displayed as a vector, defined by the impedance magnitude, $|Z|$, and phase angle, φ (phase shift), described above. The vector and the complex quantity are different representations of total impedance and are mathematically equivalent. The graph below shows the representations of EIS data as vector and complex plane for single frequency (Fig. 2.7).

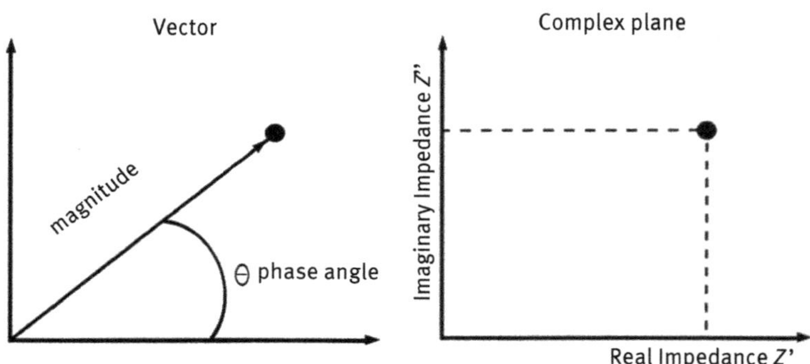

Fig. 2.7: The representations of EIS data as a vector and a complex plane for a single frequency.

Now, let us consider an ideally reversible electrode that can be represented as a simple resistor, R. Then, since in the case of a resistor there is no phase shift, the total impedance will read:

$$Z = \frac{E_t}{I_t} = \frac{E_0 \sin(\omega t)}{I_0(\sin\omega t)} = Z_0 = R \tag{2.24}$$

In complex plane Z' versus Z', it will be represented by a single point, as shown in the simulation for R = 2,000 Ohm (Fig. 2.8). The same data can be also represented as the so-called Bode plots to visualize the effect of the frequency of the applied sinusoidal perturbation on both Z' and φ (phase shift) (Fig. 2.8), respectively. All EIS graphs were obtained with help of a freeware **EIS Spectrum Analyser**, by **Genady Ragoisha**, Research Institute for Physical-Chemical Problems, Belarusian State University.

However, in the case of an ideally polarizable electrode, there will be only charge accumulation at the interface. In this case, the circuit element should be a capacitor C. Then, the expression of the total impedance should be derived as follows. First, we take eq. (2.12) defining the potential polarization, then, we take the definitions of a capacitance C, and resultant charging and discharging current, I_C, assuming the independence of C of polarization:

$$E_t = E_0 \sin(\omega t), \quad Q = C \cdot E \text{ and } I_C = \frac{dQ}{dt} \tag{2.25}$$

This will give us finally:

$$I_C = \omega C E_0 \cos(\omega t) = \omega C E_0 \sin\left(\omega t + \frac{\pi}{2}\right) \tag{2.26}$$

This equation shows that for an ideal capacitor the flow of current, I_C, is shifted to 90° with respect to the applied sinusoidal perturbation E_t. Equation (2.26), *per analogiam* to Ohm's law, can be rewritten as follows:

$$I_C = \frac{E_0}{X_C} \sin\left(\omega t + \frac{\pi}{2}\right) \tag{2.27}$$

where $X_C = 1/\omega C$ is called the ***capacitive reactance***, which, unlike the resistance, R, depends not only on the capacitance, C, but also on the angular frequency, ω, of the AC current flow through this capacitor. Now, the total impedance contains only the imaginary part:

$$Z = j \cdot Z'' = j/\omega C \tag{2.28}$$

It is obvious from the above equations that the larger the ω, the lower the impedance of an ideal capacitor. Again, this can be visualized with the help of Nyquist and Bode plots (Fig. 2.9) (C = 1 μF):

Let us now put together the above two circuit elements, R and C, in series. Then, the total impedance Z will be equal:

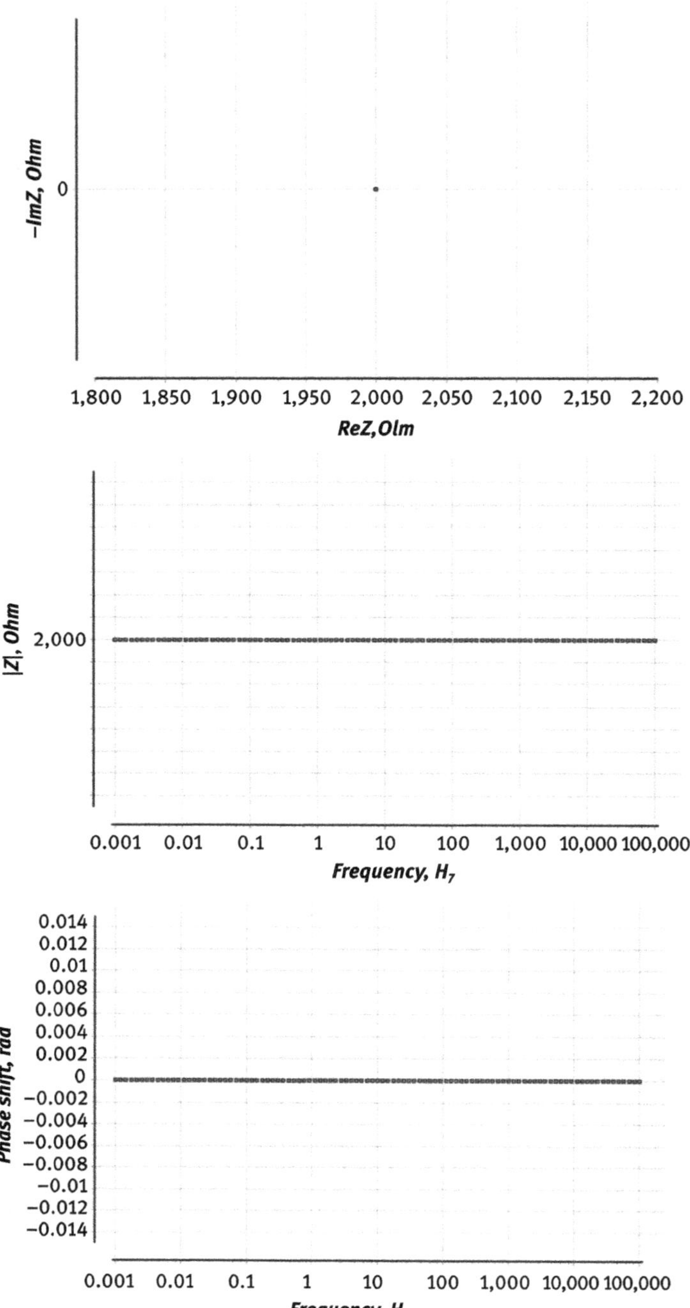

Fig. 2.8: The Nyquist and Bode graphs; simulations for the resistor (2 kOhm).

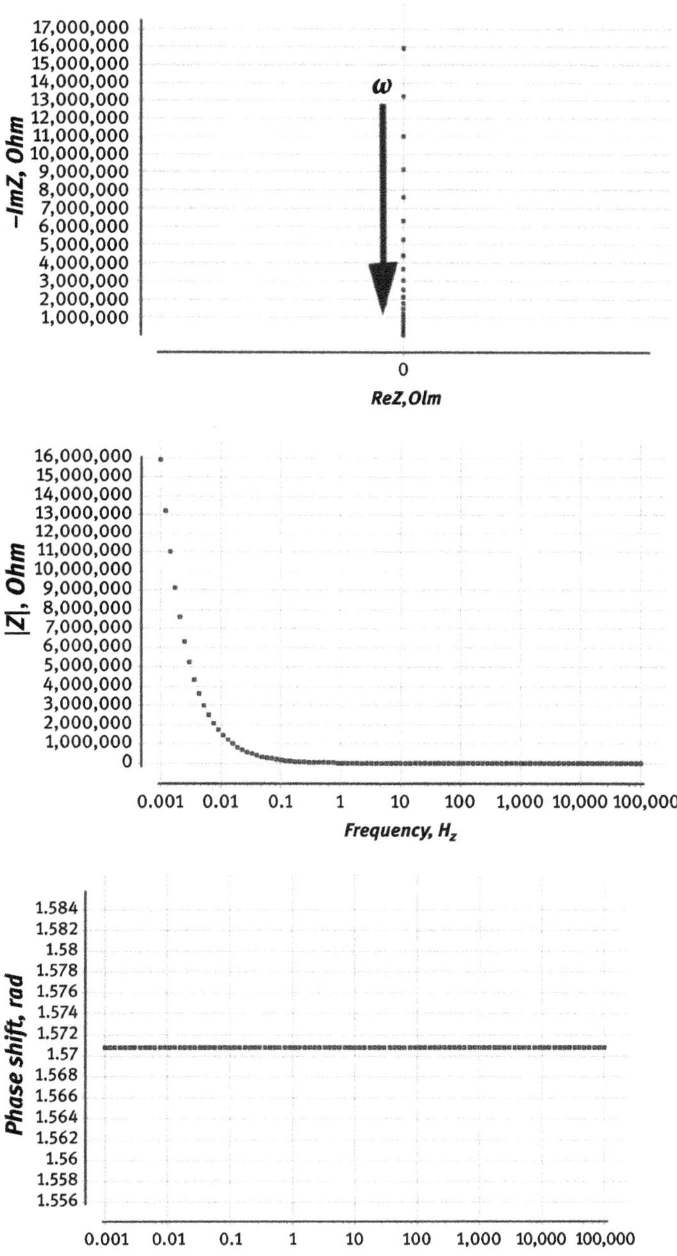

Fig. 2.9: The Nyquist and Bode plots for a single capacitance. Black arrow indicates the direction of ω increase. Please note that the *Y*-axis in the right-hand side panel is in radians, so the value of phase shift φ = 1.570796 radians corresponds to $\pi/2$ = 90°.

$$Z = R + jX_C = R + \frac{j}{\omega C} \tag{2.29}$$

The next step in our representation of an electronic equivalent circuit should be the circuitry shown in Fig. 2.1, modeling the typical electrode/electrolyte interface. Here, the interface is not only charged (ideally polarizable interface), but also some part of this charge can participate in the redox reaction and can be transferred across this interface. In other words, this reaction is partially short-circuiting the charge accumulation with a charge transfer resistance, R_{ct}. This is shown in Fig. 2.1 as a parallel mesh of a capacitor, C_{dl}, and a resistor, R_{ct}. Moreover, we should include also the ionic resistance of the solution, R_s, that does not participate in the redox reaction itself. Such a solution, usually of concentration at least 10 times larger than that of the redox species, is usually called the supporting electrolyte and its role is to maintain possibly high conductance of the solution and to keep the ionic strength constant, so the structure of the electrochemical double layer should remain unaffected by the electrode processes (Fig. 2.10).

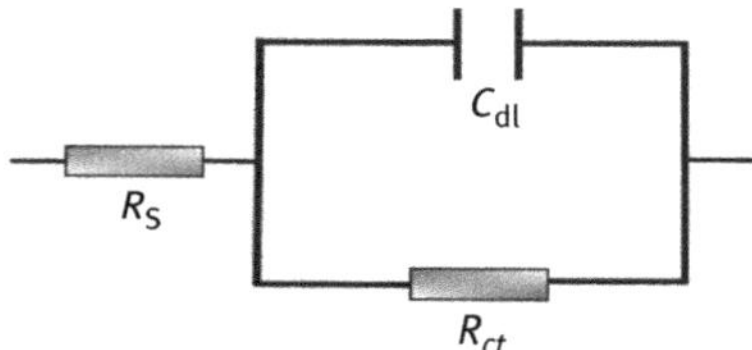

Fig. 2.10: A simplified Randles' electronic equivalent circuit.

Here, the total impedance Z of such circuit is a sum of R_S in series with the impedance, Zp, of parallel connection of C_{dl} and R_{ct}. So

For a parallel mesh:

$$\frac{1}{Z_p} = \frac{1}{R_{ct}} + j\omega\, C_{dl}, \qquad Z_p = \frac{R_{ct}}{1 + j\omega R_{ct} C_{dl}} \tag{2.30}$$

For total impedance:

$$Z = R_S + Z_p = R_S + \frac{R_{ct}}{1 + j\omega R_{ct} C_{dl}} \tag{2.31}$$

It is important to note at this point that in the case of real systems, R_{ct} should be replaced by the Faradaic impedance, Z_F, aiming to account for all possible electrode processes, such as diffusion, adsorption, surface structuring/evolution, electrodeposition, coating, and corrosion. In chapters that follow, we will exemplify some of the simplest models of electrochemical processes; however, the interested reader is directed for further readings listed at the end of this chapter. For such a circuit, called the simplified Randles circuit, the Nyquist plot is shown in Fig. 2.11, together with the Bode plots.

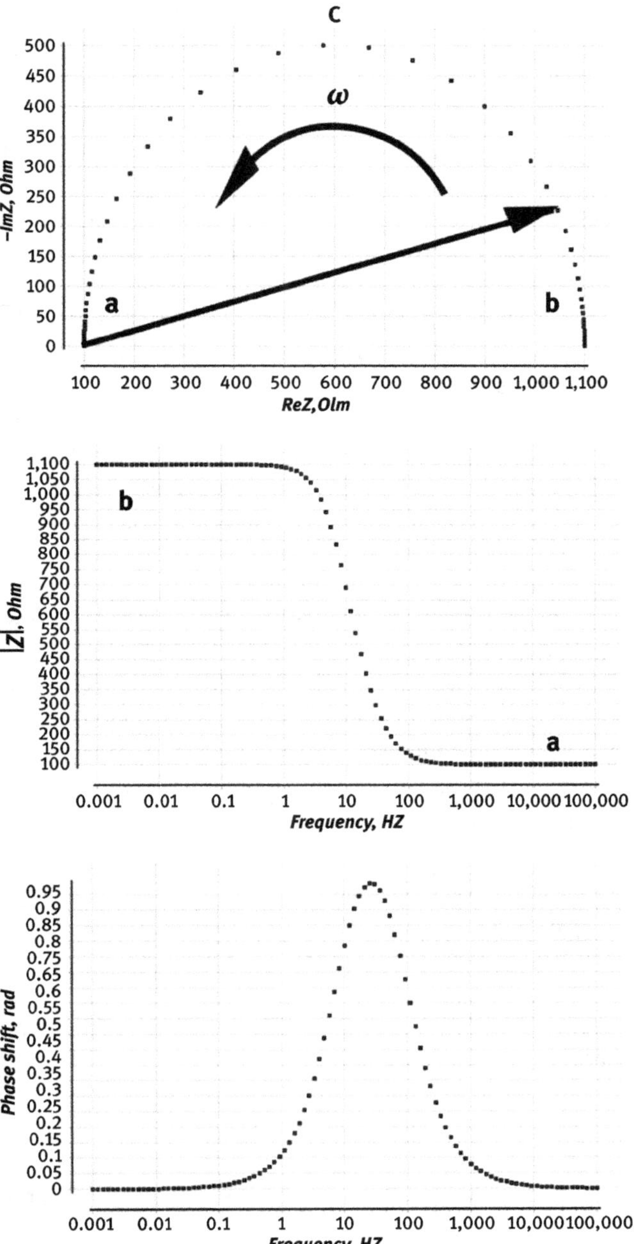

Fig. 2.11: Nyquist and Bode plots for R_S (100 Ohm) in series with parallel connection of C_{dl} (2E-5 F) and R_{ct} (1,000 Ohm), as in Fig. 2.7. Black arrow indicates the value of total impedance for frequency of approximately 2 Hz, whereas its length is the impedance magnitude, according to the Pythagorean law: $Z^2 = (Z'^2 + Z''^2)^{0.5}$. This is the vector representation of the impedance $|Z|$. The angle between this vector and

The Nyquist plot also shows three very important points, denoted **a**, **b**, and **c**. These are the characteristic points allowing us to immediately retrieve the information on R_S, R_{ct} and C_{dl}, because the impedance value at point **a** equals to R_S, at point **b** it is equal to R_S + R_{ct} (the imaginary part at both points equals zero), whereas at point **c** (the maximum of the semicircle) $\omega C_{dl} R_{ct} = 1$. The values of R_S (point **a**) and R_S + R_{ct} (point **b**) can also be found on the Bode plot of Z' versus log(frequency).

As stated earlier, in the case of real electrochemical systems, in order to account for all possible electrode processes, the charge transfer resistance, R_{ct}, should be replaced by the Faradaic impedance, Z_F. The most common (but not limited to) elements that constitute part of Z_F are the Warburg impedance, W, and the constant phase element (CPE). Warburg impedance represents a resistance to mass transfer, showing the extent of diffusion control in the investigated system. Typically, it shows a 45° phase angle (phase shift). The diffusion control has to be considered, for example, in studies of corrosion processes, where there is limited access to oxygen (discussed later in chapters that follow). The CPE approximates the "real" surface that is not an ideally flat, uniform surface with "smeared" charge uniformly distributed over the whole surface. Normally CPE shows a 70°–90° phase shift. The latter phase angle is characteristic of a perfect capacitor. Generally, the lower the phase angle, the more "imperfect" the interface is with diffusive processes taking control over the whole electrochemical process. This can happen for instance in porous electrodes, conductive polymer coatings, or lithium batteries. Figure 2.12 shows the Nyquist and Bode plots for Warburg impedance, W, in series with R_{ct}, both in parallel with C_{dl}, and the whole mesh is in series with R_S.

One may question why we presented the impedance data in both the complex plane (Nyquist plot) and Bode (Z', Z'' vs. logarithm of frequency). Therefore, let us compare the information that can be immediately obtained from both types of plots.

Nyquist plot

- Individual charge transfer processes are resolvable.
- Frequency is not obvious, apart from its increase toward the lower frequencies.
- Small impedances can be hidden by large impedances.

The two last points are the disadvantage of a Nyquist plot. Therefore, to make the response of the system more specific, the Bode plots are used.

Bode plot

- Individual charge transfer processes are resolvable as in Nyquist.
- Frequency is explicit.
- Small impedances can be identified easily even in the presence of large impedances.

Fig. 2.11 (continued)
the X-axis, commonly called the "phase angle, φ" and $\tan\varphi = \frac{Z''}{Z'}$. Black circular arrow shows the direction of the increase of angular frequency, ω.

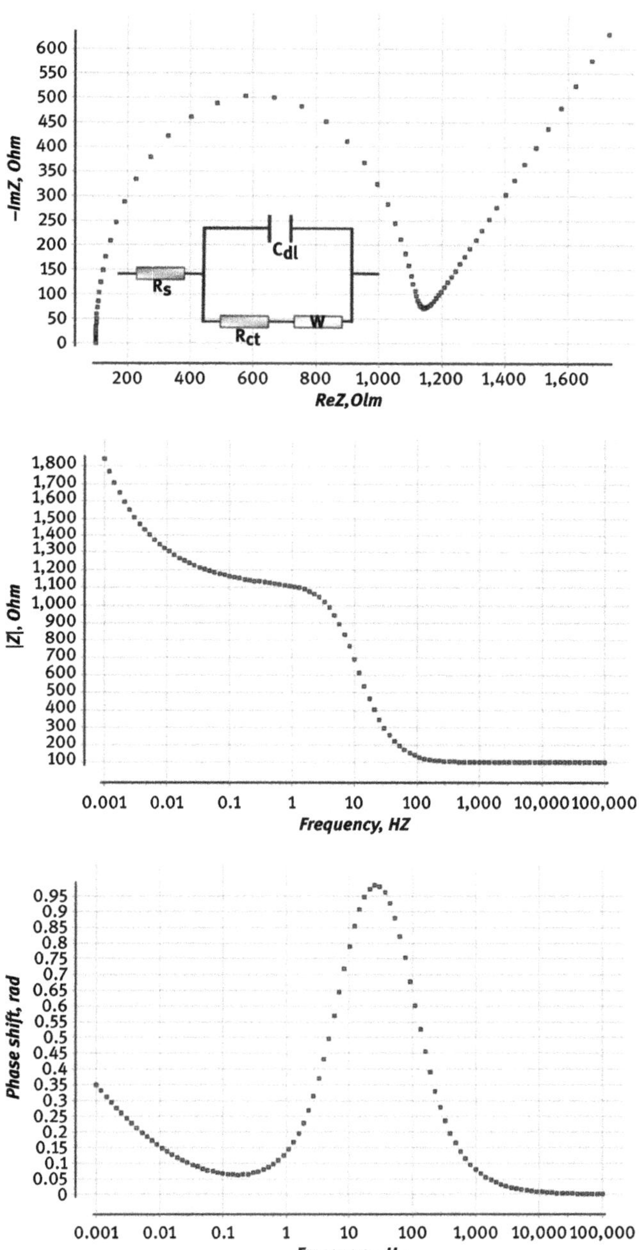

Fig. 2.12: An example of EIS simulation for the "full" Randles' equivalent circuit that includes Warburg impedance *W* (left panel – Nyquist; middle and right-hand side – Bode) for R_S = 100 Ohm, R_{ct} = 1,000 Ohm, C_{dl} = 2E-5 F, and *W* = 50 Ohm.

Finally, a word of caution should be said with respect to EIS. Even though EIS can provide very useful and precise information about the studied processes and their mechanism and therefore become more and more popular among the electrochemists, several criteria have to be fulfilled before the appropriate conclusions could be inferred from the results and model fitting:

A. The system under study
   1. The investigated electrochemical system obeys Ohm's law, that is, the value of impedance is the proportionality coefficient of potential stimulus and current response: $E = Z{\cdot}I$. Moreover, the value of Z is independent of the magnitude of potential perturbation.
   2. The system is stable during the EIS experiment and does not change in time since some of the experiments can last quite long. For instance, if a wide frequency range is used with several steady-state DC potentials chosen for each frequency range, the experiment can take up to hours.
   3. The observed response of the system is due only to the applied potentials (both DC and AC).

B. Data presentation and modeling
   1. It is a must that each element of the electrical equivalent circuit corresponds to a specific mechanism or process in the studied system.
   2. One should not simply multiply the equivalent circuit elements until a good fit is obtained, but use the simplest model that fits the data (the ***Occham's razor*** principle). This model should comply with point B.1, of course.

At the end of this section, let us now summarize what was said above in the form of a flowchart graph. We should state, however, that this flowchart, with some modifications, should be used in all applied electrochemistry experiments.

2.3 Electrochemical quartz crystal microbalance method

As we discussed in the previous section, numerous electrochemical methods are available for measurements of characteristic parameters (current, potential, and charge) of electrochemical processes and systems.

In the last years, many hybrid methods were developed for the investigations of electrochemical systems. They are "in situ" techniques that give us the possibility to examine the state of electrode surface and reaction products in real time of electrochemical processes. One such technique is electrochemical quartz crystal microbalance (EQCM), a very powerful tool for the investigations of thin films and interfacial phenomena. EQCM method gives us reach information about deposition, dissolution of metal films, redox processes in polymer films, formation of self-assembled monolayers, and changes at the electrode/solution interface resulting from adsorption or reaction. EQCM method is based

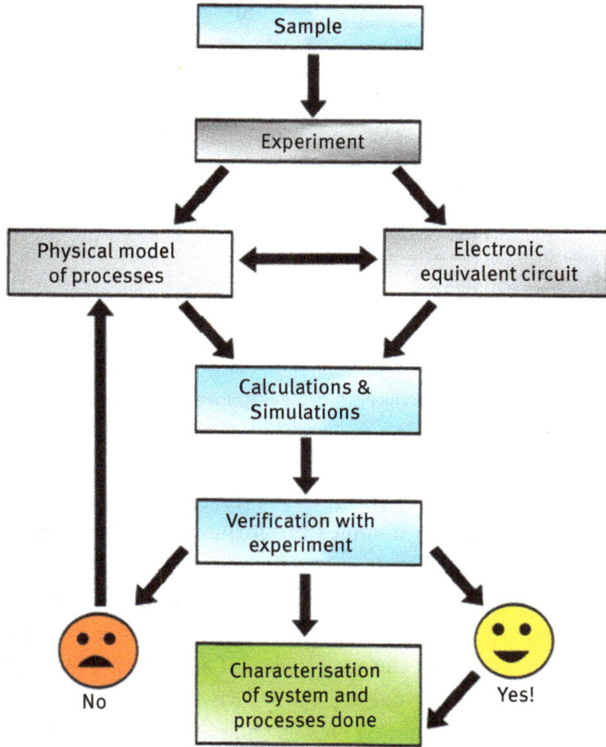

Scheme 2: Flowchart graph summarizing a critical approach to the electrochemical results obtained with different techniques.

on simultaneous detection of frequency changes and, in consequence, of mass changes on the electrode during its electrochemical response.

Figure 2.13(a) shows a schematic diagram of an EQCM apparatus. It contains three electrodes cell. Resonator (working electrode) consists of AT-cut quartz crystal (QC) in a form of a thin disk of approximately 320 µm thickness, covered on both sides by a thin layer of gold (200 nm thickness). Other coinage metals and carbon coatings are also available commercially. Figure 2.13(b) shows the top view of QC with the deposited Au layer. One side of the Au-covered crystal is in contact with the electrolyte and works as a typical electrode. The Ref electrode and C counter electrode are as usual. All equipment used, that is, potentiostat/galvanostat, oscillator, and frequency counter, are controlled by a computer. The typical experiment would involve the application of the potential waveform, or the potential step, to the working electrode, and simultaneous measurements of current flowing through the cell and the oscillation frequency of the crystal. In many cases, the oscillator and frequency counter is replaced by the impedance analyzer.

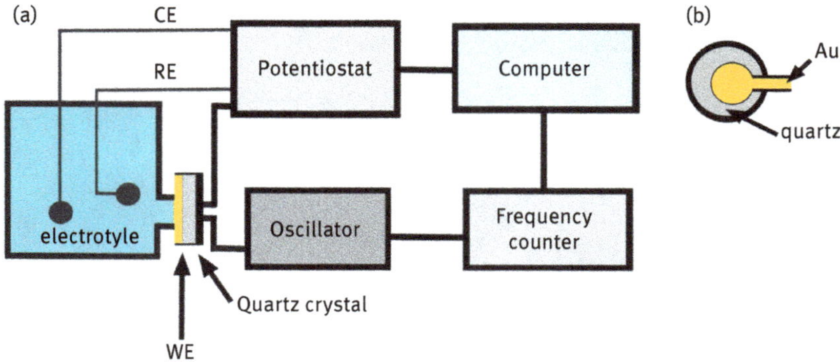

Fig. 2.13: (a) Schematic diagram of apparatus for electrochemical QCM measurements. (b) Schematic top view of quartz crystal with deposited gold electrode.

QCM and in consequence EQCM methods are based on the reverse (converse) piezoelectric effect. The primary piezoelectric effect involves the formation of the charge on a crystal surface caused by mechanical strain. In consequence, one can observe a potential drop across such a crystal that is proportional to the magnitude of strain. Such an effect is characteristic for acentric crystals, such as quartz and other crystals (topaz, sucrose). The converse effect means that the application of an external electric field across the crystal will generate the strain and crystal deformation. The application of the alternating potential across quartz crystal causes changes in the strain polarity, in consequence, the vibrational motions and further formation of acoustic transverse wave. This transverse acoustic wave propagates across the crystal depth between the two crystal surfaces (Fig. 2.14).

The standing wave conditions are attained when acoustic wavelength, λ, is equal to $2d_q$ ($\lambda = 2d_q$), where d_q is the thickness of the quartz crystal. In the resonant condition, the resonant frequency of acoustic wave f_0 is given by the equation:

$$f_0 = \frac{v_{tr}}{2d_q} = \left(\frac{\mu_q^{1/2}}{\rho_q^{1/2}}\right) : (2d_q) \tag{2.32}$$

where v_{tr} is the transverse velocity of acoustic wave in AT-cut quartz, ρ_q is the density of quartz ($\rho_q = 2.648 \mathrm{g/cm^2}$), μ_q is the shear modulus of quartz ($\mu_q = 2.947 \times 10^{10}\ \mathrm{N/m}$). Under the resonant conditions, such specially prepared quartz crystal works as a resonator. Let us suppose that on this resonator we deposited compact, uniform layer of some foreign (unknown) material with acoustic properties identical to those of quartz. In this case, the change of the thickness resulting from the deposited layer can be treated as the change in the thickness of the quartz crystal. The acoustic wave will now propagate across quartz and deposited layer, but with different frequency (lower) caused by an increase in thickness. The fractional changes in the frequency Δf caused by fractional changes in thickness Δd are described by the equation:

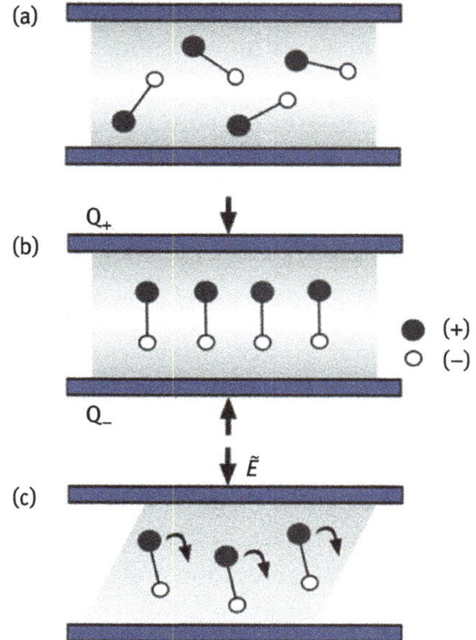

Fig. 2.14: Schematic representation of piezoelectric effects: (a) quartz crystal (QC) is not affected by external forces. The distribution of dipoles in QC is chaotic (no piezoelectric effect). (b) Quartz crystal under the influence of stress. Stress causes the orientation of dipoles resulting in the formation of electric field (primary piezoelectric effect). (c) Quartz crystal under the influence of the electric field. The field causes reorientation of dipoles resulting in the lattice strain and shear deformation of crystal (converse piezoelectric effect).

$$\frac{\Delta f}{f_0} = \frac{\Delta d}{d_q} = -2f_0 \frac{\Delta \mathrm{d}}{v_{tr}} \tag{2.33}$$

Using this equation, and taking into account that $\Delta d = \frac{\Delta m}{A\rho_q}$, where A is piezoelectrically active area, we can obtain the well-known Sauerbrey's equation:

$$\Delta f = -\frac{2f_0^2 \Delta m}{A\left(\mu_q \rho_q\right)^{1/2}} = -C_f \Delta m \tag{2.34}$$

C_f is called the integral mass sensitivity of a resonator.

Because the film is treated as an extension of quartz crystal, the Sauerbrey's equation applies only to systems in which the following conditions are met: the deposited mass must be rigid and distributed evenly over the surface, so the deposited film has a uniform thickness across the active region of resonator, and the mass of the film does not exceed 2% of the mass of quartz crystal, that is, typically lower than 20 µg/cm^2. The integral mass sensitivity has a value of 56.6 Hz·cm^2/µg for 5 MHz crystal in air, and in aqueous solution, it is lower (about 42 Hz·cm^2/µg).

In practice, depending on the measured films, there are some effects that disturb the Sauerbrey's relation. We should mention here such effects, as the changes in viscoelasticity of films due to swelling and ionic exchange. The other effects that can influence the results are microsurface roughness of QC resulting in trapping of liquid in cavities of the electrode surface, too high mass load on the electrode, temperature changes that affect density, and viscosity of the solution in which the crystal is operating.

As mentioned earlier, EQCM method is widely used in the investigations of different types of thin films. Let us consider the simple deposition of thin film of metal on a QC electrode. In this case, the frequency changes Δf result only from mass changes (Δm) caused by the reduction of metal ions, and they are related proportionally to the charge Q $(\mathrm{C/cm^2})$,

$$\Delta f = \frac{10^6 M C_{\mathrm{f}} Q}{nF} \quad \text{because} \quad \Delta m = \frac{MQ}{nF} \tag{2.35}$$

Here M is the apparent molar mass (g/mol), n is the number of electrons involved in the unit process, and F is the Faraday constant. The factor 10^6 completes the conversion from units of $C_{\mathrm{f}}\left(\mathrm{Hz}/\mu\mathrm{g\,cm^2}\right)$ to the units of M $(\mathrm{g/mol})$.

A plot of Δf versus Q will give the apparent molar mass of deposited metallic species. Any deviation from linearity will point out that the process is more complicated and can involve the exchange of solvent or ions or both between the film and solution, as it is observed for thin film of conducting polymers. Figure 2.15(a) presents the CV voltammogram and frequency change of QCM during the deposition of Ag film on Au-QC electrode.

CV scan shown in Fig. 2.15(a) begins at +0.6 V where only Ag^+ ions exist, and the frequency of resonator (electrode) is then set to zero (by the equipment circuitry). Upon Ag^+ reduction, the QCM frequency decreases as a result of Ag deposited mass increase. Upon the oxidation of metal Ag during the reverse scan, the frequency of QCM is going back to zero value. The relation between frequency changes and charge flowing during the reduction and oxidation of Ag^+, Ag species on the Au-QCM electrode is shown in Fig. 2.15(b). Based on eq. (2.35), the relation is linear and points out that there is no additional reaction at the resonator electrode. From the slope of Δf versus Q, this relation allows us to determine the molar mass of Ag. As molar mass of

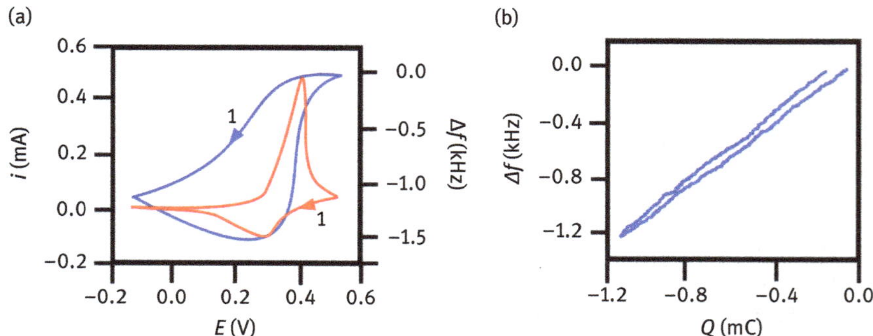

Fig. 2.15: (a) Cyclic voltammogram (red curve) and frequency change (blue curve) recorded on QCM|Au electrode in solution containing 0.002 M $AgClO_4$ and 0.1 M $HClO_4$; (b) Plot of frequency shift versus charge passed during the measurements in the solution containing 0.002 M $AgClO_4$ and 0.1 M $HClO_4$.. Unpublished data, courtesy of M. Skompska (Faculty of Chemistry, Warsaw University).

Ag is well known, the deposition of Ag on Au-QCM electrode is often used for calibration and experimental determination of C_f.

To learn more about the theory and applications of the EQCM method, an interested reader will find in refs. [5–7].

Bibliography

[1] Bard AJ, Faulkner LR. Electrochemical Methods, Fundamentals and Applications, John Wiley & Sons, Inc., 2001, 87–416.

[2] Marken F, Neudeck A, Bond AM. Cyclic Voltammetry. In: Scholtz F., ed., Electroanalytical Methods. Guide to experiments and applications. Springer-Verlag, 2010.

[3] Retter U, Lohse H. Electrochemical impedance spectroscopy. In: Scholtz F., ed., Electroanalytical Methods. Guide to experiments and applications. Springer-Verlag, 2010.

[4] Lasia A. Electrochemical impedance spectroscopy and its applications, Springer-Verlag, 2014.

[5] Gileadi E. Physical electrochemistry. Fundamentals, techniques, applications. Weinheim, Germany, Wiley-VCH Verlag GmbH & Co. KGaA, 2011, 253–264.

[6] Buttry DA, Ward MD. Measurements of interfacial processes at electrode surface with the electrochemical quartz crystal microbalance. Chem. Rev 1992, 92, 1355–1379.

[7] Buttry DA. Application of the quartz microbalance to electrochemistry. In Bard AJ, ed. Electroanalytical Chemistry: A Series of Advances. New York, USA, Marcel Dekker, 1991, 17, 1–85.

Part II: **Electrochemistry in material science – selected topics**

3 Corrosion

3.1 General remarks

Because corrosion causes spontaneous and uncontrolled degradation of materials in their environment, it is one of the largest challenges in materials science and industry. The economic costs of corrosion are extremely high and include the prevention of corrosion and replacement of corroded materials. It is estimated that approximately 5–7% of global income is spent on corrosion-related costs. Corrosion occurs naturally and reflects the tendency of a material to achieve a more chemically stable form. Even though this term can be applied to all materials, such as metals, alloys, semiconductors, conducting polymers, and even dielectrics, here we will focus on metals and their electrochemical reactions with the environment, leading to the destruction of metal construction to more environmentally stable products, such as oxides, hydroxides, or sulfides.

3.2 Corrosion – what does it mean? Mechanism of corrosion

Corrosion is caused by oxidation of metals paralleled by the reduction of electroactive species present in the environment. Both types of reactions require charge transfer; therefore, electrochemistry is a technique of choice in the investigations of corrosion processes. Corrosion reaction consists of oxidation and reduction reactions of the redox couples being different from each other. Thus, it differs from the equilibrium electrochemistry, where the redox reaction at equilibrium is described by the Nernst equations. Under this condition, the reduction (cathodic) and oxidation (anodic) currents are the same, though different in directions of electron flow, and the net current is zero. In corrosion, however, usually, more than two different redox couples are involved in the overall process. The open-circuit potential or the potential that is formed in the system is determined by the potential at which the net current is zero, as in the equilibrium electrochemistry, but the anodic current is related to the constant dissolution of metal, while the cathodic current is related to the reduction of different species from the environment.

We can classify the corrosion processes, for example, with respect to the environment:

– Atmospheric corrosion: it is caused by air and air pollutants present in the air. It depends on the presence of water (rain, dew, humidity, or melting snow) that dissolves various salts from the environment (e.g., salt used in icy roads). Atmospheric corrosion is predominant of all the other forms of corrosion and the cost of protection against atmospheric corrosion is approximately 50% of the total cost

https://doi.org/10.1515/9783111160986-003

of all other corrosion types. It affects all structures from small articles and utilities to skyscrapers, industrial plants, and bridges.
- Soil corrosion: soil corrosion affects metallic constructions (such as oil pipelines) in direct contact with soil or bedrock. It is not limited to metals, such as iron, steel, or copper, but also constitutes a hazard to concrete structures. The rate of soil corrosion depends greatly on the type of soil due to the fact that some soils are more corrosive than the others because of their composition and water content.
- (Micro)biological corrosion: because living organisms can provide local corrosive environments (such as biofilm), for instance, under the aerobic conditions, some bacteria can oxidize sulfur, producing sulfuric acid: $2\ S + 3\ O_2 + 2\ H_2O \rightarrow 2\ H_2SO_4$, which in turn can corrode ferrous structures, such as in-ground pipes or other underground steel constructions. On the other hand, some anaerobic bacteria can reduce sulfates to sulfides obtaining hydrogen from organic wastes: $SO_4^{2-} + 4\ H_2 \rightarrow 4\ H_2O + S^{2-}$. Both processes can be promoted in wet environments.

Or
- Wet corrosion (usually in an aqueous environment): it appears when two different metals located in an electrolytic solution are in contact with each other. This environment creates an electrochemical cell with potential between the two metals: the more reactive metal acting as an anode (oxidation), while the other one acts as a cathode (reduction). When metals are exposed to air at temperatures lower than 100 °C, water vapor can condense on their surface forming an aqueous layer, promoting the electrochemical corrosion reactions.

One can also classify corrosion from the point of view of its mechanism:
- Chemical – in dry gases, nonconducting liquids
- Electrochemical – in electrolyte solutions (aqueous, wet soil, humid atmosphere)

Although this chapter is devoted to the electrochemistry of corrosion, it is worth to mention also that the mechanical action can facilitate the corrosion and resulting material damage. The first type of mechanical wear (fatigue) results from mechanical stress – the stress corrosion cracking (SCC). During SCC, the initiation of small defects – cracks, pits, and their subsequent growth on metal or protective layer in an aqueous solution can enhance the destructive effect of corrosion. It concentrates at the front of the defects, stimulating the crack and pits growth and corrosion progress under the stress. We will discuss this later in the text. Other mechanisms stimulating corrosion are due to the flow of suspension of solid particles or even turbulent flow of gas bubbles crashing and destroying the rust layer or passive film, thus exposing bare metal surface to the fluid. These types of corrosion are called erosion corrosion and cavitation corrosion, respectively.

As can be seen, the classifications criteria frequently interleave; therefore, for the purpose of this handbook, we will focus on electrochemical models of corrosion.

Corrosion of metal starts when the oxidation of metal leading to the anodic dissolution and a formation of hydrated metal oxide begins. This can be described by the following reactions:

$$\mathrm{Me} \longrightarrow \mathrm{Me}^{n+} + ne^-$$
$$\mathrm{Me} + x \cdot \mathrm{H_2O} \longrightarrow \mathrm{MeO}_y(\mathrm{OH})_{x-y} + \; 2x \cdot \mathrm{H}^+ + 2x \cdot \mathrm{e}^-$$

The dissolution rate is exponentially increased with an increase in the anodic potential of the metal following the Butler–Volmer formalism discussed in Chapter 1 (eq. (1.103)).

In order to maintain neutrality condition, a reduction reaction of a substance present in the environment must occur simultaneously. In an aqueous solution, this is typically the hydrogen evolution due to the reduction of protons (acidic media) or water reduction (neutral to alkaline media):

$$2\,\mathrm{H}^+ + 2\,\mathrm{e}^- \longrightarrow \mathrm{H_2}$$
$$2\,\mathrm{H_2O} + 2\,\mathrm{e}^- \longrightarrow \mathrm{H_2} + 2\,\mathrm{OH}^-$$

Other typical reduction is due to the presence of oxygen dissolved in the aqueous solutions:

$$\mathrm{O_2} + 4\,\mathrm{H}^+ + 4\,\mathrm{e}^- \longrightarrow 2\;\mathrm{H_2O}$$
$$\mathrm{O_2} + 2\,\mathrm{H_2O} + 4\,\mathrm{e}^- \longrightarrow 4\,\mathrm{OH}^-$$

Obviously, the first reduction will take place preferentially in acidic media, whereas the latter takes place in alkaline ones. Both reactions are controlled by the diffusion of oxygen; therefore, their rates are similar. As already discussed at the beginning of this chapter, corrosion consists of oxidation and reduction reactions proceeding simultaneously of the redox couples being different from each other. The potential that appears naturally in the system, the open-circuit potential, is determined by several participating redox reactions whose net current is zero. This corrosion potential is called the mixed potential and is not the equilibrium potential determined by the equilibrium conditions of each of the reactions, but results from their kinetic relations. If only two redox couples are involved in corrosion, the open-circuit potential will be between the two equilibrium potentials, even though it is impossible to predict its precise value based on the thermodynamics.

Let us consider corrosion processes of Zn (valid also for other metals, such as Fe, but at different potential regime) in acidic media, as shown in Fig. 3.1. At thermodynamic equilibrium, the Nernst potential for zinc redox reaction (*A* – anodic) and hydrogen electrode reaction (C – cathodic), respectively, will be given as follows:

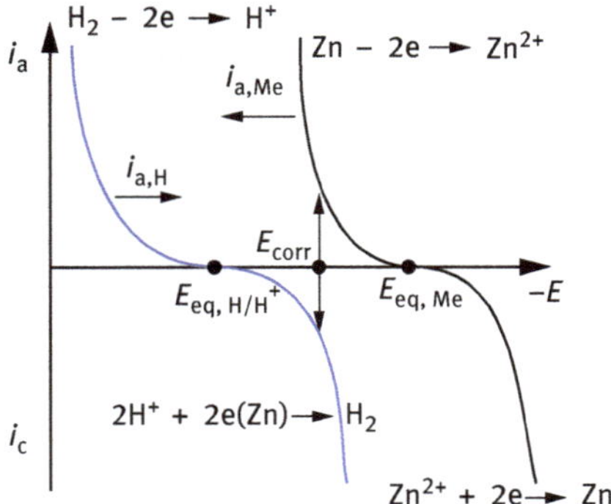

Fig. 3.1: Schematic representation of the formation of corrosion potential, E_{corr}, for two different redox couples under wet, acidic conditions (solution, Butler–Volmer formalism only for clarity). Dissolution of Me (Zn, Fe) in the oxidation (anodic) reaction coupled to the reduction of protons with hydrogen evolution in the cathodic reaction. Arrows show the shifts of the *i–V* curves until the E_{corr} is achieved with the net current equal to zero. Please note the sign of *E* on *x*-axis.
No contribution of the reduction of O_2 is considered.

$$A: E_{eq, Zn/Zn^{2+}} = E^0 + \frac{RT}{2F} \ln a_{Zn^{2+}} \tag{3.1}$$

$$C: E_{eq, H_2/2H^+} = E^{0'} - \frac{RT}{2F} \ln \frac{p_{H_2}}{(a_{H^+})^2} \tag{3.2}$$

Under equilibrium, $i_{a.\,Zn} = i_{c,\,Zn} = i_{0,Zn}$ for metal and $i_{a,H} = {}_{c,H} = i_{0,H}$.

It is immediately seen that in the corrosion (oxidation) process, the concentration (activity) of Zn^{2+} ions in the solution will increase, shifting the potential of metallic Zn toward more positive values (to the left in Fig. 3.1, shown by an arrow), whereas the hydrogen evolution (reduction) coupled to this process will decrease the concentration of H^+ ions, shifting the potential of $H_2/2\ H^+$ redox couple (redox system **2**) toward more negative values (to the right, as marked by an arrow). Preservation of charge and energy will finally equilibrate the rates of dissolution of metal and evolution of hydrogen, so the oxidative and reductive currents will equilibrate at the potential value E_{corr}, being established between the two equilibrium potentials of both redox couples:

$$i_{a,\,Me} = -\,i_{c,\,2} = i_{corr}$$

where $i_{a,Me}$ and $i_{c,2}$ are the anodic dissolution current of metal Me and cathodic current of the redox couple **2**, respectively, whereas i_{corr} is the resultant corrosion current, anodic positive, cathodic negative. At the potential E_{corr}, metal will corrode at a constant rate, providing the presence of a sufficient amount of coupled redox system **2** in the environment.

Other reactions that can participate in the overall corrosion process of Zn metal are exemplified as follows:

$$O_2 + 4\,H^+ + 4\,e \rightarrow 2\,H_2O$$
$$O_2 + 2\,H_2O + 4\,e \rightarrow 4\,OH^-$$
$$Fe^{3+} + e \rightarrow Fe^{2+}$$
$$NO_3{}^- + 3\,H^+ + 2\,e \rightarrow HNO_2 + H_2O$$

The corrosion example discussed above does not require any additional metal to be present in the environment: the corrosion will proceed only due to the environment surrounding the metal. In the above case of Zn (but also for Fe), it was an aqueous solution of acidic pH.

Now we consider other possible cases of electrochemical corrosion, introducing to the aqueous solution two metal slabs of different redox potentials such as Zn and Fe Fig. 3.2. When these metals are separated from each other in the solution, upon the open-circuit conditions, no corrosion will proceed, apart from the possible environmental corrosion. If these two slabs are brought into contact, the electrochemical potentials of electrons of one metal will equilibrate with the electrochemical potential of the second one; hence, the metal of more negative Nernst potential will act as anode and start to dissolve (Zn), whereas the metal of more positive Nernst potential will act as cathode and will decrease greatly its corrosion (Fe). The electric contact between the two metals induces a potential change in the more negative direction from the equilibrium value at Fe and in the more positive direction from the equilibrium value at Zn. The positive shift of the potential of Zn^{2+}/Zn redox couple from $E_{eq,\,Zn/Zn^{2+}}$ will result in an increase of the anodic dissolution current of Zn, which can be described by the electrochemical kinetic formalism of Butler–Volmer (eq. (1.103)). Contrary to this corrosion process, the negative shift from the equilibrium potential of Fe^{2+}/Fe redox couple from $E_{eq,\,Fe/Fe^{2+}}$ will decrease the dissolution of Fe. The Zn/Fe system is therefore used for the corrosion protection of steel structures. The zinc metal coupled with steel is called the sacrificial anode. This is illustrated in Fig. 3.2, where initially both metals corrode in acidic media with hydrogen evolution, and, after being brought into contact, the potentials shift as described earlier to finally attain the corrosion potential of Zn/Fe pair, $E_{corr,Zn/Fe}$.

The above discussion applies also to the situation depicted in Fig. 3.3, where Fe and more "noble" metal, such as Cu in contact, submerged, for example, in seawater at neutral to alkaline pH. For this Fe/Cu pair, iron will act as anode, whereas Cu as cathode, facilitating the corrosion of iron, particularly next to the Fe/Cu contact, etching Fe. This graphics illustrates also the formation of the corrosion macrocell (contact) that forms always when two metals of different redox potentials contact in a corrosion-promoting environment. Metal of more negative redox potential will always undergo anodic dissolution, "protecting" its counterpart with more positive redox potential.

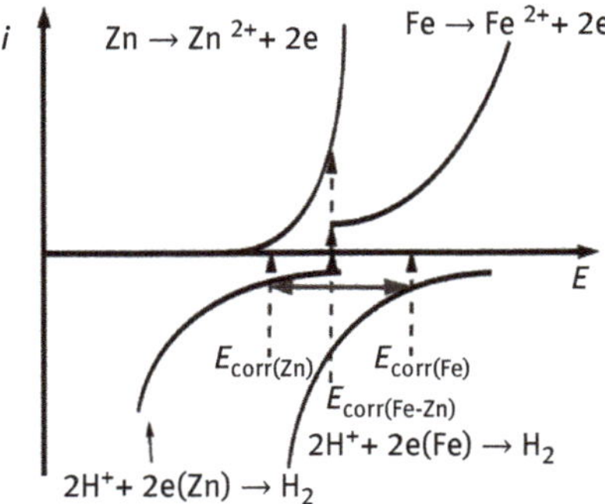

Fig. 3.2: Example of electrochemical corrosion in the case of two metal slabs of different redox potentials such as Zn and Fe immersed in an aqueous solution of slightly acidic pH. Upon electrical connection of these metals, they will tend to equilibrate their electrochemical potentials (compare Fig. 1.10) and (Zn) starts to dissolve, whereas Fe decreases its corrosion.
This shifts the potentials of both metals along the potential axis toward more positive values (Zn) and negative (Fe) toward E_{corr}(Fe–Zn).

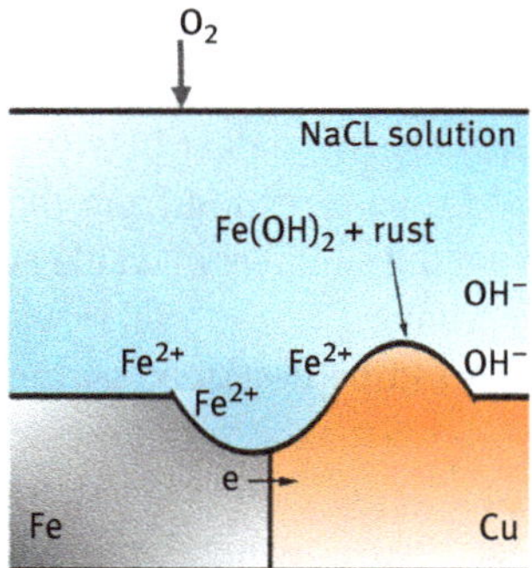

Fig. 3.3: Formation of the corrosion macrocell. Here, Fe/Cu contact pair is submerged in, for example, seawater. Iron acts as an anode, whereas Cu as a cathode, facilitating the corrosion of iron, particularly next to the Fe/Cu contact, etching Fe.

Electrochemical corrosion can also appear within the same metal or alloy having inhomogeneous grain structure, such as steel. This type of corrosion, called the intergranular corrosion, appears because the energy of the interior of grains is smaller than that of the grain boundaries (Fig. 3.4. Here Zn granular structure is exemplified). Then, the grain interior will act as multiple cathodes and boundaries as multiple anodes, forming a series of microcells. Corrosion due to the formation of microcells is generally hard to control and therefore dangerous, mainly because it is within the metallic structure and no rust is visible on the surface until the structure failure and breakdown. The above two examples of electrochemical corrosion are classified based on the spatial locations of anodic and cathodic sites (macro- and microcells) but they differ also with respect to participants: a macrocell is formed when anode and cathode are formed from different metals, while a corrosion microcell is formed when multiple cathodic and anodic sites are within the same metal or alloy due to their granular structure.

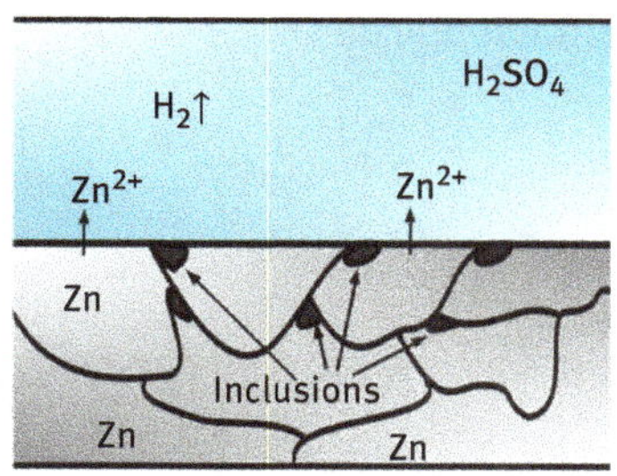

Fig. 3.4: Corrosion microcell formed with multiple cathodic and anodic sites existing within the same metal or alloy due to their granular structure.

3.3 Characterization of corrosion: corrosion potential, corrosion current

Corrosion potential (open-circuit potential) was discussed earlier. The second important parameter describing the corrosion process is the corrosion current, $\boldsymbol{i}_{\mathbf{corr}}$. The higher the corrosion current, the greater the corrosion rate. Its value depends on all participating species: their exchange currents, equilibrium potential differences, pH, and so on. Let us consider the simplest possible system with only two participants, such as corroding metal Me and the reaction counterpart – redox system **2**. Then, according to the Butler–Volmer equation (eq. (1.103)) we can write the following set of equations:

For corrosion of Me:

$$i_{\mathrm{a,Me}} = i_{0,\mathrm{Me}} \left[\exp \frac{(1-\alpha) zF\eta_{\mathrm{Me}}}{RT} - \exp \frac{-\alpha zF\eta_{\mathrm{Me}}}{RT} \right] \tag{3.3}$$

For the developing overpotential of Me:

$$\eta_{\mathrm{Me}} = E_{\mathrm{corr}} - E_{\mathrm{eq,Me}} \tag{3.4}$$

For the redox system *2*:

$$i_{\mathrm{c},2} = i_{0,2} \left[\exp \frac{(1-\alpha)\ zF\eta_2}{RT} - \exp \frac{-\alpha zF\eta_2}{RT} \right] \tag{3.5}$$

For the overpotential of *2*:

$$\eta_2 = E_{\mathrm{corr}} - E_{\mathrm{eq},2} \tag{3.6}$$

For the case of charge transfer coefficient $\alpha = 0.5$, the above set of equations will yield:

$$i_{\mathrm{corr}} = (i_{0,\mathrm{Me}} \times i_{0,2})^{\frac{1}{2}} \times \exp \left[\frac{zF\left(E_{\mathrm{eq},2} - E_{\mathrm{eq,Me}}\right)}{RT} \right] \tag{3.7}$$

This equation shows that the measured corrosion current is a function of $i_{0,\,\mathrm{Me}}$, $i_{0,\,2}$, ΔE_{eq}:

$$i_{\mathrm{corr}} = f(i_{0,\mathrm{Me}},\ i_{0,2},\ \Delta E_{\mathrm{eq}}) \tag{3.8}$$

3.3.1 Stability of materials: potential/pH (Pourbaix) diagrams – the thermodynamic aspect

In 1966, M. Pourbaix proposed a system allowing classifications of various regions for different metals in aqueous solution, depending on pH and potential versus standard hydrogen electrode, assuming a concentration of Me ions of 10^{-6} M.

$$\text{a)}\ O_2 + 4\,H^+ + 4\,e \leftrightarrow 2\,H_2O$$

$$\text{b)}\ 2\,H^+ + 2\,e \leftrightarrow H_2$$

$$\text{c)}\ Me^{n+} + ne \leftrightarrow Me$$

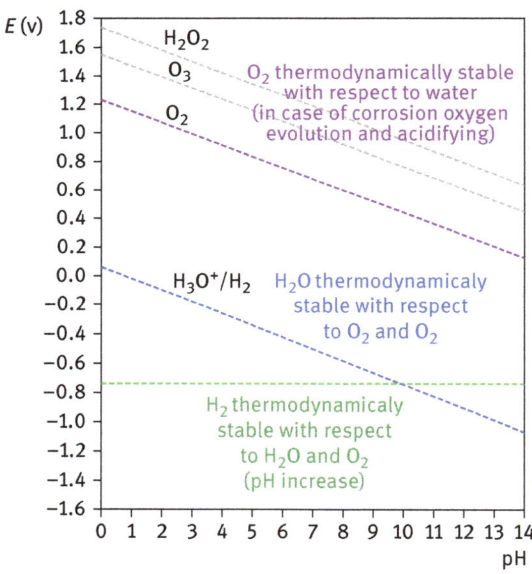

Fig. 3.5: The Pourbaix graph identifying several regions and boundaries on the potential/pH plane for the case of aqueous solution and a simple redox couple Me^{n+}/Me.

The Pourbaix graph (Fig. 3.5) distinguishes several regions on the potential/pH plane for the case of aqueous solution and a simple redox couple Me^{n+}/Me, characterized by its Nernst potential. The upper dashed line corresponds to the oxygen four-electron reduction at equilibrium (reaction a) and its slope reflects its pH dependence, according to the equation:

$$E_{eq} = E^0_{O_2/H_2O} + \frac{RT}{4F} \ln \frac{a^4_{H^+} \cdot p_{O_2}}{a_{H_2O}{}^2} \tag{3.9}$$

At a pressure of 1,013 hPa and temperature 25 °C, it will read:

$$E_{eq} = 1.23 - 0.059\text{pH} + 0.012\log p_{O_2} \tag{3.10}$$

Therefore, the resultant slope of line **a** is −59 mV per unit pH.

Above this equilibrium line, O_2 is thermodynamically stable with respect to water. If corrosion appears, it will be accompanied by oxygen evolution and acidification of the solution. Below this equilibrium line, water is thermodynamically stable with respect to H_2 and O_2.

The lower dashed line (line **b**) corresponds to the reduction of protons at equilibrium:

$$E_{eq} = E^0_{2\,H^+/H_2} + \frac{RT}{2} \ln \frac{a^2_{H^+}}{p_{H_2}} \tag{3.11}$$

At constant H_2 pressure and at 25 °C, this equation will give

$$E_{eq} = -0.059\text{pH} - 0.029\log p_{H_2} \tag{3.12}$$

Thus, this line will also have a slope of −59 mV per unit pH. Below this line H_2 is thermodynamically stable with respect to water and oxygen. If corrosion appears, it will be accompanied by hydrogen evolution and alkalization of the solution.

The last line, line **c**, is parallel to abscissa because the Nernst equation for Me reduction/oxidation is independent of pH:

$$E = E^0_{Me^{z+}/Me} + \frac{RT}{zF} \ln a_{Me^{z+}} \tag{3.13}$$

We are now ready to consider two examples of Pourbaix diagrams for the case of Zn and Fe. In the potential–pH diagrams, Pourbaix arbitrary chose 10^{-6} M for the concentration of metal ions and in the next examples, we follow this choice and T = 298 K.

In the case of zinc, keep in mind the possible amphoteric equilibria.

The four equilibrium lines 1–4 describe the following Nernstian equilibria of zinc with respect to pH:

1. $Zn^{2+} + 2\,e \leftrightarrow Zn$

$$E_{eq,1} = E^0_{Zn^{2+}/Zn} + \frac{RT}{2F} \ln a_{Zn^{2+}} \tag{3.14}$$

2. $Zn(OH)_2 + 2\,H^+ + 2\,e \leftrightarrow Zn + 2\,H_2O$

$$E_{eq,2} = E_2^0 + \frac{RT}{F} \ln a_{H^+}$$

since activities of solid $Zn(OH)_2$, metal Zn, and H_2O are equal to unity. Thus, this equation provides the equilibrium line 2 with slope −59 mV per unit pH:

$$E_2 = -0.439 - 0.0591\text{pH} \tag{3.15}$$

3. $HZnO_2^- + 3\,H^+ + 2\,e \leftrightarrow Zn + 2H_2O$

$$E_{eq,3} = E_3^0 + \frac{3RT}{2F} \ln a_{HZnO_2^-} \cdot a_{H^+} \tag{3.16}$$

$$E_3 = 0.054 - 0.0886\text{pH} + 0.0295\log a_{HZnO_2^-} \tag{3.17}$$

4. $ZnO_2^{2-} + 4\,H^+ + 2\,e \leftrightarrow Zn + 2\,H_2O$

$$E_{eq,4} = E_4^0 + 2\frac{RT}{F} \ln a_{ZnO_2^{2-}} \cdot a_{H^+} \tag{3.18}$$

$$E_4 = 0.441 - 0.1182\text{pH} + 0.0295\log a_{ZnO_2^{2-}} \tag{3.19}$$

Below all these four lines (denoted as solid, thick line), metallic zinc will be immune against corrosion. The line separating zinc stability against corrosion from the rest of this Pourbaix diagram is shown as a thick solid line in Fig. 3.6.

As one probably notice at this point that there are several pH-dependent forms of oxidized zinc such as $Zn(OH)_2$, $HZnO_2^-$, and ZnO_2^{2-}, the equilibrium lines separating the stability area of these forms are independent of potential:

$$\begin{aligned} &Zn(OH)_2 + 2\,H^+ \leftrightarrow Zn^{2+} + 2\,H_2O; && pK_a = 8.5 \\ &Zn(OH)_2 \leftrightarrow HZnO_2^-; && pK_a = 10.7 \\ &HZnO_2^- \leftrightarrow ZnO_2^{2-}; && pK_a = 13.1 \end{aligned}$$

Therefore, they are marked as lines parallel to the ordinate (thin, solid lines). Thus, this Pourbaix diagram distinguishes, apart from the stability region of Zn metal, additional four different stable Zn oxidation products, depending on pH and E_{eq}.

As more important, but also a more complicated diagram, taking into account various possible equilibria, we introduce now the diagram for iron (Fig. 3.7). For soluble ionic species, there are Fe^{2+}, Fe^{3+}, and also various oxoacid ion species in high pH, alkaline solutions. The two iron oxides, Fe_2O_3 and Fe_3O_4, are also considered. Nevertheless, having the appropriate data, it is not difficult to draw the equilibrium lines corresponding to the reactions taking place in the Fe/H_2O system.

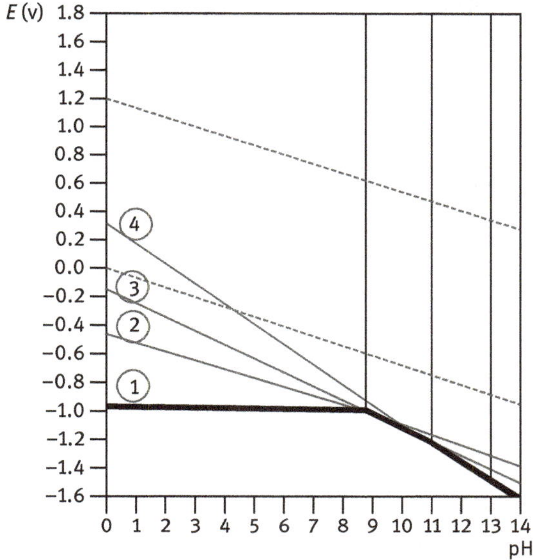

Fig. 3.6: Construction of the Pourbaix diagram in the case of redox reactions of Zn in aqueous media at different pH.

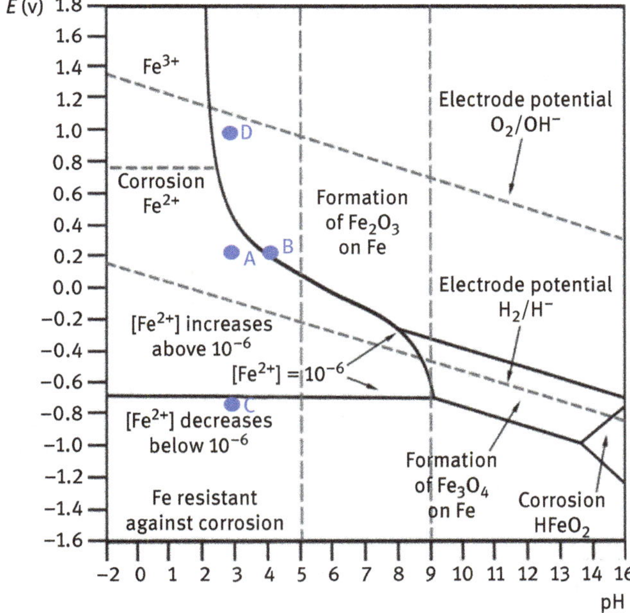

Fig. 3.7: The Pourbaix diagram of iron in aqueous media at different pH.

In this diagram there are three distinct regions:

1. Passivation, immunity region of Fe against corrosion: the potential is sufficiently low, and bare iron is stable.
2. Corrosion region: Fe^{2+} and Fe^{3+} are stable, continuous corrosion of metallic iron.
3. Passivity region: stable iron oxide films are formed, inhibiting corrosion by separating bare iron from the environment.

As mentioned earlier, Pourbaix assumed the concentration of iron ions equal to 10^{-6} M. If it is higher, iron undergoes corrosion; if lower, iron is in the anticorrosive state. This is shown in Fig. 3.7.

Pourbaix diagrams can be used to predict corrosiveness of the metal. Let us use the above diagram to predict qualitatively the corrosion of Fe in aqueous solution. To do this, the following data will be required:

1. pH of the solution
2. Corrosion potential, E_{corr}
3. Redox potential E_{Redox} of the solution, measured with the help of the so-called redox electrode (Pt, glassy carbon electrode) with respect to the saturated hydrogen electrode as a reference electrode.

From the value of E_{corr} and solution pH, we can judge in which of the regions iron metal (as electrode) is localized. If E_{corr} is more negative than that of E_{redox} of the solution with environmental redox species, then the corrosion of Fe (anodic dissolution) can be predicted to proceed to a great degree, and vice versa. In other words, when the E_{corr} of iron sample is within the corrosion region of the Pourbaix diagram, a high corrosion rate of such a sample should be expected. Then, one should consider the protection of this sample. Consider, for example, point **A** on the diagram. If pH of solution is weakly acidic, then iron is in the corrosion region. There are several ways to protect this sample. One is obvious but hardly practical, namely, we should change the solution pH to a more alkaline, resulting in shifting our sample into the passivity region (point **B**). But as we said, this method is unsuitable in reality. Another possibility is to shift the potential of iron in a negative direction to place it inside the immunity region. This method is called cathodic protection and is effected by contacting directly our Fe sample with other metals that can act as sacrificial anode, such as Zn, Al, and Mg. This will shift the potential of Fe to a more negative value (point **C**). The mechanism of such protection was described earlier in the text. Another possible way to shift the potential of iron toward the immunity region is to polarize it negatively from an external source (point **C**). The third mechanism is called the anodic protection and this protection is achieved by polarizing the iron sample in a positive direction in order to place it in the passivation region (point **D**). In the case of anodic protection by a passivating layer, attention should be paid to the existence of defects, even pinhole type in the continuous layer. Their existence in the presence of aggressive ions in the solution, for example,

Cl^-, will lead to localized corrosion. The passive layer will be etched by adsorption of these ions forming small pits of large (few millimeters) penetrating depths.

3.3.2 Stability of materials: current–potential (Evans) diagrams – the kinetic aspect

The potential–pH (Pourbaix) diagrams are based on the thermodynamic equilibria or redox reactions of metals (oxidation) and environmental counterparts (reduction) participating in corrosion reactions. As we said earlier, the corrosion potential does not correspond directly to the equilibrium redox potentials of participants. The actual corrosion potential is a result of a mixed potential system and its value is between the equilibrium potentials of all redox systems present. At this potential, the anodic positive current of metal dissolution is coupled to the cathodic, negative current of environmental counterparts. From the electrochemical measurements of the current–potential relationship, one can estimate the corrosion current of metal under study. Keeping in mind that the current values are directly related to the corrosion rate, let us plot the potential–current diagrams showing the effects of the exchange current, i^0_{Me}, and the equilibrium potential of corroding metal, $E_{eq,Me}$, on the resultant corrosion current $\boldsymbol{i_{corr}}$ and corrosion potential, E_{corr}. The initial system is presented in Fig. 3.8(a), in which ordinate shows the potential values: equilibrium for Me and environmental redox system **2**, with resultant corrosion potential in between the two. Abscissa presents the corresponding current values in the logarithmic scale, resembling the Tafel plots (Chapter 1, Fig. 1.16) for the evaluations of the exchange current and transfer coefficient characteristics of single redox couple in the solution. We will leave unchanged the electrochemical properties of the environmental couple **2**, for clarity.

Figure 3.8(a) shows the values of Nernst potential of Me, $E_{eq,Me}$, its exchange current, i^0_{Me}, as well as similar values for the environmental redox couple **2**. The point where the anodic dissolution current of Me is equal to the cathodic current of **2** marks the corrosion potential, E_{corr}, and corrosion current, i_{corr}. Now, consider the effect of replacing metal Me, with more noble metal Me′, with its equilibrium potential $E'_{eq,Me'}$, more positive than that of Me, that is, being closer to $E_{eq,2}$ (Fig. 3.8(b)). This will result in a drastic (logarithmic scale) decrease of corrosion current, i'_{corr}. Figure 3.8(c) shows the effect of lowering the exchange current of corroding metal by, for example, modification of its surface properties, such as by introduction of a passivating layer. In this case, without changing the Nernst equilibrium potential, we introduce a kinetic barrier against the electron transfer, decreasing the resultant i_{corr} and shifting the E_{corr} toward the equilibrium potential of the environmental redox counterpart **2**.

In the previous cases, we kept unchanged the electrochemical behavior of the environmental couple **2**, assuming its kinetic-type redox properties. However, in neutral pH, the reduction of dissolved oxygen gas can be considered as the preferential cathodic reaction coupled to the anodic corrosion reaction of iron. Since the oxygen re-

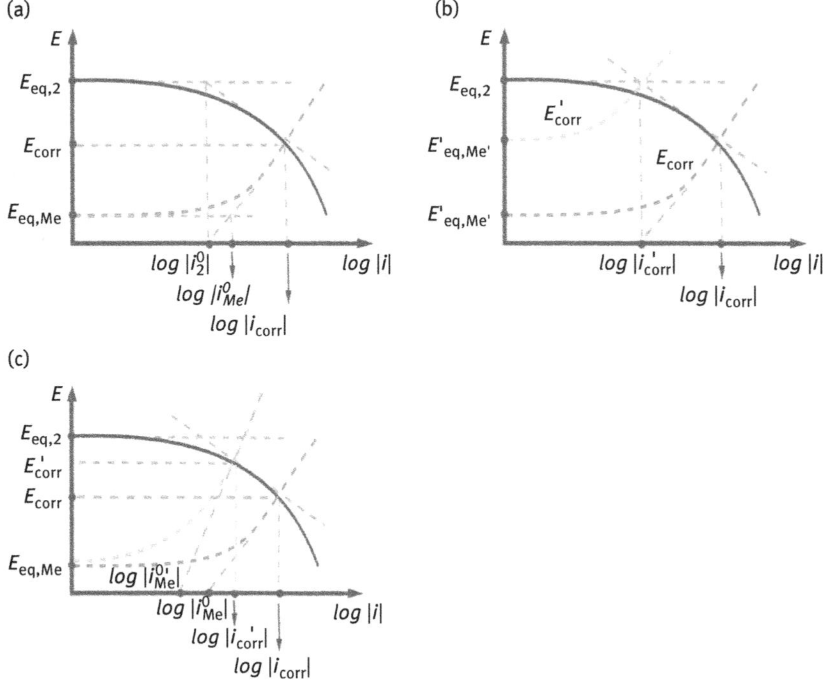

Fig. 3.8: The Evans kinetic diagram for metal: (a) Me and environmental redox system **2**; (b) more noble Me′ replaces Me (e.g., being electroplated on Me), being now in contact with **2**. This results in a decrease of corrosion current. (c) A decrease of the exchange current of Me also decreases the corrosion rate. See text.

duction current is usually diffusion-controlled one in the stagnant solution in equilibrium with oxygen, it reaches the limiting current value. Figure 3.9 shows an example of an oxygen-rich solution after its concentration was limited (e.g., by deaeration or changing the environmental conditions).

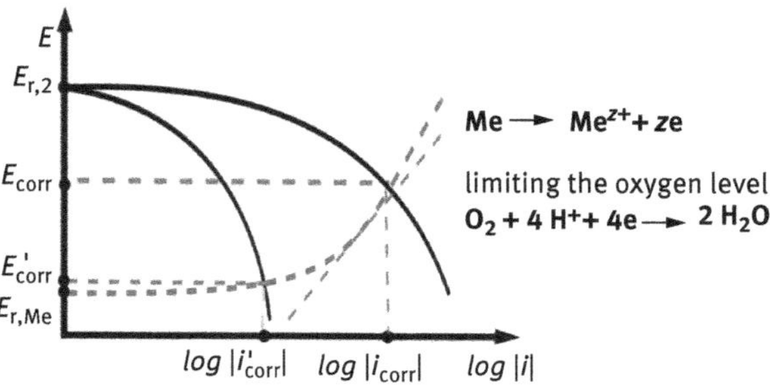

Fig. 3.9: The effect of decreasing the activity of environmental redox couple **2**, for example, depletion of oxygen in the previously oxygen-rich environment. Such situation can occur near the bottom of a stagnant water.

Let us now consider the Pourbaix diagram (potential–pH) of iron in light of the Evans kinetic diagram (current–potential). To do this we will plot current measured in the external circuit with respect to the polarization potential of an iron electrode in a deaerated solution. We will go along the constant pH line at pH = 5 and pH = 9 (Fig. 3.7, dashed lines pH = 5 and 9). Here, we will explain in more detail potential–current behavior for solution pH = 5. Similar considerations can be carried out at pH = 9.

Initially, at sufficiently low potential $E_{eq,Fe}$ = −0.7 V (for concentration of Fe^{2+} lower than 10^{-6} M), iron will be in its immunity state, resistant to corrosion at pH 5. Corrosion begins at a potential more positive (anodic) with rapid exponential rise of corrosion current until it is equalized by the reduction of environmental couple **2**, to achieve E_{corr} and i_{corr}. Further positive polarization of Fe up to approximately 0.2 V results in moving to the passivation region, when the surface is covered with iron oxide film. At this potential, called the passivation potential, roughly corresponding to the formation potential of Fe_2O_3, the corrosion current drops to very small values, constant within the passivity region of Fe. Subsequent positive polarization of iron causes the breakdown of passivity and steep rise of current. In general, three regions can be distinguished: active (corrosion), transition, and passive regions, as shown in Fig. 3.10.

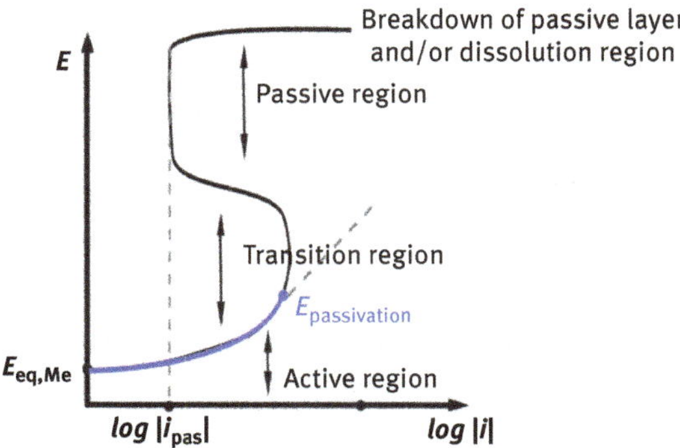

Fig. 3.10: Potential–current diagram of Fe identifying different corrosion processes.

3.4 Evaluation of corrosion rate from electrochemical measurements

By combining the electrochemical kinetics with the corrosion process, we have now tools to estimate the corrosion current and the corrosion mechanism.

3.4.1 Linear scan voltammetry

The simplest way to evaluate the corrosion process is to measure the value of the open-circuit voltage E_{OCP} that can be approximated as the corrosion potential, E_{corr}, and then, to measure the so-called polarization resistance, R_p, defined as the resistance of the metal to oxidation during the application of an external potential in a given corrosive condition. The R_p value is typically measured by polarizing linearly the sample in a very narrow potential window around the E_{corr}, typically within +/−5 mV to +/−10 mV range. Data analysis should provide a line in the i_{meas} versus E_{pol} coordinates, as shown in Fig. 3.11.

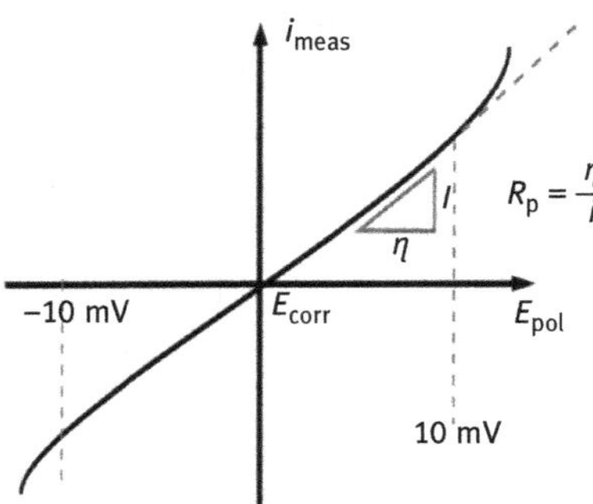

Fig. 3.11: Evaluation of corrosion parameters from the linear scan voltammetry at narrow potential range around E_{corr}.

You may want to recall now the similarity of this approach to the evaluation of the electrode polarization discussed in the case of the Butler–Volmer theory of electrode kinetics for low overpotentials η values.

Next, the corrosion current can be used to calculate the corrosion rate with the help of Faraday's law. At the corrosion potential, E_{corr}, both the cathodic reaction in the environment and the anodic dissolution of metal will have the same corrosion current, i_{corr}. At the corrosion potential, the metal is "self"-polarized to E_{corr}, and the overpotential of the metal for this situation is

$$\eta = E - E_{eq,\,Me} = E_{corr} - E_{eq,\,Me}$$

The dissolution current will follow the Butler–Volmer kinetic equations (see Figs. 1.16 and 1.17, and eq. (1.108)). Thus, the anodic dissolution of a metal electrode can be written as follows:

$$i_{A,\,Me} = i^0_{Me} \times \exp\frac{(1-\alpha)zF\eta}{RT} = i^0_{Me} \times \exp\frac{(1-\alpha)zF\left(E_{corr} - E_{eq,\,Me}\right)}{RT} \tag{3.20}$$

We now introduce a small perturbation ΔE to this overpotential in the range of +/−5 mV to +/−10 mV, as discussed earlier, for the evaluation of polarization resistance R_p. Thus, due to this perturbation, a new anodic dissolution current will flow:

$$i_{A,\,Me} = i^0_{Me} \times \exp\frac{E_{corr} - E_{eq,\,Me} + \Delta E}{b_A} \tag{3.21}$$

and, after some rearrangement:

$$i_{A,\,Me} = i^0_{Me} \times \exp\frac{E_{corr} - E_{eq,\,Me}}{b_A} \times \exp\frac{\Delta E}{b_A} \tag{3.22}$$

where $b_A = \frac{RT}{(1-\alpha)zF}$. Finally, we will have

$$i_{A,\,Me} = i_{corr} \times \exp\left(\frac{\Delta E}{b_A}\right) \tag{3.23}$$

the same time, on the cathode:

$$i_{C,2} = i_{corr} \times \exp\left(\frac{\Delta E}{b_C}\right); \quad b_C = \frac{RT}{\alpha zF} \tag{3.24}$$

As already discussed, the current measured in such an experiment will be the difference between the anodic current and the cathodic one; therefore:

$$i = i_{A,\,Me} - i_{C,2} = i_{corr}\left[\exp\frac{\Delta E}{b_A} - \exp\left(-\frac{\Delta E}{b_C}\right)\right] \tag{3.25}$$

In the case of small ΔE, we can develop the exponential function into a series and taking only the first term, we will get

$$i = i_{corr} \times \Delta E \times \left(\frac{b_C + b_A}{b_C \times b_A}\right), \quad \text{where } \frac{\Delta E}{i} = R_{pol} \tag{3.26}$$

and, finally, we will have a simple relation between the polarization resistance R_{pol} and the corrosion current i_{corr}:

$$i_{corr} = \frac{1}{R_{pol}} \times \left(\frac{b_C + b_A}{b_C \times b_A}\right) \tag{3.27}$$

where b_A and b_C are slopes of plot $\log|i|$ versus applied potential E, for E larger than approximately 0.1 V and −0.1 than E_{corr} for anodic and cathodic polarization, respectively. The value of current i can be measured by electrochemistry in a classical three-electrode cell, with metal of interest as an anode. Even though the i_{corr} (Fig. 3.12) cannot be directly measured, it can be indirectly estimated with the help of the Tafel plot (see Chapter 1, eq. (1.108)).

3.4.2 Electrochemical impedance spectroscopy in corrosion

Electrochemical impedance spectroscopy is very well suited to study corrosion. In particular, it is a very powerful tool to study various coatings used to prevent corro-

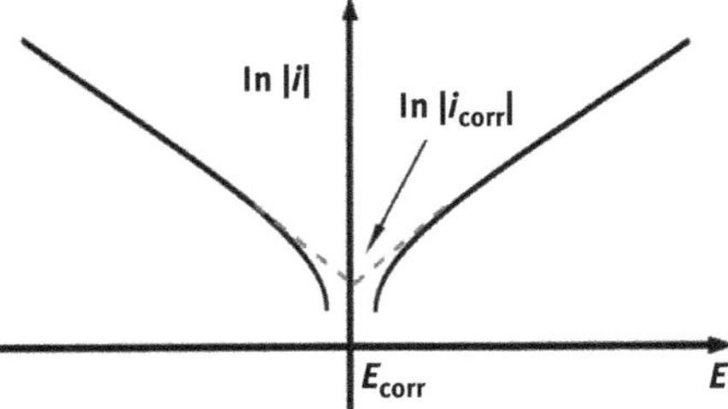

Fig. 3.12: Tafel plot allowing for the evaluation of i_{corr}.

sion. An ideal metal surface resistant to corrosion (ideal polarizable electrode), as judged for instance from the Pourbaix diagram, is a pure capacitor, C, in series with the solution resistance, R_S. Capacitance of such capacitor is related to the electrical double layer capacitance, discussed in Section 2.2. The equivalent electrical circuit, corresponding to this approach, was also shown there. However, the real metal/solution interfaces rarely behave in an ideal manner; they undergo corrosion – oxidation with charge transfer across the interface. Then, the equivalent circuit has to be modified to include the charge transfer resistance, R_{ct}, or polarization resistance, R_P, short-circuiting the interfacial capacitance, C. This is shown in Fig. 2.10 as simplified Randles model.

As in the case of plots in Section 2.2, the EIS Nyquist and Bode plots presented in this chapter were simulated using freeware **EIS Spectrum Analyser,** by **Genady Ragoisha**, Research Institute for Physical–Chemical Problems, Belarusian State University. If we now include the diffusion contribution to the corrosion process, for example, oxygen diffusion, then we will have to add the Warburg coefficient to the circuit. Thus, we obtained the Randles full equivalent circuit of mixed kinetic and diffusion process (Fig. 2.12).

In the real corrosion system, the metal is often already covered with corrosion products, such as an oxide layer (may be passive), the adsorption layer of inhibitors, protective passivating layer, or paint. In time, the protective layers may deteriorate and cracks or other types of defects may be formed. Therefore, we now consider an even more complex system, which involves the passivation or protection layer, for example, iron oxide layer, again in reference to the Pourbaix diagram (Fig. 3.7) at various stages of integrity: from the totally passivating one to that containing defects. First, let us consider ideally passivating/protective coating thin layer on a metal surface. Intuitively, the equivalent electric circuit for such a system should be represented by a solution resistance, R_S, and the capacitance of metal/coating/solution interface, C_C, in series with R_S. The value of dielectric capacitance coating, well adhered to the metallic surface is usually one to two orders of magnitude lower than that of the e.d.l. capacitance, C, so now even though the circuit looks identical to that describing the ideally polarizable electrode, it will differ in the capacitance value, for the same solution resistance. You may not see this difference in the Nyquist plot, but it will become evident if the data are plotted in the form of Bode graphs that separate the effects of imaginary and real parts of the data (capacitance and resistance, for short). During the long-term exposure of the coating to solution, corrosion is triggered in the cracks and pinholes. In the beginning, it is localized but can spread around in time. Thus, the circuit parameters will change in time, too, providing means of controlling corrosion progress. A pinhole or crack penetrating through the coating produces a double-layer capacitance of the clean metal interface, C_{edl}, greater than that of the coating: $C_{edl} > C_C$. The Faradaic reaction that proceeds is represented by the resistance, R_{PC}. And so, now we have the following equivalent circuit: R_S in series with a mesh containing C_C, parallel to R_{PC} that is in series with C_{edl} short-circuited with the polarization resistance, R_P. This circuit is shown in Fig. 3.13, with its Nyquist and Bode representation for the following data: $C_C = 1 \times 10^{-8}$ F, $C_{edl} = 1 \times 10^{-6}$ F, $R_S = 100$ Ohm, $R_p = 10$ kOhm, R_P 5 kOhm.

The complex plane (Nyquist) plot of this circuit displays two semicircles with low-frequency impedance decreasing with further deterioration of the protective coating. Electrochemical impedance spectroscopy has proven to be ideally suited for studies of the quality of protective coatings (passivating or organic layers), resulting in the publication of ISO norms. Other electrochemical studies discussed above are not sensitive enough, although much simpler and easy to use. For the case of EIS, however, one has to assure that the experiments and modeling are carried out correctly. Also, the equipment has to be capable of measuring at high frequencies and resistances that are required for thin dielectric films [10].

3.5 Localized corrosion: pits, crevices, intergranular corrosion – oxygen reduction as accompanying cathodic reaction

Localized corrosion is observed in small, restricted, local areas on metallic surfaces. It is dangerous in transferring pipelines and storage tanks, particularly in chemical processing plants. Contrasting to general corrosion affecting whole areas of metallic constructions and equipment, localized corrosion penetrates the metal very rapidly, with rates often several orders of magnitude higher than the corrosion rates for general corrosion of the very same metal or alloy. It is also very difficult to localize in advance, causing a substantial threat to life and damage to the installations and engineering systems. Localized corrosion is mainly due to heterogeneities and defects in metal, metal coatings, or environment and takes place wherever the anodic and cathodic sites on a metal surface can form corrosion microcells. Three main types of localized corrosion can be distinguished, such as pitting corrosion, crevice corrosion, and intergranular corrosion; the latter was described earlier, in the microcell corrosion section.

Pitting corrosion is caused by local dissolution or defects of passive film on the metal surface. Cavities are formed in this process, surrounded by an intact passivated surface [5, 6]. Pitting can be initiated by various factors and the most common are:

– Localized chemical or mechanical damage to the protective oxide film. Acidity of aqueous solution, low dissolved oxygen concentration or high concentrations of

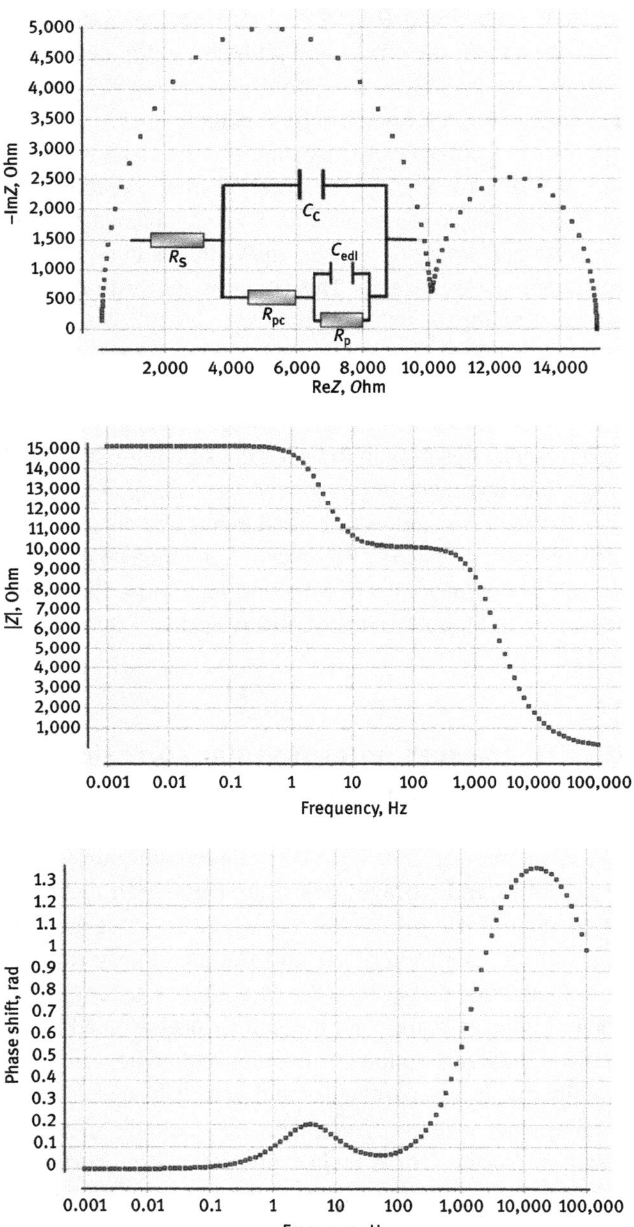

Fig. 3.13: Electrochemical impedance spectroscopy with simulated results for a metallic surface covered with protective (passivating) coating with a small pinhole/crack penetrating through the passivating layer. The electric equivalent circuit and its simulated response are shown in the form of Nyquist and Bode plots. Circuit parameters are described in the text.

chloride (such as in seawater) or other aggressive compounds found in the chemical processing plants can account for majority of pitting damage.
- Localized, poor application of protective coatings.
- Localized nonuniformities and inclusions in the metal surface.

Pitting corrosion is most difficult to detect, predict, and protect against, because, apart from a very small area of corrosive attack, corrosion products may often cover the pits with semipermeable “plugs” that do not form passivating layer. Pitting corrosion can also be harmful due to the formation of stress areas with fatigue and stress cracking that may be stimulated at the bottom of corrosion pits.

Crevice corrosion appears in confined areas of the metal surface – crevices, where access to the environment is restricted [6]. This type of corrosion is usually associated with a stagnant solution in the microenvironment of the site under attack. Such stagnant surrounding, where mass transport is limited, typically appears in narrow crevices (e.g., under gaskets, insulation material and slightly detached coating layers) or under deposits. Crevice corrosion is usually initiated by changes in local chemistry within the crevice, such as depletion of oxygen, inhibitor, changes in pH, and increase of concentration of aggressive ions (e.g., Cl^- and S^{2-}). For example, metal ions produced from the anodic reaction can accumulate in the crevice. The increase of the metal ion concentration in the stagnant region promotes, in turn, the hydrolysis reaction between the metal ions and water. The progress of this reaction will gradually lower the local pH in the crevice, facilitating further corrosion and increasing its rate. The most common form is oxygen depletion cell corrosion. It occurs because the stagnant solution inside the crevice has lower oxygen content as compared to the surface of a metal or alloy immersed in an aqueous environment. Thus, the crevice site forms an anode at the metal surface, whereas the metal surface in contact with a normally aerated aqueous environment forms a cathode. The oxygen differentiation corrosion is not limited to crevices. It appears everywhere where the concentration gradient of oxygen is formed. All metal constructions that are designed for seawater, or penetrating wet soil, are susceptible to the differential oxygen concentrations, as illustrated in Fig. 3.14.

3.6 Hydrogen evolution as accompanying reaction – role in corrosion: embrittlement and cracking

Hydrogen evolution may occur during the various steps of manufacturing of metal structures and constructions (e.g., cathodic protection, electroplating, and welding) or proton reduction process, accompanying the primary corrosion reaction. Hydrogen may also appear over time through environmental exposure, such as soil, salt, water, or gradual corrosion of coatings. In the practical use of iron-derived material, such as steels, atmospheric corrosion plays an important role in hydrogen generation and

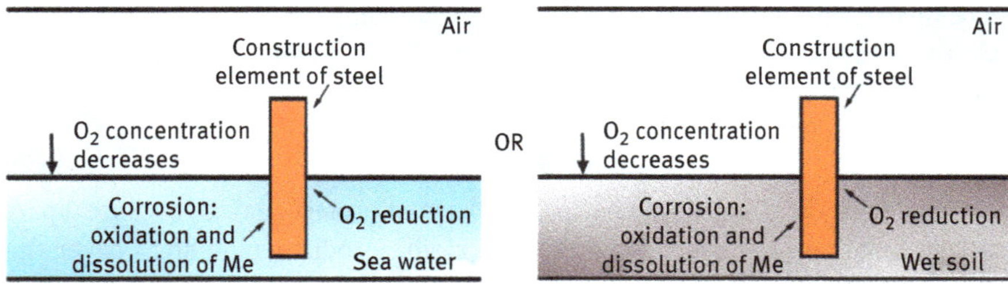

Fig. 3.14: A scheme of metal corrosion due to the difference in oxygen concentration.

partition into the corroding construction, even though the primary cathodic reaction is the reduction of oxygen. The cathodic reaction of hydrogen evolution results from the combination of several elementary steps. In acidic media the following reactions will proceed:

Volmer reaction:	$H_3O^+ + e \leftrightarrow H_{ad} + H_2O$
Tafel reaction:	$H_{ad} + H_{ad} \leftrightarrow H_2$
Heyrovsky reaction:	$H_{ad} + H_3O^+ + e \leftrightarrow H_2 + H_2O$

Whereas in alkaline solutions:

Volmer reaction:	$H_2O + e \leftrightarrow H_{ad} + OH^-$
Heyrovsky reaction:	$H_{ad} + H_2O + e \leftrightarrow H_2 + OH^-$

Obviously, the source of electrons is the anodic dissolution (corrosion) of the metal.

Most of the hydrogen atoms present at the surface of a metal desorb from the surface as H_2 formed in Tafel and Heyrovsky reactions. However, small part of hydrogen atoms being adsorbed, H_{ad}, produced during Volmer reaction become absorbed, H_{ab}, by the surface layer of the metal and can diffuse through the metal structure, accumulating at the grain boundaries, defective microscopic voids, creating pressure from within the metal, for instance, in iron, the specific volume of the absorbed hydrogen, V_H, can reach approximately 2.6 cm^3/mol corresponding to the pressure build-up to approximately 10^5–10^7 atm. This pressure can easily reduce metal strength up to the point where it cracks – the so-called **hydrogen embrittlement** or hydrogen-assisted cracking. Hydrogen embrittlement is the process by which metals become brittle and susceptible to fracture due to the introduction and subsequent diffusion of hydrogen into the metal. Metals become more susceptible to hydrogen embrittlement under stress: cracking will initiate faster when the metal structure is subjected to stress. Although hydrogen atoms embrittle a variety of substances, including steel and other metals and alloys, hydrogen-assisted cracking of high-strength steel is of the most importance.

Hydrogen embrittlement can be prevented by several methods, mainly relying on the minimization of the contact between the metal and hydrogen, particularly during fabrication and use. Often preheating or postheating can be applied to force hydrogen to diffuse out of a structure before the latter can be used. However, one has to consider that if high temperature is applied, hydrogen can react with carbon atoms present in steel, forming methane. Methane molecules will not diffuse out of steel but accumulate in tiny voids and grain boundaries at high pressures, also initiating cracking of the material.

Much more about corrosion fundamentals, as well as about some other topics described in the text here, the interested reader can find in [1–9].

3.7 Protection against corrosion

Protection against corrosion is a must nowadays because anything made of iron or steel is a prime target of corrosion processes. Therefore, we will focus now on corrosion inhibitors. They are chemical substances used as additives in very small quantities to the environment. These substances can inhibit and either minimize or prevent corrosion. An efficient inhibitor should be environmentally friendly, economical for application, and producing stable effect even if used in small concentrations. The efficiency of inhibitors, P, is evaluated by making a direct comparison of corrosion rate with and without the presence of the inhibitor:

$$P = (v_0 - v/v_0) \cdot 100$$

where v_0 is the corrosion rate in the absence of inhibitor, and v is the corrosion rate in the same environment in the presence of inhibitor.

A qualitative classification of inhibitors is shown in Fig. 3.15, yet for the sake of clarity and to stay within the scope of this book, we will discuss on cathodic and anodic inhibition; however, it can be seen that, for example, cathodic inhibitors can also be classified as scavengers, decreasing oxygen concentration (hydrazine – N_2H_2) or increasing pH (Sb_2O_3) (Fig. 3.15).

Let us now recall eq. (3.7), relating the corrosion current with the exchange currents of metal, the $i_{0,Me}$, environmental couple, $i_{0,2}$, and the equilibrium potential difference of both participating couples, $E_{eq,2} - E_{eq,Me}$:

$$i_{corr} = (i_{0,Me} \times i_{0,2})^{\frac{1}{2}} \times \exp\left[\frac{zF(E_{eq,2} - E_{eq,Me})}{RT}\right]$$

Close inspection of this equation provides us immediately with tools that rely on the decrease of the product of the exchange currents and can be used to protect the metal against corrosion.

1. *Corrosion inhibitors I.* Inhibition of corrosion by the addition of cathodic inhibitors; decrease of $i_{0,2}$ no change of $i_{0,Me}$.

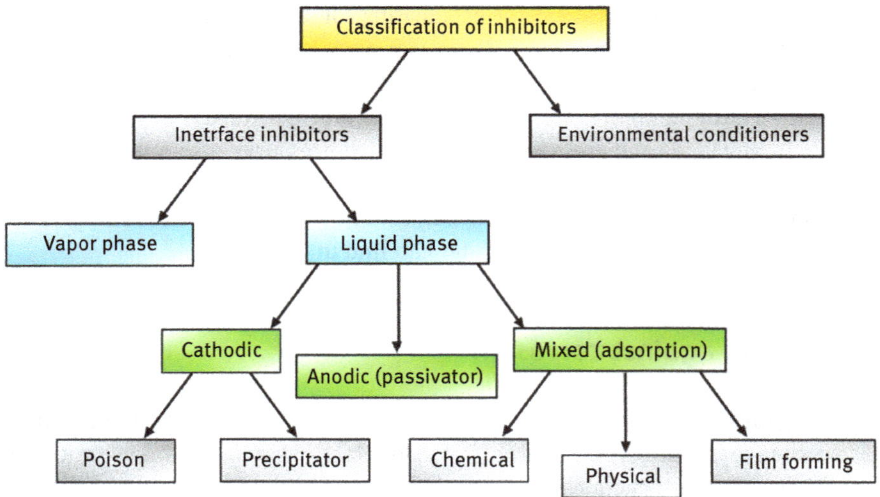

Fig. 3.15: Classification of corrosion inhibitors.

This type of cathodic inhibition can be achieved in two ways, all of them lead to a shift of E_{corr} toward the equilibrium potential of metal:

– by inhibiting the reduction of H^+ as cathodic reaction, for example:
  by adding Sb_2O_3. Then, the following reaction will proceed:

$$Sb_2O_3 + 6\,H^+ + 6\,e \rightarrow 2\,Sb + 3\,H_2O$$

– by inhibiting oxygen reduction as cathodic reaction, for example:

$$2\,N_2H_2 + 6O_2 \rightarrow 4\,NO_3^- + 4\,H^+$$

or

$$2\,SO_3^{2-} + O_2 \rightarrow 2\,SO_4^{2-}$$

Graphically, this type of protection can be presented on Evans diagram (Fig. 3.16).

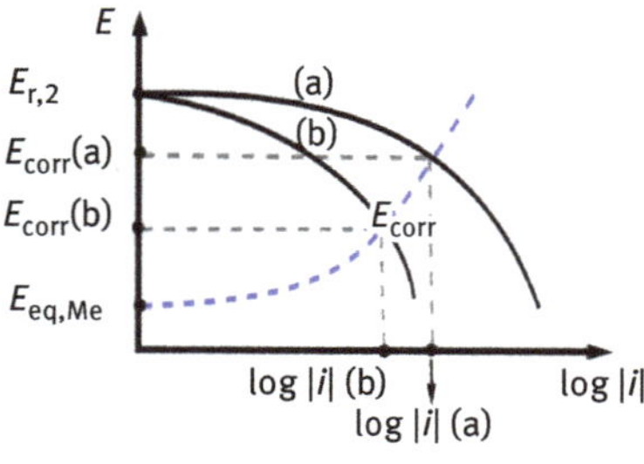

Fig. 3.16: Evans diagram in the case of the presence of a cathodic inhibitor.

In this diagram, E_{corr}(a) and E_{corr}(b) stand for corrosion potentials before and after inhibitor was added, respectively, and i(a) and i(b) are corrosion currents before and after the inhibition, respectively.

2. *Corrosion inhibitors II.* Inhibition of corrosion by the addition of anodic inhibitors; decrease of $i_{0,Me}$ no change of $i_{0,2}$. E_{corr} increases toward $E_{eq,2}$.

Anodic inhibitors are chemical substances that alter the anodic reaction of metals, forcing their surface into the passivation region (see Pourbaix diagram). They also form the protective oxide films on the surface of metal; therefore, they are frequently referred to as passivators. In the kinetic Evans diagram, the oxidation reaction current diminishes (from curve a to curve b in log scale), shifting also the corrosion potential E_{corr} toward the equilibrium, Nernstian potential of the environmental counterpart, $E_{eq,2}$, as shown in Fig. 3.17.

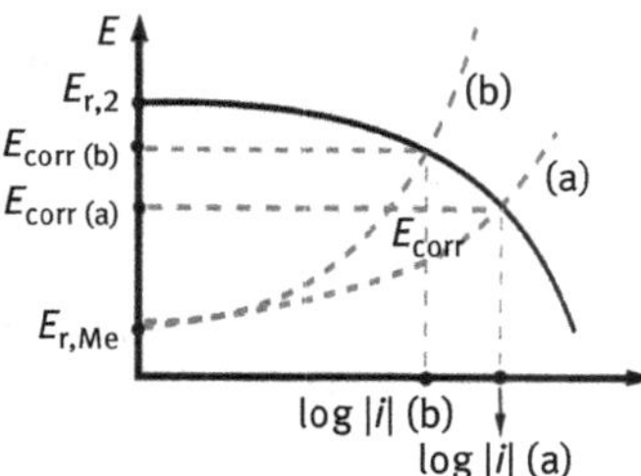

Fig. 3.17: Evans diagram change induced by an anodic inhibitor.

Examples of anodic inhibitors include
- chromates,
- molybdates,
- tungstates,
- nitrites,
- orthophosphates,
- organic compounds containing heteroatoms (S, O, N), double bonds, and/or polar functional groups.

Due to the chemical characteristics of anodic inhibitors, they are considered environmentally unsafe.

Most authors suggest that synthetic organic inhibitors work by being adsorbed onto the metal surface through electrostatic interactions or interactions with empty *d*-orbitals of iron atoms. Nevertheless, there exist still major drawbacks of such compounds related to their toxicity, non-biodegradability, or production costs. Therefore, to overcome these problems, the researchers turned to cheap, eco-friendly, green corrosion inhibitors, derived from plant extracts – seeds, fruits, roots, barks, and leaves. These extracts contain a large number of phytochemicals with their green anticorrosion activity due to the structural presence of nitrogen, oxygen, or sulfur heteroatoms, as well as unsaturated organic rings and double bonds which can adsorb on the surface of the metal and provide effective inhibition against the corrosive environment. Moreover, the source plants for such inhibitors are renewable, of relatively high availability, and reasonably simple and low-cost extract processes, utilizing environmentally safe solvents like water or alcohol. However, this approach is still far from being commercialized. Much more on the principles, design, and utilization of eco-friendly corrosion inhibitors can be found in [11].

3.7.1 Electroplating

Electroplating is the method that uses electrochemical deposition of a metal to form a thin, coherent metal coating layer on another metal that serves as an electrode. Electroplating can be used to protect iron and iron-derived metals (e.g., steel) against corrosion, but more generally, it can be used to change the surface properties (functional and/or decorative) of metal, such as wear resistance, lubricity, abrasion, or even the aesthetic qualities. As it forms an additional layer of metal, it may also be used in electroforming to build up forms on a model or to add size on a metallic substrate. The metal to be plated is connected as the cathode in a DC circuit, whereas the metal to be electrodeposited can be dissolved in the electrolyte or can be the anode that is sacrificed by its electro-oxidation. The choice of metal for the protective layer depends on its purpose:
- functional/decorative, or
- sacrificial coating.

A functional or decorative layer is used primarily to enhance the functionality of the metal or its attractiveness. Common metals used for such a purpose include nickel, copper, chromium, zinc, and tin; less frequently, for special purposes (e.g., electronics and jewelry) these will be gold, silver, platinum, ruthenium, rhodium, palladium, and iridium. Let us present some of the electroplating examples below.

Nickel electroplating is widely used for its corrosion-resistant properties. Such coverage is an excellent replacement for chrome electroplating because the overall procedure of nickel electroplating is less harmful to the environment. Additionally, the nickel layer is aesthetically pleasing. The layer of nickel also provides good lubricity.

The silver coating layer in engineering is done mainly to increase the electronic and heat conductivity of the base metal, a feature necessary for electronic devices, as well as for its excellent corrosion resistance. One has to consider also the enhancement of the aesthetic look of the product coated with polished silver layer. This feature is due to the very good light reflectance of silver.

We now address the so-called sacrificial coating, where the coating layer serves not only as a mechanical barrier against corrosion of underlying iron or steel but participates actively in its protection. The metal used for such coating is sacrificial, that is, it is used up in the reaction. Common examples are zinc and cadmium, the latter being gradually prohibited because of the environment. The remarkable effectiveness of zinc electroplating and relatively low cost of such procedure have made it the most common and widespread choice for protecting surfaces in all types of processes. Almost one-third of all zinc metal is used for such protective coatings. When zinc is electroplated onto the surface of ferrous metals, it is exposed in most cases to atmospheric oxygen and humidity, producing zinc hydroxide:

$$2\,Zn + O_2 + 2\,H_2O \rightarrow 2\,Zn(OH)_2$$

Then, in the presence of atmospheric carbon dioxide (greenhouse gas), the hydroxide combines to produce an insoluble layer of zinc carbonate:

$$Zn(OH)_2 + CO_2 \rightarrow ZnCO_3 + H_2O$$

This insoluble, gray layer of zinc carbonate adheres strongly to the underlying zinc, sealing it from further rusting. This way also the iron or steel construction covered with a protective zinc layer is safe from corrosion.

Finally, electroplating creates a much stronger bond with the surface of the material to protect it from corrosion when compared with other methods, such as painting.

Bibliography

[1] Revie RW, Uhlig HH, Corrosion and corrosion control. An Introduction to corrosion science and engineering. John Wiley & Sons, Inc., 2008.

[2] Landolt D. Corrosion and surface chemistry of metals. CRC Press, Taylor & Francis Group, LLC, Boca Raton, FL, 33487, 2007, first edition, EPFL Press.

[3] Pourbaix M. Lectures on electrochemical corrosion. Plenum Press, New York-London, 1973.

[4] Pedeferri P. Corrosion Science and engineering. Engineering Materials book series. Lazzari L, Pedeferri MP, Springer Nature Switzerland AG, 2018.

[5] Szklarska-Smialowska Z. Pitting corrosion of metals. Natl. Assn. Corr. Eng.. 1986.

[6] Szklarska-Smialowska Z. Pitting and crevice corrosion. Houston Texas, NACE International Corrosion Soc., 2005.

[7] Tait WS. Electrochemical corrosion basics. In: Kutz M., ed., Handbook of environmental degradation of materials. 3rd ed., William Andrew Applied Sci. Publishers., Elsevier Inc. 2018, Chapter 5, pp. 97–115.

[8] Popov BN, Lee J-W, Djukic MB, Hydrogen permeation and hydrogen-induced cracking. In: Kutz M., ed., Handbook of environmental degradation of materials. 3rd ed., William Andrew Applied Sci. Publishers., Elsevier Inc. 2018, Chapter 7, pp. 133–162.

[9] Ohtsuka T, Nishikata A, Sakairi M, Fushimi K. Electrochemistry for corrosion fundamentals. Springer, 2018.

[10] Lasia A. Electrochemical impedance spectroscopy and its applications, Springer-Verlag, 2014.

[11] Guo L.,Verma Ch., Zhang D., (Eds.), Eco-Friendly Corrosion Inhibitors. Principles, Designing and Applications, Elsevier, 2022.

4 Electrocatalysis

4.1 General remarks

Electrocatalysis is a type of heterogeneous catalysis that results in the increase of the rate of a reaction occurring at the electrode–electrolyte interface. In this type of catalysis, the electrode surface or electrode surface covered with catalytic species (e.g., metal particles and bimetal particles) play the role of catalyst. The catalyst assists in transferring the electrons between the electrode and reactants or/and facilitates the chemical transformation of adsorbed intermediate formed during reaction. Before going further we will briefly characterize the catalysis, looking for differences in both types.

Catalysis is the process in which the addition of some substance increases the rate of chemical reaction. This substance named catalyst is not consumed and not changed chemically in the overall reaction. Catalyst reacts with one or more reactants to form an intermediate, which subsequently gives the final reaction product and regenerated catalyst:

$$A + B \rightarrow P, \quad A + K \leftrightarrow A \cdots K, \quad A \cdots K + B \rightarrow K + P$$

where A, B – reactants, P – product, K – catalyst, $A \cdots K$ – intermediate.

In the presence of a catalyst, chemical reactions occur faster because the catalyst provides an alternative reaction pathway with lower activation energy than in the case of a noncatalyzed reaction. Figure 4.1(a, b) illustrates the changes in energy and reaction pathway during catalyzed reaction.

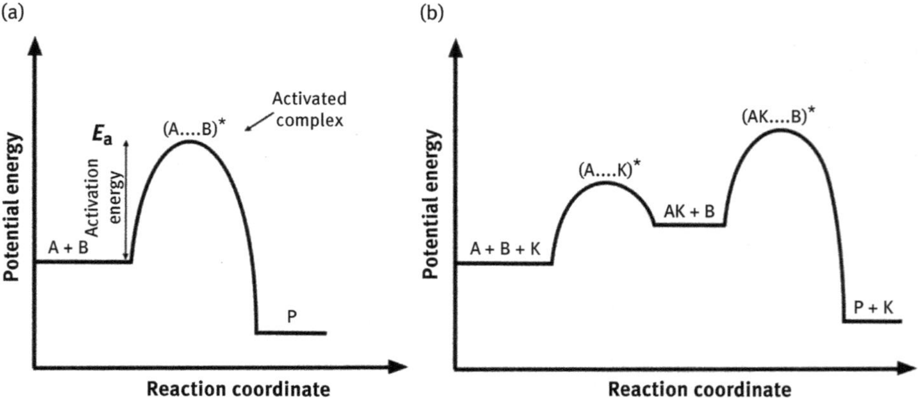

Fig. 4.1: Schemes showing the changes in energy and reaction pathways between the noncatalyzed (a) and catalyzed (b) chemical reactions. A, B – reactants; P – product; K – catalyst.

https://doi.org/10.1515/9783111160986-004

Catalysts have no effect on the chemical equilibrium and do not change the equilibrium constant $K = \frac{k_1}{k_2}$, since the rate constant of both the forward k_1 and the backward k_2 reactions are affected in the same way. In consequence, the catalysts cannot change the standard Gibbs free energy of reaction $\Delta_r G^0$. Two types of catalysis can be distinguished – homogeneous and heterogeneous one, depending on whether a catalyst is in the same phase as the substrate or not. Homogeneous catalysis means that catalyst and reagents exist in the same phase. Such kinds of electrocatalysis can be found when, for instance, soluble organometallic compounds are used as catalysts in the solution together with reactants. In heterogeneous catalysis, the solid-state catalysts are used and reactions take place on their surface. The reactions occur when the reagents are strongly adsorbed, chemisorbed on the catalyst surface, and some bonds are formed between a catalyst and adsorbed molecules, resulting in intermediate formation. On the other hand, if the strength of bonds becomes too great, the activity of catalyst deteriorates because the adsorbed molecules block the active sites of the catalyst. The activity of the catalysts is indicated by the "volcano" curve and will be discussed later. In heterogeneous catalysis, the total surface area of the catalyst has a great impact on its activity. Because of that, some materials with very large surfaces are used as the carriers for catalyst species (e.g., zeolites or other mesoporous materials). The other way to improve the catalytic activity and increase the reaction rate is to minimize the size of catalyst particles.

The electrocatalysis has many common features with classic heterogeneous catalysis. Many catalytic materials used in both types of catalysis are the same such as Pt, Pd, Ni, or oxides. In both cases, the rate of reaction depends on the catalyst properties such as chemical composition and surface area of catalyst. Also in both cases, the catalyst increases the rate of reaction by the acceleration of the slow step or by changing the reaction pathway. What does distinguish the electrocatalysis from chemical catalysis? The main difference is the dependence of the electrocatalytic reaction rate on electrode potential, more exactly on the potential difference across the interface, which can be modified by an external source of voltage. Following equations show the differences in the rates (v) of chemical and electrochemical reaction and roughly in the catalytic and electrocatalytic reactions of first order, as well:

$$\text{chemical catalysis:}\quad \mathrm{A} \xrightarrow{k_1} B,\quad v_r = c_A k_1,\qquad v_r = c_A P \exp\left(-\frac{\Delta G^{0\#}}{RT}\right) \tag{4.1}$$

$$\text{electrocatalysis:}\quad \mathrm{A}^{n+} + ne \xrightarrow{k_2} \mathrm{B},\qquad v_r = nFc_A k_2,\quad v_r = \frac{j}{nF}$$

$$v_r = c_A P \exp\left(-\frac{\Delta G^{0\#}}{RT}\right)\exp\left(-\frac{\alpha nFE}{RT}\right) \tag{4.2}$$

where $\Delta G^{0\#}$ is the Gibbs free energy of activation (activated complex theory), P is the preexponential factor, E the electrode potential, j the current density, and α the transfer coefficient.

From eq. (4.2), it is clearly seen that small changes in electrode potential cause the tremendous increase of reaction rate, even by several orders of magnitude.

The other features that differentiate the electrocatalytic and catalytic reactions are: (i) the participation of electrons in the electrocatalytic reaction (they must be supplied to or withdrawn from the catalyst surface), (ii) the presence of nonreactive species in solution such as solvent and ions, (iii) the electric field existing at the electrode–solution interface can significantly change the surface properties of catalyst and affect the rate of reaction, and (iv) temperature of electrocatalytic reactions is much lower than the chemical catalytic reaction.

4.2 How to compare the activity of catalysts in electrochemical reactions?

In practice, the problem that arises is the selection of well-determined quantitative parameters characterizing the activity of electrocatalyst. Generally in electrochemical experiments, the relation current–potential is measured. To eliminate the influence of the differences in the surface area of electrodes, the results are presented as current density–potential plots. In many cases, the geometric area of the electrode is used for current density calculations, but in the case of electrocatalysis, we have to use the "true" (real) working surface area. This real surface area can be determined by in situ or ex situ methods, such as hydrogen adsorption, voltammetry, BET (Brunauer S, Emmet P, Teller E isotherm), and porosimetry. If metal nanoparticles are used as catalyst on conducting substrate, it is possible to determine their amount by electrochemical quartz microbalance (see Section 2.3).

Let us consider the same oxidation reaction occurring with electrocatalysts 1 and 2. The current density in both cases is described by the Volmer–Butler equation (Chapter 1):

$$j_1 = j_{0,1}\exp\left(\frac{(1-\alpha_1)nF(E-E_{eq})}{RT}\right) \tag{4.3a}$$

$$j_2 = j_{0,2}\exp\left(\frac{(1-\alpha_2)nF(E-E_{eq})}{RT}\right) \tag{4.3b}$$

$$\eta = E - E_{eq}$$

If the pathway of reaction is the same in both cases, we may assume that transfer coefficients α are equal: $\alpha_1 = \alpha_2$. Then the ratio of current density is expressed by the ratio of exchange current densities:

$$\frac{j_1}{j_2} = \frac{j_{0,1}}{j_{0,2}} \tag{4.4}$$

Since the exchange current density is related to the standard rate constant (Chapter 1), we may also use this parameter for comparison of different electrocatalysts. The higher the exchange current density value, the better the electrocatalyst for reaction is. We can also fix the current density values, for example, 50 mA/cm^2 and compare the overpotential values at which such current density is reached. The lower the overpotential value, the better the electrocatalyst is. In practice, if the reaction is irreversible or very complicated (E_{eq}, η are not determined), we use working Tafel expression $E = a + b\log|j|$ and compare the catalyst at chosen potential or current density. It is illustrated in Fig. 4.2(a, b) for the same reaction occurring at two different catalysts (electrodes).

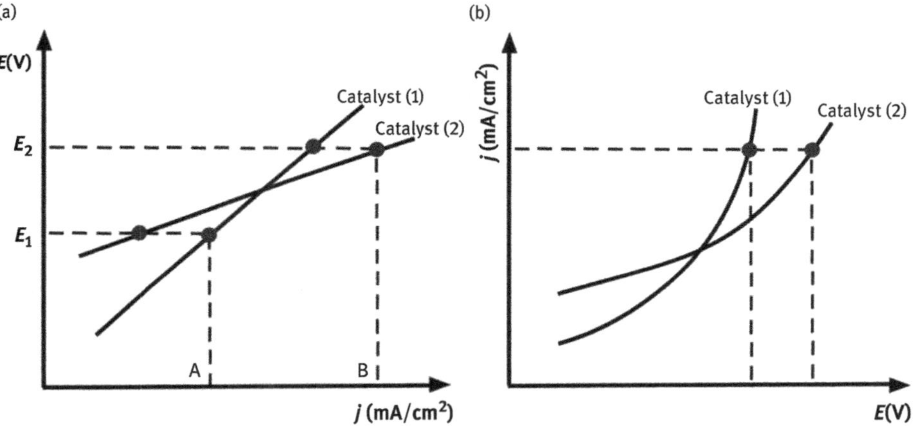

Fig. 4.2: Schemes illustrating how to compare the activity of catalysts in an electrochemical reaction. Scheme (a) presents potential–current density curves for two catalysts. The best catalyst has a higher value of current density at the same potential value. Scheme (b) presents current density–potential curves for two catalysts. The best catalyst has a lower value of potential at the same current density value.

It is clearly seen that in the case (a), in the region (A) the first (1) and in the region (B) the second (2) catalyst are more active showing a higher value of current density. At constant current density (case b), this catalyst is better, for which the same current value is attained at lower potential (catalyst 1) [1, 2].

4.3 Electrocatalysts

The optimum electrocatalytic material containing the catalyst species and the support should be characterized by large, catalytically active surface area with active catalytic sites distributed uniformly across the support surface. The support should prevent the agglomeration of catalytic species, exhibit the activating interaction with them, and provide good electrical conductivity to catalytic centers. In heterogeneous electrocatalysis, plenty of materials of various compositions and sizes are applied as catalysts. Below we will briefly describe some materials used as catalysts and supports.

4.3.1 Metals, alloys, and oxides

The metal electrodes are widely used in electrochemical laboratories and in technological processes: electrolysis, electroplating, as sources of energy. Their usefulness is only limited by the potential range in which they are electrochemically stable, and do not undergo dissolution or oxidation. The catalytic activity of metals is mostly governed by the electron structure of their atoms, more exactly by the existence of free unpaired electrons in d-orbitals and their ability to form bonds with adsorbed reactants. Catalytic activity of metals varies from metal to metal and from reaction to reaction. The high catalytically active metals belong to subgroup VIIIB of elements. These are irons: Fe, Ru, Os; cobaltous: Co, Rh, Ir; platinums: Ni, Pd, Pt. Platinum is the most universal catalyst and is stable in wide potential range in acidic and alkaline solutions. Platinum and the metals of its group are used in many reactions such as hydrogen evolution (HER), oxygen reduction (ORR), and organic compound reduction, for example, nitrobenzene and benzofuroxan.

All these materials are very expensive and their amount is limited and because of that the scientific works went and still continue in two directions: (i) the formation of bi- or multicomponent catalysts with very high activity and (ii) to minimize the catalyst amount by decreasing particle size of a catalyst (nanoparticles, ad-atoms are used) or by application of thin films as electrode. Catalysts containing two metals were examined in many reactions taking place in fuel cells, such as oxidation of formic acid (FA), oxidation of methanol, and HRE. Bimetallic catalysts can be prepared in many ways. The most straightforward way is to prepare a bulk alloy by metallurgy, that is, melting of two components together, followed by homogenization. The other methods used are chemical precipitation and further annealing or the electrochemical deposition (Chapters 5 and 6). Some surface alloys were prepared by using layer-by-layer deposition in the UHV (ultra-high vacuum) system and their catalytic activity was tested in the oxidation reaction of CO. The subject of intensive studies is the difference between the bulk and surface composition of alloys [3]. It is well documented that the annealed surface of alloy is enriched in Pt amount, when Pt bulk alloys composition is also Pt rich, for example, Pt_3Me (Me – Sn, Fe, Co, Ni, Ru, Ag). The most popular and highly investigated catalyst is Pt–Ru alloy. In the late 1960s, the strong synergy effect was found for this system. The increase of three order magnitude in catalytic activity was observed at anodic oxidation of methanol in comparison with pure Pt, whereas pure Ru is inactive for this reaction. This alloy–catalyst has also another feature; in contrast to pure Pt, it is not poisoned by CO_{ads} species. The mechanism currently accepted is the adsorption of methanol molecules at Pt sites of the surface and their partial oxidation:

$$Pt + CH_3OH \rightarrow Pt(CH_3OH)_{ads}$$

$$Pt{-}CH_3OH_{ads} \rightarrow Pt{-}(CO)_{ads} + 4H^+ + 4e$$

The oxidation of CO_{ads} is accelerated by molecules containing oxygen atom and adsorbed on Ru sites close to Pt sites:

$$Ru + H_2O \rightarrow Ru{-}OH_{ads} + H^+ + e$$

$$Ru{-}OH_{ads} + Pt{-}\,(CO)_{ads} \rightarrow Pt + Ru + CO_2 + H^+ + e$$

The catalytic activity Pt–Ru depends on the surface concentration of Ru. The effect is significant even at very low Ru concentration. For these pathways, the quantitative model was developed connecting the oxidation rate of methanol with a surface concentration of Ru in alloy. Generally, the methanol oxidation reaction may be described by the so-called bifunctional mechanism.

The other organic compound with potential application in the fuel cells is FA. The mechanism of electrooxidation of FA on Pt and other Pt-group metals was widely investigated. Dual pathway of reaction was proposed:

$$HCOOH \xrightarrow{\text{dehydrogenation}} CO_2 + 2H^+ + 2e$$

$$HCOOH \xrightarrow{\text{dehydration}} CO_{ads} + H_2O$$

The mechanism of FA electrooxidation on Pt electrode and Ru differs. In the first case, the overall reaction of HCOOH on Pt is dehydrogenation reaction with the dehydration reaction being the blocking reaction. On Ru electrode, oxidation of HCOOH proceeds via dehydration and formation of adsorbed CO species that are finally oxidized to $CO_2(CO + H_2O \rightarrow CO_2 + 2H^+ + 2e)$. The rate of HCOOH electrooxidation on Ru electrode is about 20 times lower than on the Pt electrode. The application of Pt–Ru alloy increases a few times the rate of FA oxidation in comparison with Pt as being a result of the increase of CO_{ads} oxidation rate. On the Pt–Ru alloy, two pathways are included in the oxidation reaction. The best results were obtained for the alloy with surface Ru concentration of 50%. To know much more about bimetallic electrocatalysts, interested readers will find in refs. [2, 3].

The surface of many metals is covered with spontaneously formed oxides, and other more noble metals become covered during anodic polarization. The composition and properties of oxides obtained electrochemically depend on the potential and electrolyte used. In their structure, vacancies can be found as a result of nonstoichiometric formation and lack of oxide or metal ions in a crystal lattice or on the surface. Many of oxides are semiconductors characterized by energy band >2.5eV and low intrinsic conductivity. The special doping by donor or electron acceptors increases the conductivity and changes the type of conductivity from intrinsic to n or p (Chapter 10). The semiconducting oxides are widely used in the photocells for water splitting and in photocatalysis (Chapters 11 and 12). The application of oxides as electrodes is limited. They are not stable, undergoing the reduction during cathodic reactions and often dissolving in acidic solution. The best-known oxide applied in batteries is PbO_2 (Chapter 8). The electrocatalytic properties have nickel oxide widely used in the oxidation of organic compounds. The oxide Co_3O_4 was tested as a catalyst in ORR [2]. In the last decade, the mixed oxides are of scientific interest as catalysts in chemical and electrochemical catalysis. The term mixed oxides is applied to solid ionic compounds containing the oxide anion O^{2-} and two or more types of cations. Typical examples are $FeTiO_3$ (a mixed oxide of iron (Fe^{2+}) and titanium (Ti^{4+}) cations) and compounds with spinel structure such as cobaltites MCo_2O_4 or perovskite structure RTO_3, where R is a rare earth element (La) and T is the transition metal (e.g., Ni, Co, and Mn). Since the perovskites are not conducting they have to be doped (usually with BaO) before being used as the electrode. The perovskites catalyze the oxygen evolution reaction from alkaline solutions. Many of the perovskites were tested in the reaction of oxygen reduction for future application in fuel cells [4, 5].

4.3.2 Carbon catalysts and supports

There are plenty of carbon materials that are applied in heterogeneous chemical catalysis as well as in the electrocatalysis [2]. Materials such as black carbon, activated carbon, and Vulcan carbon have a very large specific surface area (from 200 for Vulcan to 3,000 m^2/g for activated C), but the high electrical resistance limited their usefulness in electrochemistry. However, when mixed and sintered with conducting carbons (graphite powder), they are often applied as a support for catalyst particles. Glassy carbon (GC) often named vitreous carbon has large conductivity, almost smooth ideal surface, and the high values of overpotentials of water oxidation and reduction. The potential window of GC is large, the material is stable in acidic and alkaline solutions, because of that GC is widely used as an electrode material in electrochemical investigation. Vitreous carbon produced as the foam is called reticulated vitreous carbon. It is a good conducting material with developed surface and large volume and is often used in electrochemistry as three-dimensional electrode for nanoparticles, conducting polymer deposition. Graphite is another carbon material used in the form of paste as electrode material.

Graphite is a good conductor, is stable, and its surface area is large but unfortunately should be determined before experiments. Highly oriented pyrolytic graphite (HOPG) is an analog of metal single crystal (in carbon chemistry) with preferential crystallographic orientation. HOPG is often used for the in situ determination of adsorption or underpotential deposited (UPD) on adatom structure by means of AFM (atomic force microscopy) and scanning tunneling microscopy (STM). In recent years, the most intensively investigated material is graphene (Nobel price 2010, Andre Geim, Konstantin Novoselov), a two-dimensional carbon nanomaterial. The graphene (G) exhibits interesting and reach features that the perspectives of application are very wide. Structurally, graphene is a one-atom-thick planar sheet of sp^2 bonded carbon atoms closely packed in a honeycomb lattice. This material has unique physicochemical properties including excellent electrical and thermal conductivity, fast charge transfer mobility, mechanical flexibility, and very large specific surface area (theoretically 2,630 m^2/g for the single layer). Besides graphene, some of its derivatives can be obtained during synthesis such as graphene oxide (GO) and reduced graphene oxide (RGO). They contain oxygenated functional groups: – O, = O, – OH, or others, which can be involved in various chemical reactions. The graphene, GO, and RGO can be applied in electrocatalysis in two ways: as the catalytic support or as carbocatalyst. Graphene, RGO-supported Pt, Au particles, binary alloy composed of Pt, Au, Co, Ni, and oxides such as Co_3O_4, MnO_2, and Cu_2O were explored as ORR catalyst for fuel cell application [6].

Recently, nitrogen (N)-doped carbon nanomaterials have been developed. It was found that the incorporation of nitrogen in graphene (NG) or its derivatives results in enhancement of catalytic activity of metal particles due to improvement in metal–graphene binding. What is interesting is that NG itself exhibits the enhanced electrocatalytic activity toward the reduction of oxygen in comparison with undoped graphene. Recently, efforts were made to apply carbocatalyst (metal-free catalyst) such as graphene and graphene doped with N or other elements (e.g., B, S, Se, and P) as catalyst for oxygen reduction. However, the application of other elements than N has a small impact on the catalytic activity of graphene. The rich literature on this topic can be found in the review [6]. The other carbon nanomaterials that were used as a support for catalysts are carbon nanotubes (CNTs) and carbon nanofibers (CNFs).

4.4 Catalyst activity

Why some metals have catalytic properties? If we compare the electron structure of Pt, Ru, Ir, Co, and Ni atoms, we see common features. All these atoms have unpaired electrons in d-orbitals. Atoms of Pt and Ru have also unpaired electrons in s orbitals: Pt – $5s^2p^6d^96s^1$ and Ru – $4s^2p^6d^75s^1$. Such atoms can interact with species that are sources of electrons for creating the electron pairs. This interaction results in catalytic activity in the formation of intermediates. One may expect that we can predict the catalytic behavior of a metal and its catalytic activity on the basis of the electron structure of atoms, we can – but only to some extent. The large effort was made to find the correlation between bulk properties of a catalyst such as an electron work function (Chapter 1), heat of sublimation, melting temperature, or other physicochemical parameters and the activity of a catalyst. We will briefly describe here the electron work function (simply, work function) effect and after that, since the heterogeneous electrocatalytic reaction is a surface process, we will concentrate on the adsorption and size effect [1, 2, 7].

4.4.1 Electron work function effect

In the electrochemical reaction, electrons are transferred from the metal to reactants in the solution, so the rate of reaction should increase with lowering of the work function. But it is not the case. Two work functions should be distinguished: one term defines the work that must be done by the external force in transferring the electron from the metal into vacuum Φ^{M} and other term means the work of electron transferring from the metal to the contacting electrolyte Φ^{E}. It was shown [2] that work function Φ^{E} is independent of metal nature, when it is determined at the same electrode potential. However, this is true if some simple redox reaction is considered such as $Fe^{3+} + e \rightarrow Fe^{2+}$. The measurements of activation enthalpy $\Delta H_{a}^{0\#}$ (activation heat) carried out for this reaction on various electrodes have shown independence of $\Delta H_{a}^{0\#}$ of Φ^{M} [1, 2]. It means that the activation enthalpy of the reaction that involves only charge transfer is independent of the electron structure of the electrode. The decisive role in electrocatalytic reactions plays the surface process – adsorption.

4.4.2 Adsorption impact

The extent of the adsorption effect is dependent on the bond strength between catalyst and adsorbed reactant or intermediate species. Therefore, in consequence, it is dependent upon the Gibbs free energy of adsorption ΔG_{ads}^{0}, degree of surface coverage, θ, and influence of potential on adsorption parameters. If we consider, very roughly, few catalysts and an increase of bonding strength between them and adsorbed species, we should expect both the increase of θ and increase of the rate of the formation of reaction products. If the increase of θ in a series of catalysts is continued, their free surface for adsorption decreases causing also the decrease of reaction rate. It is the clash of opposite tendencies; in consequence, the reaction rate will increase until it reaches its maximum and then it decreases. The plots of the rate of reaction (i.e., exchange current density) against the enthalpy of adsorption (the heat of adsorption is connected with the strength of adsorption bond) for different catalyst and the same reaction will show the volcano (bell) shape (Fig. 4.3) [8, 9].

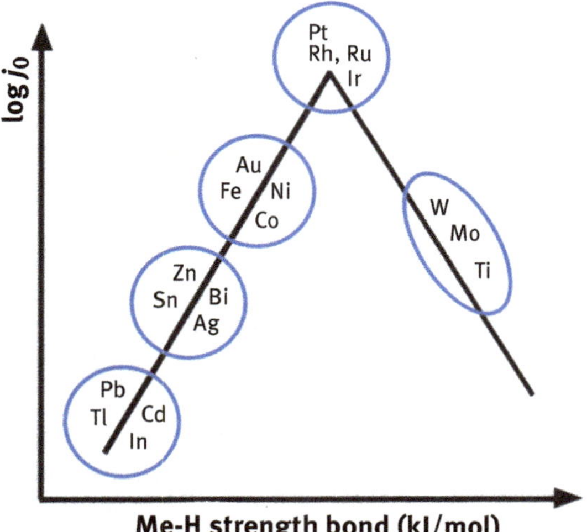

Fig. 4.3: Volcano plot for the hydrogen evolution reaction, showing exchange current density dependence on metal–hydrogen strength bond. Adapted from [9].

Let us consider in more detail the simple two-step reaction A → I → B with moderate adsorption of intermediate I, so the equilibrium between adsorption and desorption can be established. Further, we assume that (i) the surface of catalyst (electrode) is homogeneous, which means that the Gibbs free energy of adsorption is coverage-independent, (ii) the active sites are well defined with one molecule occupying a single site, and (iii) the active sites are energetically equal and there is no interaction between the adsorbed species. At such assumptions, the Langmuir isotherm is applicable:

$$\theta = \frac{k_{\mathrm{ads}} c_{\mathrm{A}}}{k_{\mathrm{des}} + k_{\mathrm{ads}} c_{\mathrm{A}}}, \qquad \text{or in the form} \qquad \frac{\theta}{1-\theta} = K_{\mathrm{ads},0} c_{\mathrm{A}} \tag{4.5}$$

where $k_{\mathrm{ads}}, k_{\mathrm{des}}$ are rate constants of adsorption, desorption, respectively, c_i is bulk concentration of reactant; in this case c_{A}, $K_{\mathrm{ads},0} = \frac{k_{\mathrm{ads}}}{k_{\mathrm{des}}}$ is the equilibrium constant for adsorption and is related to the standard Gibbs free energy of adsorption $\Delta G^0_{\mathrm{ads}}$,

$$K_{\mathrm{ads},0} = \exp\left(-\frac{\Delta G^0_{\mathrm{ads}}}{RT}\right) \tag{4.6}$$

In steady state, the rate of reaction: A → I → B, $v_{\mathrm{r}} = k_{\mathrm{des}}\theta$,

$$v_{\mathrm{r}} = \frac{k_{\mathrm{des}} k_{\mathrm{ads}} c_{\mathrm{A}}}{k_{\mathrm{des}} + k_{\mathrm{ads}} c_{\mathrm{A}}} \tag{4.7}$$

Keeping in mind that rate constants can be described by the Arrhenius equation and depend upon the activation energy of adsorption (or desorption), E_{a}, or on the enthalpy of activation (heat of activation) of adsorption $\Delta H^{0\#}_{\mathrm{a}}$, one can write the equations:

$$k_{\mathrm{ads}} = C\exp\left(\frac{\Delta H_{\mathrm{a}}^{0\#}}{RT}\right) \tag{4.8a}$$

$$k_{\mathrm{des}} = C'\exp\left(-\frac{\Delta H_{\mathrm{a}}^{0\#}}{RT}\right) \tag{4.8b}$$

$$\Delta H_{\mathrm{a}}^{0\#} = E_{\mathrm{a}} - RT$$

Putting k_{ads}, k_{des} expressions from eqs. (4.8a) and (4.8b) in eq. (4.7), we obtain the volcano shape dependence between the rate of reaction and parameters characterizing energetically the adsorption [2].

Langmuir isotherm was derived under assumptions that adsorption sites are equivalent, independent from each other and there is a lack of interaction between the adsorbed species. There are many experimental results where the Langmuir isotherm is not valid or is valid in a limited range of θ. The other isotherms were introduced, which take into account the inhomogeneity of catalyst surface (Temkin isotherm) and lateral interaction between adsorbed molecules (Frumkin isotherm) [7]. Both these isotherms were delivered under assumption that $\Delta G_{\mathrm{ads}}^{0}$ depend linearly on θ, more exactly the $\Delta H_{\mathrm{ads}}^{0}$ is θ dependent:

$$\Delta G_{\mathrm{ads},\theta}^{0} = \Delta G_{\mathrm{ads}}^{0} + r\theta \tag{4.9}$$

The r parameter in Temkin isotherm is connected with inhomogeneity factor f, whereas in Frumkin isotherm, parameter $r = f_{\mathrm{int}}RT$, where f_{int} is the interaction factor. This factor has a positive value when repulsion interaction is involved or negative for attraction interaction between the adsorbed species. Applying the expression $\Delta G^{0} = -RT\ln K$ and eq. (4.9) one can easily derive the equation describing the equilibrium constant $K_{\mathrm{ads},\theta}$ at coverage θ:

$$K_{\mathrm{ads},\theta} = K_{\mathrm{ads},0}\exp\left(-\frac{r\theta}{RT}\right) = K_{\mathrm{ads},0}\exp(-f_{\mathrm{int}}\theta) \tag{4.10}$$

going further Frumkin's isotherm (1928) has the form:

$$\frac{\theta}{1-\theta} = K_{\mathrm{ads},\theta}c_i \tag{4.11a}$$

or

$$\frac{\theta}{1-\theta} = K_{\mathrm{ads},0}c_i\exp(-f_{\mathrm{int}}\theta) \tag{4.11b}$$

while Temkin's isotherm (1939) can be written as follows:

$$\theta = \frac{1}{f}\ln K_{\text{ads},\theta} + \frac{1}{f}\ln c_{\text{i}} \tag{4.12a}$$

or

$$\exp(f\theta) = K_{\text{ads},\theta} c_{\text{i}} \tag{4.12b}$$

If θ is small, the difference $1-\theta \approx 1$ and Frumkin's isotherm becomes Temkin's isotherm, however, with different meanings of f factor.

The coverage of catalyst surface depends not only on the bond strength, electron structure of catalyst, and surface inhomogeneity but also on the applied polarizing potential. The influence of potential upon coverage is shown below for Frumkin and Temkin isotherms, and the expressions are derived by using basic equations (Chapter 1):

$$\Delta G = \Delta G^{\#0} \pm \Delta G^{\text{E}}, \quad \Delta G^{\text{E}} = -nFE$$

$$\text{Frumkin isotherm:} \quad \frac{\theta}{1-\theta}\exp(f_{\text{int}}\theta) = K_{\text{ads},0} c_{\text{i}} \exp\left(\frac{nFE}{RT}\right) \tag{4.13}$$

$$\text{Temkin isotherm:} \quad \exp(f\theta) = K_{\text{ads},0} c_i \exp\left(\frac{nFE}{RT}\right) \tag{4.14}$$

Both isotherms show a linear dependence of coverage on potential:

$$\theta = \frac{2.3}{f}\log K_{\text{ads},0} c_{\text{i}} + \frac{nF}{fRT}E \tag{4.15}$$

However, Frumkin isotherm can be used in this form only in such coverages where the assumption $\frac{\theta}{1-\theta} \approx 1$ is valid.

4.4.3 Size effect

Except for the surface energy and adsorption, the surface factors that influence the activity of catalyst are the surface composition, the orientation of crystals creating the surface, the surface defects, and catalyst size and shape. All these factors may be changed during catalyst preparation and surface pretreatment; therefore, the surface state should be carefully controlled before catalyst application by spectroscopic and microscopic methods.

In the heterogeneous electrocatalysis, the catalysts are used in various forms, such as planar electrodes, thin solid films, particles, nanoparticles, adatoms, monolayers, and others. All these forms vary in size. The application of high-resolution microscopy for surface characterization points out that they are nanostructured materials. Even the surface of the planar electrode is covered by many defects of nanoscale size. Some experimental results point out that the size of catalyst particles influences the catalytic activity causing its increase or decrease. This effect is referred to as size effect [2, 10, 11]. The investigations of size effect is an important goal because the minimization of catalyst size and its dispersion on the substrate to form very high surface area decrease the cost of manufactured batteries, fuel cells, or other reactors where catalyst are used. The investigation results are often unexpected. For instance: (i) Pt particles larger than 5 μm cause the oxygen reduction by 4e direct pathway to H_2O, while for particles lower than 5 μm, the dominant product of reduction is H_2O_2 (see Section 4.5.1); (ii) CO is oxidized to CO_2 at polycrystalline Au electrode with high overpotential of 0.7 V as a result of AuO formation, 3 nm Au particles are much more active, and oxidation of CO takes place at lower overpotential 0.3 V; (iii) at platinized platinum, the rate of H_2 evolution is about order magnitude lower than at Pt electrode; however, there are some examples in literature pointing out to the opposite effect. The problems with the comparison of catalytic activity of particles of various sizes are caused by difficulties in the preparation of uniform particles and by the interaction between particles and support.

The mechanisms called bifunctional catalysis and other called "third body effect" are connected with size impact on catalysis [3, 10]. The oxidation of methanol is an example of a bifunctional mechanism (see the beginning of this section). This mechanism corresponds to the reactions where the rate of reaction is related to the two-component of catalyst (e.g., Pt–Ru) selectively chemisorbing the species taking part in the slowest step of reactions.

Up till now, the "third body effect" is not well understood. As an example, the oxidation of FA at Pt modified by UPD (Chapter 6) metals of groups III, IV, or VA can be considered.

Let us assume that the processes involving the organic substances occur in two parallel ways (e.g., dehydrogenation and dehydration of FA):

$$\mathrm{HCOOH} \xrightarrow{\text{dehydrogenation}} \mathrm{CO_2} + 2\,\mathrm{H^+} + 2\,\mathrm{e}$$

$$\mathrm{HCOOH} \xrightarrow{\text{dehydration}} \mathrm{CO_{ads}} + \mathrm{H_2O}$$

The adsorbed CO_{ads} inhibited the catalytic reaction on Pt. To improve the rate of reaction, the foreign adatoms were deposited, called third body, which selectively suppressed CO adsorption by blocking neighboring adsorption sites [12].

Not only the size of crystallites forming catalyst surface or particles but also crystallographic surface structure influenced the catalytic activity. Many results on the influence of face structure on catalytic activity are contradictory. It was found that the order of decreasing catalytic activity for H_2 evolution at different faces of gold and wolfram tungsten is as follows: (110) > (100) > (111) and (111) > (110) > (100), respectively. The influence of crystallographic face depends also on the solution. It was reported that for Pt single crystals, the catalytic activity against ORR changes in order (110) > (111) > (100) in $HClO_4$ (110) > (100) > (111) in H_2SO_4 [2].

The investigations of disperse catalyst are of special interest [2]. They contain the crystallites and crystalline aggregates of different shapes and sizes. Disperse catalyst can be obtained on foreign supports, for example, various C, oxides, zeolites, conducting polymers, and others. The catalytic particles can be also deposited on a substrate consisting of the same metal, for example, platinized platinum or on foreign metals like UPD metal adatoms. The important parameter for dispersed catalyst is specific surface area (true working surface area), which can be determined by measurements of low-temperature argon adsorption (by BET method). Experimentally, it was found that, in some cases, the specific catalytic activity (referred to the true working surface) of electrodes with metal disperse catalyst is lower than the catalytic activity of the same electrode without the disperse catalyst of the same metal. The reasons for such facts are the porous structure of disperse catalyst resulting in the increase of concentration overpotential, loses connected with boundaries of particles grains, and with the boundaries of particles and substrate. However, as the total surface area of dispersed catalysts increases very much, the overall rate of reaction is larger in such systems but not so much as expected in comparison with compact electrodes.

Nowadays, there is a new field, which is developing very widely in chemical catalysis and electrocatalysis. It is the so-called single-atom catalysis. Single atom catalyst contains isolated metal atoms singly dispersed on a support. This topic is a subject of reviews [13].

4.5 Electrocatalysts in hydrogen–oxygen fuel cells

From the mid-twentieth century, the huge development of investigations of new electrochemical sources of energy is observed. Among these sources, hydrogen–oxygen fuel cells play the most important role. There, the electrical energy is produced as the result of a conversion of chemical energy of the oxidation of H_2 (HOR) at the anode and reduction of O_2 (ORR) at the cathode. A very important reaction from the technological point of view is the electrochemical production of pure H_2 (HER), the fuel used in the hydrogen–oxygen cell. The reader can find much more about the electrochemical energy converters in Chapter 8. Below we will mostly concentrate on the ORR and electrocatalysts improving its efficiency. Also, we will very briefly describe the reaction of oxidation (ionization) and the evolution of hydrogen.

4.5.1 Oxygen reduction reaction

In acidic aqueous solution oxygen reduction occurs by two overall pathways: a direct four- electron reduction and a peroxide two-electron, which involve H_2O_2 intermediate formation:

$$O_2 + 4H^+ + 4e \rightarrow 2H_2O, \quad E^0 = 1.299\,\text{V vs SHE}$$

$$O_2 + 2H^+ + 2e \rightarrow H_2O_2, \quad E^0 = 0.69\,\text{V vs SHE}$$

Peroxide can further undergo reduction or decomposition

$$H_2O_2 + 2H^+ + 2e \rightarrow 2H_2O, \quad E^0 = 1.776\,\text{V vs SHE}$$

$$2\,H_2O_2 \rightarrow 2H_2O + O_2$$

In alkaline solution, four- or two-electron reduction of O_2 occurs with ions formation

$$O_2 + 2H_2O + 4e \rightarrow 4OH^-, \quad E^0 = 0.401\,\text{V vs SHE}$$

$$O_2 + H_2O + 2e \rightarrow HO_2^- + OH^-, \quad E^0 = -0.065\,\text{V vs SHE}$$

The HO_2^- ion is unstable and can undergo further reactions

$$HO_2^- + H_2O + 2e \rightarrow 3OH^-$$

$$2HO_2^- \rightarrow 2OH^- + O_2$$

The standard potential E_0 values of these reactions were obtained from thermodynamic data. For the first reaction, the equilibrium potential is not established because of the very small value of exchange current density (10^{-8}–10^{-7} mA/cm^2). The measured open-circuit potential (OCP) in the presence of oxygen at the Pt electrode is about 1 V versus NHE (normal hydrogen electrode), which is 0.2–0.3 V more negative than the equilibrium potential. This difference and a very small value of exchange current density pointed out that the reaction of O_2 direct reduction is slow even on Pt electrode, catalyst of the reaction. One of the reasons is the stability of O – O bond in O_2 molecule, as the dissociation energy of this molecule is 494 kJ/mol. The mechanisms of oxygen reduction on Pt were intensively investigated. They are described in [14] and references there. According to the experimental and theoretical studies, the OR reaction on Pt is usually explained by two mechanisms: dissociative and associative [15]. They consist of chemical and electron transfer steps as shown below.

Dissociative mechanism:

$$2\,Pt + O_2 \rightarrow 2PtO$$

$$2\,PtO + 2\,H^+ + 2\,e \rightarrow 2PtOH$$

$$2\,PtOH + 2\,H^+ + 2\,e \rightarrow 2Pt + 2\,H_2O$$

Associative mechanism:

$$Pt + O_2 \rightarrow PtO_2$$

$$PtO_2 + H^+ + e \rightarrow Pt - O_2H$$

$$Pt - O_2H + H^+ + e \rightarrow \left(\frac{H_2O + PtO}{Pt + H_2O_2} \right)$$

$$Pt - O + H^+ + e \rightarrow Pt - OH$$

$$Pt - OH + H^+ + e \rightarrow Pt + H_2O$$

If the OR reaction proceeds according to the dissociative mechanism, the formation of H_2O_2 is not possible, because the O–O bonds of adsorbed molecules are broken. However, if the oxygen bonds are not broken, H_2O_2 can be produced undergoing further decomposition to water (associative mechanism). The decomposition reaction is very quick so the formation of H_2O_2 is not observed experimentally at Pt electrode. The first step in ORR is the adsorption of O_2 on the electrode surface. Three models for oxygen adsorption on metal surface are considered: (i) O_2 interacts by two bonds with the single substrate atoms (Griffiths model), (ii) O_2 interacts by one bond with a single substrate atom (Pauling model), (iii) O_2 interacts by two bonds with two substrate atoms (Yeager bridge model, Fig. 4.4).

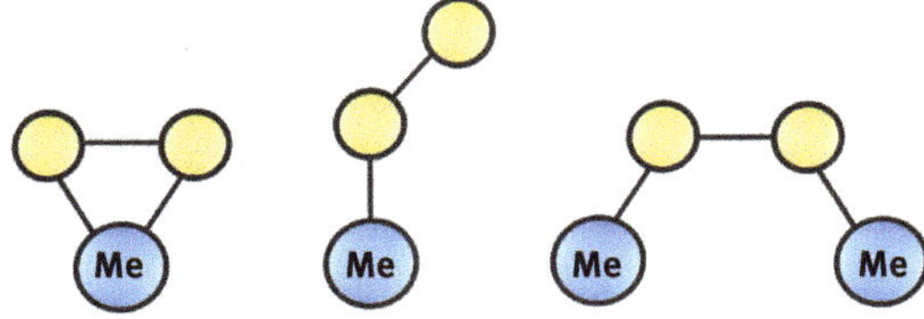

Fig. 4.4: Models for O_2 adsorbed on metal active sites. Adapted from [14].

The last model was proposed for the reaction on Pt metals. It appears that dual adsorption sites more likely are involved in the dissociation of O_2, while the single adsorption sites are preferable in hydrogen peroxide formation. The direct four-electron pathway is characteristic for Pt and other Pt group metals. The peroxide pathway was found for mercury and graphite electrodes that are inactive in the catalytic decomposition of H_2O_2. At other metal electrodes, the ORRs occur in parallel following two pathways, one of which dominates based on the experimental conditions such as surface treatment and electrolyte composition.

There are plenty of experimental results obtained for different electrodes mostly in acidic media but also in alkaline solution. They are described and discussed by Adžić [14]. As was mentioned, the ORR is slow and requires suitable catalysts eliminating the peroxide pathway, having an impact on catalytic efficiency. Catalysts should also have a large surface area and be stable chemically and electrochemically. Pt is the only commercially available catalyst with sensible activity and stability. However, the cost of this catalyst is high and huge effort is made to replace Pt by other materials or minimalize the amount of it by applying nanoparticles. Except Pt other metals from group VIII B such as Pd, Ru, Ir, and Rh have been tested as cathode materials. These metals in terms of activity in ORR follow the trend Pt > Pd > Ir > Rh > Ru. Pt alloys with transition metals such as Fe, Ni, Co, and Cr have been extensively studied on different carbon substrate. The amount of catalyst was minimalized by applying Pt and Pt alloy nanoparticles. Although bimetallic catalysts (Pt–Fe, Ni, Co, etc.) show improvement in catalytic activity in comparison with Pt, there is unfortunately a drawback, since these metals are soluble in acidic solution in the potential range from 0.3 to 1.0 V versus NHE. To overcome this problem, some multicomponent catalysts were synthesized on the Pt basis such as PtTiCo, PtTiFe, PtCuCo, and PtCuCoNi [15]. Metal oxides such as NiO, ZrO_2, TiO_2, and SnO_2 were tested as a support for metal particles and as catalysts in both acidic and alkaline solutions. Many metal oxides are unstable in acidic media; therefore, the layers of conducting polymer–polypyrrole were used for protection. Apart from metal and metal alloy nanoparticles, the nanostructured carbon materials were applied as support and catalyst as well. CNFs, single-walled CNTs or multiwalled CNTs have been studied extensively as a support of metal nanoparticles, due to their very high surface area and high electric conductivity. They were used for Pt or Pd nanoparticle dispersions and shown enhanced catalytic activity as compared with commercial Pt/C. The application of graphene and its derivatives in ORR catalysis is described earlier. The noble metals can be replaced as catalysts also by transition metal macrocyclic complexes. The role of ligand type, ligand structure, and type of central metal cation was determined in the catalytic activity of O_2 reduction. For instance, the catalytic activity toward oxygen reduction decreases in this sequence [14]: Co phthalocyanine (Pc) > Co tetraphenyl porphyrin > Co tetramethyl porphyrin. The four-electron reduction of O_2 has been found for Fe–Pc and Mn–Pc, while Co–Pc two-electron reduction was observed.

4.5.2 Hydrogen evolution and oxidation reactions

The cathodic evolution reaction of H_2 (HER)

$$2\,H_3O^+ + 2\,e \rightarrow H_2 + 2\,H_2O$$

is the most common and popular electrochemical reaction. It takes place in water electrolysis in the production of chlorine. It is important in the corrosion of metals (Chapter 3) and in metallurgy. In the reverse reaction, the anodic oxidation of molecular H_2 (HOR) is utilized in hydrogen–oxygen fuel cells.

Let us consider both reactions in equilibrium: $2\,H_3O^+ + 2\,e \leftrightarrow H_2 + 2\,H_2O$. The equilibrium potential $E_{eq} = E^0 + \frac{RT}{F}\ln\frac{a_{H+}}{p_{H_2}}$ is established in acidic solution, if platinized Pt is used as an electrode. For Pt-platinized electrode, the standard potential E^0 is accepted to be 0 in all temperatures under conditions: p = 1,013 hPa and acid solution concentration = 1 N. The Pt|H_2, H^+ electrode is commonly used as the reference electrode called NHE and all the potentials may be referred to it. Note that in literature you can find another hydrogen electrode applied as reference: standard hydrogen electrode (SHE) and reversible hydrogen electrode. The standard potential of SHE is zero, but in this case, Pt is dipped in an ideal 1 M acid solution ($a_{H^+} = 1$). The electrochemical potential of electrons for SHE is $-\ 4.5 \pm 0.1\mathrm{eV}$. We will use this standard electrode (SHE) in Chapters 10 and 11 for the correlation of energy levels of semiconductors and electrolytes containing redox systems. The reversible hydrogen electrode does not require the proton concentration to be unity $E_{RHE} = -0.059\mathrm{pH}\ \ (\mathrm{vs\,SHE})$, at 298 K.

For the equilibrium reaction, the exchange current density is high, which is equal to 10^{-2}–10^{-4} A/cm^2 and depends on the solution composition and its purity. As Pt and metals of its group are sensitive to impurities, the OCP often differs from equilibrium potential. There are some metals such as Hg, Pb, Bi, Cd, and In, known as "low exchange current density metal," at them the equilibrium potential is not established. Cathodic hydrogen evolution is a two-electron reaction occurring through the intermediate steps. Some of them are referred to after the name of the scientist who suggested the type of rate-determining step (rds) in the overall reaction. The steps that govern the reaction: $2\,H_3O^+ + 2e \rightarrow H_2 + 2\,H_2O$ are:

(1) the transport of the hydronium ion to the electrode,

(2) the discharge of the ion by reaction:

$$H_3O^+ + e \leftrightarrow H_{ads} + H_2O \qquad (2a)\ \ (\text{Volmer reaction, discharge step}),$$

or by the Heyrovsky reaction (electrochemical desorption step)

$$H_{ads} + H_3O^+ + e \rightarrow H_2 + H_2O \qquad (2b),$$

(3) the recombination of adsorbed hydrogen atoms via the Tafel reaction

$$2\,H_{ads} \leftrightarrow H_2 \qquad (3)\quad (\text{Tafel step}).$$

Further steps are (4) desorption of hydrogen molecules from the electrode surface to the solution through the electrical double layer and, finally, (5) the transport of H_2 from the electrode via diffusion or gas evolution. The main step in anodic oxidation

$$H_2 + 2H_2O \rightarrow 2H_3O^+ + 2e$$

is adsorption of H_2 molecule on the electrode surface followed by dissociation and adsorption of hydrogen atoms (Tafel reaction). The other important step is the ionization of adsorbed H atoms and (H_3O^+) formation (Volmer reaction). The rates of individual steps 2a, 2b, and 3 are dependent on the bond energy of the hydrogen adsorbed on the metal surface. If this energy is low or intermediate, the reaction of the discharge of hydronium ion (Volmer reaction) followed by electrochemical desorption (Heyrovsky reaction) is the rds. Such a route was found for metals such as Hg, Cd, In, Pb, Ag, and Au. For metals such as W, Mo, and Nb, the discharge of hydronium ions is quick, and the rds is electrochemical desorption (Heyrovsky reaction). For Pt and metals of the Pt group, the rds is recombination resulting from the large value of Me–H energy bonds. In hydrogen oxidation, the slowest step is the dissociation of H_2 molecules adsorbed on the electrode surface. The common catalysts for HOR is Pt. Unfortunately, Pt is very sensitive for impurities in the solution and in H_2 gas. Commercially applicable H_2 is mainly produced by steam reforming of hydrocarbons and may contain CO. Carbon monoxide adsorbs very strongly on Pt poisoning the catalyst surface. To overcome this problem, Pt alloys (Me–Ru, Rh, Ni, or others) were tested [3]. It was found that the second compound decreases CO adsorption or catalysis the oxidation of adsorbed CO. The best catalyst used in HOR is Pt–Ru (Section 4.3). Much more about hydrogen evolution, processes, and catalysts you can find in review [16] and in references there.

Bibliography

[1] Bocris JO`M, Reddy AKN. Modern electrochemistry. V2 New York, USA, Plenum Press, 1970.
[2] Bagotsky VS. Fundamentals of electrochemistry. Hoboken, USA, J Willey & Sons Inc., 2006.
[3] Ross Jr. PN, The science of electrocatalysis on bimetallic surfaces. In Lipkowski J, Ross PN ed. Electrocatalysis, New York, USA, Wiley-VCH, 1998, 43–74.
[4] Hwang J, Rao RR, Giordano L, Katayama Y, Yu Y, Shao-Horn Y. Perovskites in catalysis and electrocatalysis. Science 2017, 358, 751–756.
[5] Risch M. Perovskite electrocatalysts for the oxygen reduction in alkaline media. Catalysts 2017, 7, 154–185.
[6] Zhou X, Qino J, Yang L, Zhang J. A review of graphene-based nanostructural materials for both catalyst support and metal-free catalyst in PEM fuel cell oxygen reduction reaction. Adv. Ener. Mater. 2014, 1301523, 1–25.
[7] Gileadi E. Physical electrochemistry. Fundamentals, techniques and applications. Weinheim, FRG, Willey-VCH Verlag GmbH&Co, 2011.
[8] Quaino P, Juarez F, Santos E, Schmicler W. Volcano plots in hydrogen electrocatalysis – uses and abuses. Beilstein J. Nanotechnol. 2014, 5, 846–854.

[9] Bockris JO`M`, Reedy AKN, Gamboa-Aldeco M. Modern electrochemistry 2A. Fundamentals of electrodics. Second Edition. New York, USA, Kluwer Academic. Plenum Publisher, 2000.

[10] Petrii OA, Tsirlina GA. Size effects in electrochemistry. Russ. Chem. Rev. 2001, 70, 285–298.

[11] Wieckowski A, Neurock M. Contrast and synergy between electrocatalysis and heterogeneous catalysis. Adv. Phys. Chem. 2011, ID 907129 18p.

[12] Leliva E, Iwasita T, Herrero E, Feliu JM. Effect of adatoms in the electrocatalysis of HCOOH oxidation. A theoretical model. Langmuir 1997, 13(23), 6287–6293.

[13] Su J, Ge R, Dong Y, Hao F, Chen L. Recent progress in single-atom electrocatalysts: concept, synthesis, and applications in clean energy conversion. J. Mater. Chem. A 2018, 6, 14025–14043.

[14] Adzic R. Recent advances in the kinetics of oxygen reduction. In Lipkowski J, Ross PN ed. Electrocatalysis. New York, USA, Wiley-VCH, 1998, 197–242.

[15] Khosteng L. Oxygen reduction reaction. In Ray A, Mukhopadhyay I, Pati RH ed. Electrocatalysis for fuel cells and hydrogen evolution. Theory to design. London, United Kingdom, Intech Open, 2018, 25–50.

[16] Dubouis N, Grimaud A. The hydrogen evolution reaction: from material to interfacial description. Chem. Sci. 2019, 10, 9165–9181.

5 Electrodeposition

5.1 General remarks

Plenty of electrolytic processes are applied in industry for manufacturing many of reagents and materials such as chlorine, sodium hydroxide (electrolysis of NaCl), hydrogen (electrolysis of water), and metals by electroextraction from the proper salt (e.g., Zn, Cd) or by electrorefining [1, 2]. Electrorefining means anodic dissolution of contaminated metals and further cathodic deposition of purified metals. By using electrolytic deposition (shortly named electrodeposition), we can obtain not only metals but also alloys, semiconductors, and conducting polymers. All these materials can be deposited on the surface of conducting substrate in the form of layers, thin films, or nano-objects. Electrodeposited thin layer of metal (electroplating) are widely used as protective or decorating coatings. Thin layers of semiconductors and conducting polymers found application in the construction of solar cells, alloys can be used as catalyzers in fuel cells, and all these materials are recently applied in microelectronic and nanotechnology. Comparing the electrodeposition with vapor deposition one clearly sees some advantages of previous one, such as (i) the equipment for electrodeposition is more simple and not so expensive, (ii) the electrodeposition can be easily controlled by electrode potential or current density, and (iii) the electrodeposition occurs at low temperature.

In practice (industry), the electrodeposition is carried out in the two-electrode cell. The material is deposited on the cathode during the reduction of ions in the case of metals, alloys, or on the anode during oxidation of the electrode or oxidation of monomers resulting in the formation of oxides, conducting polymers. The deposition can be carried out at constant potential or constant current from an electrolytic bath. Bath contains solvent with appropriate species (like metal ions, metal complexes, and organic monomers) and some salt to increase the conductivity of the solution. Additionally, buffers for pH stabilization and some additives for improvement of deposition kinetics and morphology of deposits can be used.

Generally, the electrolysis and thus also electrodeposition processes are governed by two Faraday's laws. They described (i) the relation between the amount of deposited mass m and the electrical charge Q passing during deposition reaction (first law), and (ii) the relation between the mass of deposited material and its equivalent weight (second law). The equivalent weight of a substance is its molar mass divided by electrons transferred per ion in the reaction (M/nF).

Faraday's laws may be summarized in the equation:

$$m = \frac{M}{nF} Q \tag{5.1}$$

https://doi.org/10.1515/9783111160986-005

If the current density j changed with time, the charge is described by the integral $Q = \int_0^t \eta_F jAdt$; if the current density, surface area A, and Faradaic efficiency η_F is not time t dependent, then

$$Q = \eta_F jAt = \eta_F it \tag{5.2}$$

The thickness of deposited layer d can be estimated by using the equation:

$$d = \frac{m}{\rho A}, \qquad d = \frac{\eta_F Mit}{\rho AnF} \tag{5.3}$$

ρ is the density of deposit and i is current.

For a proper determination of the thickness of the deposit, it is necessary to take into account the reactions, which occur simultaneously with the main reaction of deposition (side reactions). In such a case, the Faradaic efficiency of reaction is lower than 1.

The investigations of both deposition reactions and mechanisms of deposit formations are carried out in the three-electrode cell, where the working electrode can be cathode or anode depending on the measured system. The electrochemical transient techniques are used for this purpose, mainly step potential method (Section 2.1).

5.2 Electrocrystallization: nucleation and growth

During the electrodeposition, the formation of a new solid phase and crystal growth take place at electrode–liquid interface. In literature, the term electrocrystallization is often used for steps involved in these phenomena [3–5]. The considerations below deal with metal deposition; however, some of them are so general that they may be used for other materials such as alloys, semiconductors, and even conducting polymers. Schematically some electrocrystallization steps are shown in Fig. 5.1.

Let us consider the solvated (hydrated) metal ion (Me^{n+}) which will be reduced at the electrode surface. In the first step, the solvated ion should get close to the electrode to permit an electron transfer from the electrode. Then the intermediate species are formed, which have a partial charge and some solvent molecules, so they can be considered as a kind of adsorbed ions, the ad-ions. These ad-ions have higher energy than atoms incorporated into metal lattice; therefore, they cannot remain on the surface too long. They move from the surface site, where charge transfer took place and where they were created, to the surface defects with a lower energy of incorporation of new atoms. Ad-ions mostly integrate with the crystal lattice at kink sites and step sites, the process is finished when the edge of the surface is reached. This step-growth mechanism is schematically shown in Fig. 5.2.

The other mechanism is associated with screw dislocations and leads to spiral growth of crystal. The screw dislocations result from a shift of one atom with respect to a perfect

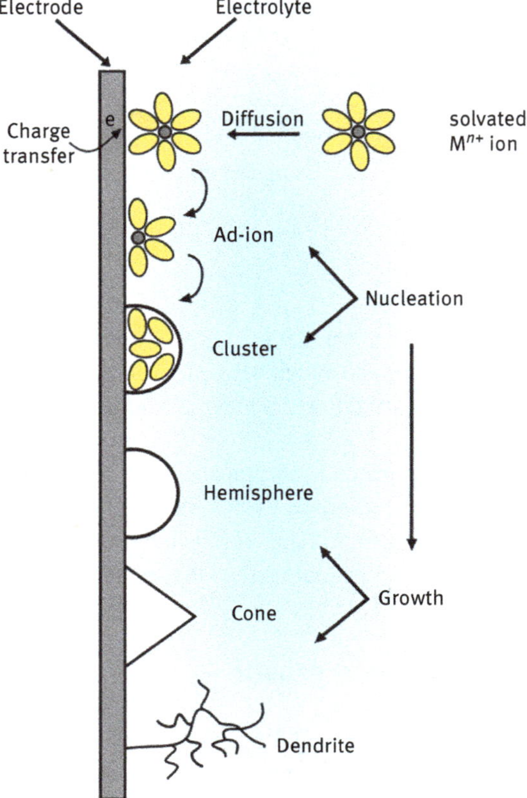

Fig. 5.1: Simplified scheme showing some steps involved in the electrocrystallization of metal on foreign substrate. Adapted from [3].

arrangement of others. This shift causes the formation of the edge at which new atoms (ad-ions and ad-atoms) can be stacked [2].

The ad-ions have to move from the place of their formation to incorporation sites. The concentration gradient of ad-ions between the place of their formation and elimination at defects is a driving force for the ad-ions motion. Such type of transport is named surface diffusion. The electrodeposition of metal or semiconductors mostly occurs at the potential more negative than the equilibrium potential (Nernst equation) of the investigated system. The difference between these two potential (η – overpotential) is the driving force for the electrocrystallization.

If the electrode surface is planar with a small amount of defects or deposition is carried out on the substrate foreign to deposited metal, the ad-ions accumulate to form a stable nucleus (crystal). The nucleus is stable and continues to grow farther if it reaches a critical size.

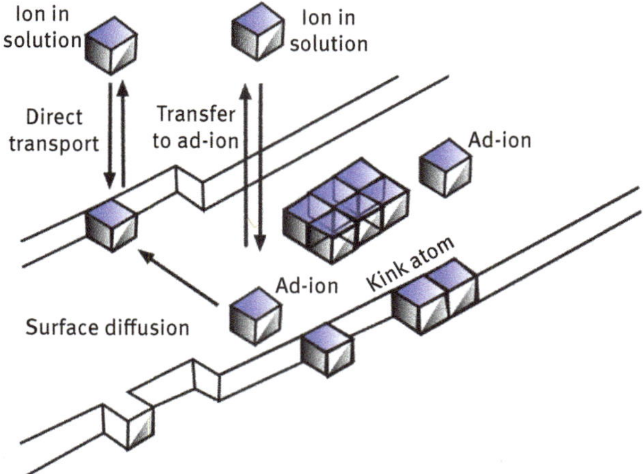

Fig. 5.2: Scheme illustrating the consecutive stages involved in incorporation of ad-ion and lattice building. Adapted from [6].

The process of nuclei formation is named nucleation and it is the beginning of the formation of the condensed phase. From nuclei further crystal growth proceeds, the relation of the rates of both processes influence the morphology of the deposit. The nuclei can be two or three dimensional (D) and their dimensionality depends on the interaction between substrate and deposited species, as well as on the relation between crystal lattice of a substrate and that of deposit. The nucleus is stable and can further grow only when it reaches a critical size.

5.2.1 Critical size of nuclei

In the classical approach, we assume that electrocrystallization has many features in common with the growth of the solid phase from its supersaturated vapor. Based on the thermodynamic description of homogeneous nucleation from the vapor [7], by analogy we can obtain the expression for the Gibbs free energy of electrochemical condensed phase formation:

$$\Delta G = \frac{nF\eta\rho}{M} \tag{5.4a}$$

M is molecular weight, ρ is the density of deposit, and η is overpotential.

The total Gibbs free energy of the formation of a disk shape nucleus (2D) with radius (r) and height (h) is the sum of two Gibbs free energies of formation, the bulk ΔG_{bulk} and the surface ΔG_{surf}:

$$\Delta G_T = \Delta G_{\text{bulk}} + \Delta G_{\text{surf}} = \frac{\pi r^2}{M} h\rho nF\eta + 2\pi rh\gamma \quad (5.4b)$$

γ is molar surface free energy (interfacial tension, Chapter 1).

For spherical nucleus (3D), this equation is written as follows:

$$\Delta G_T = \Delta G_{\text{bulk}} + \Delta G_{\text{surf}} = \frac{4\pi r^3}{3M} \rho nF\eta + 4\pi r^2\gamma \quad (5.4c)$$

Differentiations of these equations with respect to r show that free energy of nucleus formation ΔG_T passes through a maximum at the critical radius r_c. The nuclei with radii $< r_c$ are therefore unstable, whereas those with radii $> r_c$ are stable and can grow further. The free energy in maximum is called critical Gibbs free energy ΔG_c of nucleus formation and has the meaning of activation energy of homogeneous nucleation. In both cases, 2D and 3D nuclei, the critical free energy and critical radius depend on overvoltage:

2D nuclei
$$r_c = -\frac{\gamma M}{nF\rho\eta} \quad (5.5a)$$

$$\Delta G_c = -\frac{\pi hM\gamma^2}{nF\rho\eta} \quad (5.5b)$$

3D nuclei
$$r_c = -\frac{2M\gamma}{nF\eta} \quad (5.6a)$$

$$\Delta G_c = \frac{16\pi M^2\gamma^3}{3\pi\rho^2 n^2 F^2 \eta^2} \quad (5.6b)$$

note that $\eta < 0$.

The above expressions pointed out that an increase of overvoltage causes a decrease in the critical radius. It means that at the higher overpotentials we should expect a growth of small nuclei forming a fine-grained polycrystalline deposit.

5.2.2 Instantaneous and progressive nucleation

The formation of critical nuclei does not depend only on overpotential but also on thermal fluctuations, so the rate constant of nucleation (B) is related to the activation energy of nucleation ΔG_c:

$$B = F \exp\left(-\frac{\Delta G_c}{kT}\right) \quad (5.7)$$

F is a preexponential factor.

In another approach, the formation of clusters from ad-atom or ad-ions is considered as a growth center, and $\Delta G_{c,n}$ is Gibbs critical energy of clusters formation of *n*-atoms. In contrary to vapor deposition, where the homogeneous nucleation occurs, the nucleation is a heterogeneous process, since the nuclei form on the active sites of the electrode surface.

In the simple approach to the kinetics of nucleation, it is accepted that the process is the first-order reaction [3, 7]. As nucleation proceeds, the number of nuclei changes with time:

$$N(t) = N_0\ [1 - \exp(-Bt)] \tag{5.8}$$

N_0 is the initial number of active sites (kinks, steps, etc.), and B is the nucleation rate constant.

There are two limits to this equation:

i) If $Bt \gg 1$, then $\exp(-B\ t) \to 0$ and $N(t) = N_0(t)$. This equation points out that the number of nucleation sites remains constant at its initial value. This type of nucleation is referred to as **instantaneous nucleation.**
ii) if Bt is very small, $\exp(-B\ t) \approx 1 - Bt$; therefore, $N(t) = N_0\,Bt$, we can write further $N(t) = B't$, where B' is the modified nucleation rate constant. In this case, the number of nuclei increases in time. This type of nucleation is referred to as **progressive nucleation.**

The processes of nucleation and growth of 2D and 3D deposits were widely investigated by using potential step experiments. The experimental current–time curves show characteristic features like the current density maxima. Their time position t_m depends on the potential (overvoltage) (Fig. 5.3).

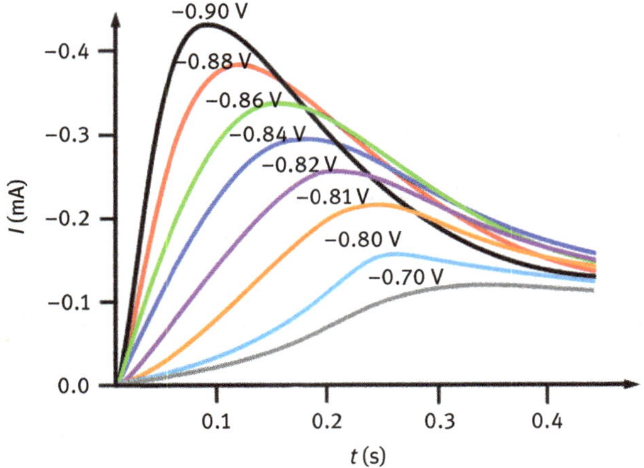

Fig. 5.3: Deposition of Cd on glassy carbon electrode at different potentials from 0.01 M $Cd(ClO_4)_2$ + 0.1 M $LiClO_4$ solution. Transients show maxima at different potential values. Unpublished data: M. Osial, K. Jackowska (Faculty of Chemistry, Warsaw University).

The quantitative analysis of current transient curves can provide information concerning the mechanism and kinetics of both nucleation and growth processes.

5.2.3 Analytical approach to experimental data

There are plenty of books and review articles dealing with the theoretical and practical aspects of electrodeposition. Interested readers will find some of them in [3–9]. Such studies deepened our understanding of electrocrystallization processes and their influence on the morphology of deposits.

All theoretical equations given below regarding $i-t$ curves are based on the assumption that the nucleation and growth are the slowest steps of deposit formation. It is convenient to consider two limiting cases of growth of 2D (disks and cylinders) and 3D (hemispheres). Let us concentrate on the models describing 2D (cylinders) growth and formation of a monolayer. It is assumed that the incorporation of ad-atoms (ad-ions) occurs at the edges of growing nuclei. The current i flowing to the single nucleus before their overlapping into a monolayer is described:

$$i = nFkA \tag{5.9a}$$

k is the rate constant of incorporation (or growth parallel to the plane of substrate), A is a surface area, $A = 2\pi rh$ for cylinders, so the current and the charge for the formation of single cylinder can be written as

$$i = 2\pi rhnFk \tag{5.9b}$$

$$Q = \frac{\pi r^2 hnF\rho}{M} \tag{5.9c}$$

The derivative of this equation leads to expression:

$$i(r,t) = \frac{2\pi hnF\rho r(t)}{M}\frac{dr(t)}{dt} \tag{5.9d}$$

As $\frac{dr(t)}{dt} = \frac{Mk}{\rho}$, the integration gives us $r(t) = \frac{Mk}{\rho}t$.

Substituting $r(t)$ into eq. (5.9d) and assuming the instantaneous nucleation of nuclei $N(t) = N_0$ we obtain the equation for current:

$$i = \frac{2N_0\ \pi hMnFk^2}{\rho}t, \qquad \text{2D instantaneous} \tag{5.10}$$

If the nucleation of nuclei (growth center) is progressive, then $N(t) = N_0\ Bt$, and the equation is given as follows:

$$i = \frac{2BN_0\ \pi hMnFk^2}{\rho} t^2, \qquad \text{2D progressive} \tag{5.11}$$

Both equations predict that current increases for all times, which is unacceptable and does not agree with experimental curves recorded in the potential step experiments. These equations were obtained under the assumption that nuclei grow independently. It may be true in a very early stage of nucleation when a low number of nuclei occur. When the number of nuclei increases and growth proceeds, neighboring nuclei will come in contact and overlap. The overlap problem has been considered theoretically by applying the concept of "extended area" A_{ex}. The extended area means the total area of nuclei without an overlap. Figure 5.4 illustrates how the overlap of nuclei changes the real surface area A and how the extended area may be presented.

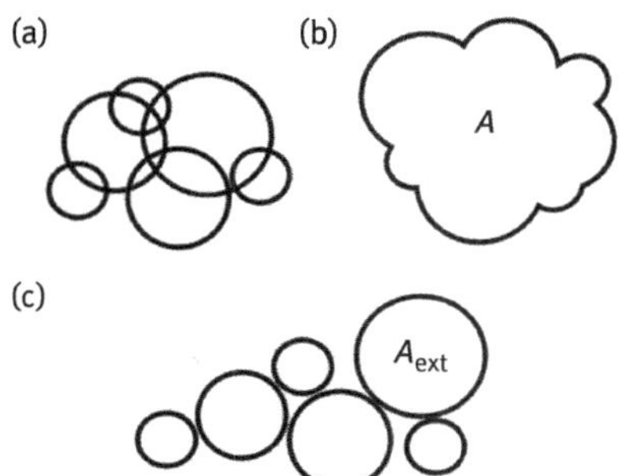

Fig. 5.4: Scheme presenting overlapping of growth centers and influence on surface area.

The relation between A and A_{ex} is given by Avrami theorem:

$A = 1 - \exp(-A_{ex})$, when A_{ex} is very large $A \to 1$, and all surface is fully covered with 2D deposit.

The current–time equations for the overlap case [3] are as follows:

$$i = \frac{2N_0\ \pi hMnFk^2 t}{\rho} \exp\left(-\frac{\pi N_0 M^2 k^2 t^2}{\rho^2}\right), \qquad \text{2D instantaneous, overlap} \tag{5.12}$$

$$i = \frac{2BN_0\ \pi hMnFk^2 t^2}{\rho} \exp\left(-\frac{\pi M^2\ BN_0\ k^2 t^3}{3\rho^2}\right), \qquad \text{2D progressive, overlap} \tag{5.13}$$

Considering the mathematical form of these equations it is clearly seen that a maximum should be observed on $i - t$ curves and that the current should go to zero. The latest is understandable as during growth all active sites (edges) are filled and the growth stops once the layer (monolayer) is completed. The current–time curves illustrating such features are shown in Fig. 5.5(a).

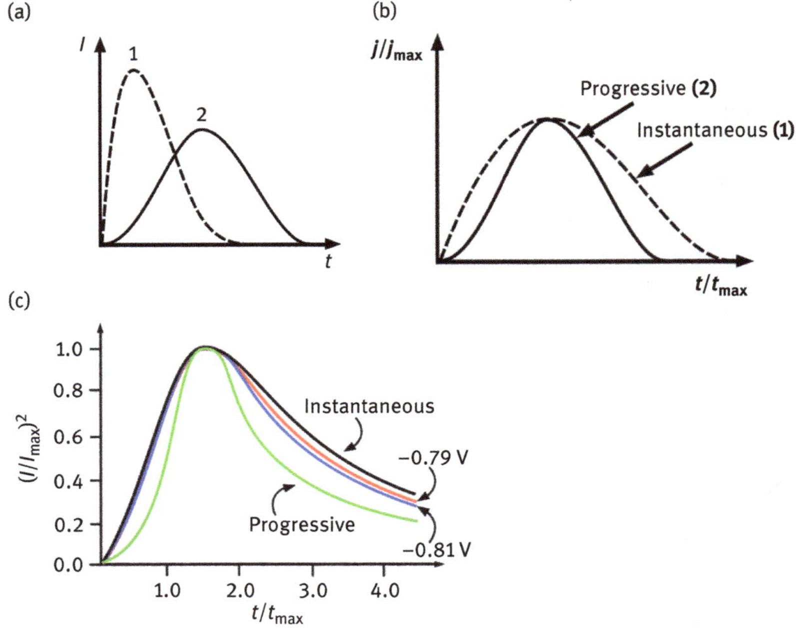

Fig. 5.5: (a) Plot illustrating the changes in the shape of transient curves. Curve 1 presents 2D instantaneous growth with overlap (eq. (5.12)) and curve 2 presents 2D progressive growth with overlap (eq. (5.13)). (b) Dimensionless plots illustrating differences between 2D instantaneous and 2D progressive growth (eqs. (5.14) and (5.15)). (c) Dimensionless plots presenting a comparison between calculated and experimental data. Experimental data were obtained for Cd deposition on a graphite paste electrode. Unpublished data: A. Frydrychewicz, M. Osial, K. Jackowska (Faculty of Chemistry, Warsaw University), G. A. Tsirlina, (Faculty of Chemistry, M.V. Lomonosov Moscow State University).

It is convenient to compare the experimental curves with calculated ones using nondimensional quantities $\frac{i}{i_{max}}$ and $\frac{t}{t_{max}}$, and i_{max}, t_{max} mean the values in maximum. The expressions in terms of nondimensional parameters are:

$$\frac{i}{i_{max}} = \frac{t}{t_{max}} \exp\left(-\frac{(t^2 - t_{max}^2)}{2t_{max}^2}\right), \qquad \text{2D instantaneous} \tag{5.14}$$

$$\frac{i}{i_{max}} = \frac{t^2}{t_{max}^3} \exp\left(-\frac{2(t^3 - t_{max}^3)}{3t_{max}^3}\right), \qquad \text{2D progressive} \tag{5.15}$$

In Fig. 5.5(b), dimensionless plots according to the above equations are presented for monolayer deposit (2D).

The comparison of calculated and experimental curves gives us the possibility to distinguish between these two mechanisms (Fig. 5.5(c)). It should be noted that for such comparison the theoretical current values have to be recalculated to current density values.

Let us describe very briefly the 3D growth of nuclei in the form of cones [3,6]. It is assumed that the nuclei may grow simultaneously in two directions parallel to the plane (substrate) and perpendicular to plane. When at short times the number of nuclei is small and they do not overlap the equations describing current–time dependence as follows:

$$i = nF\pi N_0 \frac{k_2 M^2 k_1^2}{\rho^2} t^2, \qquad \text{3D instantaneous} \tag{5.16}$$

$$i = nF\pi \frac{k_2 M^2 k_1^2 B}{3\rho^2} t^3, \qquad \text{3D progressive} \tag{5.17}$$

k_1 and k_2 are rate growth constants parallel to substrate (plane) and perpendicular, respectively. The coalescence and overlapping of nuclei lead to the expressions:

$$i = nFk_2 \left[1 - \exp\left(-\frac{\pi M^2 N_0 k_1^2}{\rho^2} t^2\right)\right], \qquad \text{3D instantaneous, overlap} \tag{5.18}$$

$$i = nFk_2 \left[1 - \exp\left(-\frac{\pi M^2 B_1 k_1^2}{3\rho^2} t^3\right)\right], \qquad \text{3D progressive, overlap} \tag{5.19}$$

In both cases for a long time, the current reaches the value of nFk_2, which means that perpendicular growth is only permitted. All these theoretical equations were derived under the assumption that nucleation and growth of nuclei are controlled by kinetics. In many cases, however, diffusion may be the slowest step in the growth of nuclei. Linear and hemispherical diffusion were considered. It is assumed that in most cases the further growth of crystallites (3D established nuclei, beyond their initial formation) is under diffusion control with hemispherical symmetry. The current–time equations for instantaneous and progressive nucleation and independent growth of crystallites are as follows:

$$i = \frac{nF\pi M^{1/2} N_0 (2Dc)^{3/2}}{\rho^{1/2}} t^{1/2}, \qquad \text{3D instantaneous} \tag{5.20}$$

$$i = \frac{2nF\pi M^{1/2} B N_0 (2Dc)^{3/2}}{3\rho^{1/2}} t^{3/2}, \qquad \text{3D progressive} \tag{5.21}$$

D is the diffusion coefficient of electroactive ions, c is its concentration, and other symbols have the meanings as above. The slope of j (current density) versus $t^{1/2}$ gives the nuclei number density N_0, if the Dc product is known.

Because during deposition the number of nuclei increases, their diffusion zone also increases and finally overlaps. The effect of overlapping of diffusion zones and its influence on $i-t$ relationship was considered by Scharifker et al. [10, 11]. Avrami theorem was adopted in the form: $\theta = 1 - \exp(-\theta_{ex})$, where θ_{ex} is a fraction of the area covered by the diffusion zone without overlap. The equations derived by Scharifker are:

$$i = \frac{nFD^{1/2}c}{\pi^{1/2}t^{1/2}}\left[1 - \exp(-N_0\pi KDt)\right], \qquad \text{3D instantaneous, overlap} \tag{5.22}$$

$$K = (8\pi cM/\rho)^{1/2}$$

$$i = \frac{nFD^{1/2}c}{\pi^{1/2}t^{1/2}}\left[1 - \exp\left(-\frac{BN_0K^*Dt^2}{2}\right)\right], \qquad \text{3D, progressive, overlap} \tag{5.23}$$

$$K^* = (4/3)(8\pi cM/\rho)^{1/2}$$

In both cases, the current passes a maximum and then approaches the limiting current for the diffusion planar to the electrode surface.

The time and current in maximum may be extracted from these equations by setting $\frac{di}{dt} = 0$.

3D instantaneous:

$$t_{max} = 1.26/N_0\pi KD, \qquad i_{max} = 0.638nFDcK^{1/2}N_0^{\ 1/2}$$

3D progressive:

$$t_{max} = (4.67/BN_0\pi K^*D)^{1/2}, \qquad i_{max} = 0.462nFD^{3/2}c(K^*)^{1/4}(BN_0)^{1/4}$$

The product $i^2_{max}t_{max}$ does not depend on K, K^*, B, N_0; therefore, it can be used as a diagnostic criterion of the type of nucleation.

The behavior for 2D nuclei is qualitatively similar to that shown by 3D nuclei. However, in 2D growth, metal deposits in the form of 2D monolayer. This type of nucleation is expected when metal deposition takes place on foreign metal substrate, exhibiting a strong metal–substrate interaction. It can lead to underpotential deposition described in Chapter 6.

The above equations were applied mostly for the description of metal deposition; however, rarely they were also used to determine the mechanism of nucleation and growth of semiconductors and conducting polymers.

5.3 Deposit morphology

As pointed earlier, the crucial electrochemical parameter influenced nucleation and crystal growth is overpotential. Of course, there are also other factors having an impact on morphology such as differences in the reaction mechanism, side reactions, methods applied in electrodeposition (steady-state method and pulse method), and conditions of deposition. Let us concentrate on the overpotential influence. At low overpotentials, the deposition takes place mostly at lower energy sites of surface (e.g., kinks). The energy for nucleation is not sufficient, so the formation of new nuclei occurs very rarely. During growth, the large crystals are formed on the surface and deposits are mostly in the form of layers and blocks. As the overpotential increases, the concentration of ad-ions increases too, so crystal growth is possible at less energetically favorable sites (e.g., steps). Deposits are formed as the bunched layers and ridges. When the overpotential is high enough and further increases, the dominant process is the nucleation leading to a large number of crystal grains. Grains coalesce forming the deposit with some defects present. The granularity of deposit (the number of crystals) is determined by the relation between nucleation and growth rates. If the nucleation rate is low compared to the rate of crystal growth, one will obtain the deposit with a small number of large grains. On the contrary, high nucleation rates lead to a fine granular deposit. At higher overpotentials, the mass transport of active species from the bulk of solution to the electrode surface becomes important. At mass transport control, deposits often grow in the form of powders and dendrites. In Fig. 5.6, the relation between deposit morphology, electrochemical parameters (j, E), and various regions of kinetic and mass transport control are shown [2, 9].

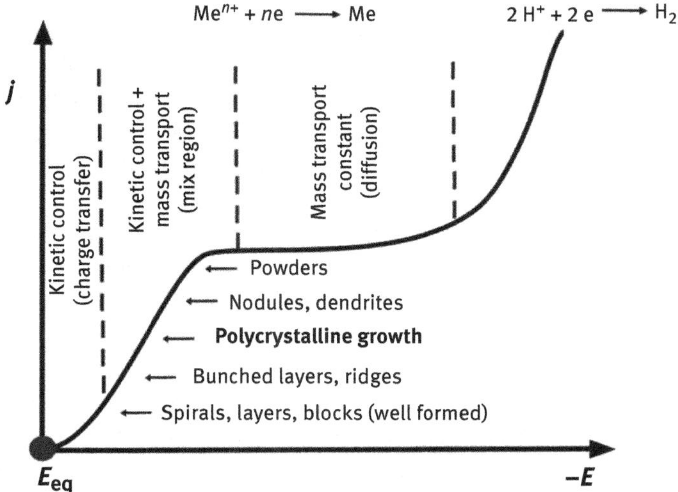

Fig. 5.6: Scheme presenting relations between deposit morphology, electrochemical parameters, and various regions of kinetic and mass transport control. Adapted from [9].

5.4 Practical aspects of electrodeposition

The industry of electroplating was developed for decades. The plating of metals and alloys plays an important role in the deposition of protective or decorative coatings and deposition of circuits in microelectronic.

Above, we briefly described mechanisms of early stages of deposition and the influence of overpotential on morphology of metal deposits. But in practice, not only morphology is important but also adhesion, mechanical properties, uniformity of coating, and its thickness. Good quality deposits are formed at current densities below the mass-transport limit, at a fraction of limiting current. Let us consider other factors that can influence the properties of deposits, such as side reactions, the additives in the plating bath, and distribution of current density [2, 12]. We will concentrate first on the side reaction. This reaction does not contribute to the main deposition reaction and decreases the Faradaic efficiency η_F. When the reaction of metal deposition is carried out from aqueous solution, the reduction of hydrogen ions is the side reaction, resulting in H_2 evolution at the cathode. Possibility of H^+ ions reduction and H_2 evolution during metal deposition depends on the relation between standard potentials of H^+/H_2 (E^0 = 0 V) and Me/Me^{n+} couples. Metals such as Pd, Ag, and Cu with positive standard electrode potential (E^0 = +0.91 V, E^0 = +0.8 V, E^0 = +0.34 V vs SHE, respectively) can be deposited even from acidic solution with 100% efficiency. Metals such as Cd, Fe, and Zn having the standard electrode potential more negative than SHE (E^0 = −0.4 V, E^0 = −0.44 V, E^0 = −0.76 V) can be deposited also, but with much lower efficiency than 100%. There are some metals (Al and Mg), which cannot be deposited from aqueous solution.

The hydrogen bubbles produced during metal deposition may cover parts of the electrode surface leading to a nonuniform deposit. To improve and make the deposited layer smooth or even bright and sheen, different compounds are used as an addition to the plating bath. Additives may influence the deposition process in a variety of ways. They may form a complex with metal ions and change the deposition potential, as well as reaction mechanism. The organic additives may adsorb strongly on the electrode surface resulting in a decrease in the rate of metal deposition. They may alter the growth of nuclei by blocking the growth sites, changing the concentration of nuclei, the rate of surface diffusion, and so on. According to their main function, the additives are referred to as carriers, brighteners, and wetting agents. The carriers influence the nucleation and growth of the forming phase, refining the deposit grains and producing brighter coating. The main role of brighteners is to make the deposition preferential on defects of the surface, resulting in the smoothing of the surface. The wetting agents decrease the surface tension, facilitating the detachment of gas bubbles generated by hydrogen evolution. There are plenty of additives used in the plating bath, such as coumarin, dextrin, saccharin, thiazole, uric acid, sodium sulfonate, and benzaldehyde.

If there is no side reaction, and only metal deposition takes place, the variations in the thickness of the deposit are connected with no uniformity of current distribution on/at an electrode during deposition [2, 12]. The uniformity of current distribution is described by the Wagner number W_a defined as

$$W_a = \left(\frac{\delta\eta/\delta j}{\rho L}\right) = \frac{R_F}{R_s} \tag{5.24}$$

ρ is the specific resistivity of the solution (Ωcm), L is a characteristic length, R_F is Faradaic (kinetic) resistance, and R_s is a resistance of solution.

The choice of L is not obvious. It may be, for instance, the difference between two electrodes: cathode (working electrode) and anode or counter electrode. The Faradaic resistance (Chapter 1) at potential near to the equilibrium potential is

$$R_F = \frac{RT}{nF}\frac{1}{j_0} \tag{5.25}$$

The large values of Wagner number correspond to more uniform current distribution and a more uniform thickness of deposit. Deposit uniformity increases with solution conductivity increase and decreases with current density increase. Therefore, in practice, the plating is often carried out initially at a lower current density to obtain uniform current distribution at thin layer coating (to increase Wagner number) and after that, the higher current density is used to achieve the thickness need. The concept connected with Wagner number is the "throwing power" T of plating bath. It is a measure of bath ability to produce coating of uniform thickness on samples having complicated geometry. In the bath with a higher T, more uniform deposit can be obtained. To quantify the throwing power of a bath, the Haring–Blum cell is often used. The cell is shown in Fig. 5.7. The cell consists of an anode and two cathodes, one on either side of the anode. Only the cathode surfaces facing anode are active. The cathodes are placed at different distances from anode x_1 and x_2.

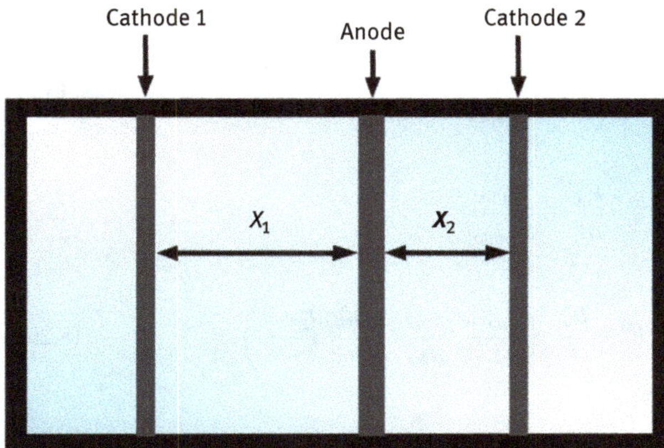

Fig. 5.7: Scheme of Haring–Blum cell.

Throwing power can be determined using the equation:

$$T(\%) = \frac{100(K - D)}{K + D - 2} \tag{5.26}$$

where $K = \frac{x_1}{x_2}$ is the ratio of cathode distances from anode, and $D = \frac{w_1}{w_2}$ is the weight ratio of deposits on the two cathodes.

The best uniformity of deposit is reached when the weight of the deposits on both electrodes is the same $(D = 1)$. It corresponds to 100% of T and is possible when $W_a \gg 1$.

5.5 Electrodeposition of binary alloys and semiconductor compounds

In this section, we will briefly describe some features of electrodepositions of binary alloys and semiconductor compounds [6, 12–17]. If the metal A is electrodeposited from a solution containing A^{n+} ions on the same substrate (A), the deposition is going at a potential slightly higher than the equilibrium potential of A/A^{n+} couple, $E_{eq} = E^0 + \frac{RT}{nF}\ln a_{A^{n+}}$ (a_A is assumed 1) .

On a foreign substrate, the deposition potential (onset potential) of A can be significantly more negative than the redox potential of A/A^{n+} couple. It results from higher overpotential η needed for the formation (called sometimes nucleation overpotential) of a new phase A. The situation is more complicated when equilibrium is established between metallic ions A^{n+}, B^{n+} in solution and binary A–B alloy. Then we can write the equations describing the equilibrium potential of systems A_{alloy}/A^{n+} and B_{alloy}/B^{n+}. Since these two potentials are equal, we can derive the equilibrium potential of alloy A–B:

$$E_{eq}(A) = E^{0,A} + \frac{RT}{nF}\ln\frac{a_{A^{n+}}}{a_{A,\,alloy}} \tag{5.27}$$

$$E_{eq}\ (B) = E^{0,B} + \frac{RT}{nF}\ln\frac{a_{B^{n+}}}{a_{B,\,alloy}} \tag{5.28}$$

$$E_{eq}(A - B) = E^0 + \frac{RT}{F}\left(\frac{x_A \ln a_{A^{n+}} + x_B \ln a_{B^{n+}}}{n x_A + n x_B}\right) \tag{5.29a}$$

$$E^0 = -\frac{\Delta G^0_{mix}}{F} + \frac{x_A E^0_A + x_B E^0_B}{x_A + x_B} \tag{5.29b}$$

where ΔG^0_{mix} is standard Gibbs free energy of the alloy mixing, and x_A, x_B are the molar fractions of the components.

If alloy forms a stable solid solution, the ΔG_{mix} is negative. It causes a positive shift of the redox potentials of A and B. The phenomenon is referred to as underpotential codeposition and was observed, for instance, during the Au–Cu alloy electrodeposition. The presence of two metal ions (or more) during the alloy formation affects also the kinetics, causing the variations in the exchange currents and limiting diffusion currents of both metals. These and the differences in the concentration of A, B ions in the solution may result that one component is deposited under mass control while the other is under activation control. In consequence, we may obtain a deposit of no uniform thickness and no uniform composition. Let us consider two cases: the first one in which the reduction reactions of A^{n+} and B^{n+} ions are under kinetic control (under control of activation overpotential). By using Volmer–Butler equation we obtain the expressions for current densities of individual reactions:

$$j_A = j_{0,A} \exp\left(- \frac{\alpha_A F(E - E_{eq,A})}{nRT} \right) \tag{5.30}$$

$$j_B = j_{0,B} \exp\left(- \frac{\alpha_B F(E - E_{eq,B})}{nRT} \right) \tag{5.31}$$

For simplicity, we assume further that number of electrons taking part in reactions $(n) = 1$ and transfer coefficients (α_A, α_B) are equal to each other $\alpha_A = \alpha_B = \alpha$, then the ratio of current densities is

$$\frac{j_A}{j_B} = \frac{j_{0,A}}{j_{0,B}} \exp\left(- \frac{\alpha F(E_{eq,A} - E_{eq,B})}{RT} \right) \tag{5.32}$$

It is clearly seen that the amount of metals in alloy A–B will depend on their exchange current density and difference of their equilibrium potentials. The higher the value of $j_{0,A}$ and more positive $E_{eq,A}$ in comparison with the same parameters of metal B, the greater the amount of metal A in the alloy (Fig. 5.8).

In the second case, we assume that the reduction of A^{n+} ions is under mass transport control (i.e., diffusion), while the B^{n+} ions reduced under kinetic control. The equations describing the current density of both processes are as follows:

$$j_A = j_{L,A} = - \frac{nFD_A c_A^0}{\sigma} \tag{5.33}$$

$$j_B = j_{0,B} \exp\left(- \frac{\alpha_B F(E - E_{eq,B})}{nRT} \right) \tag{5.34}$$

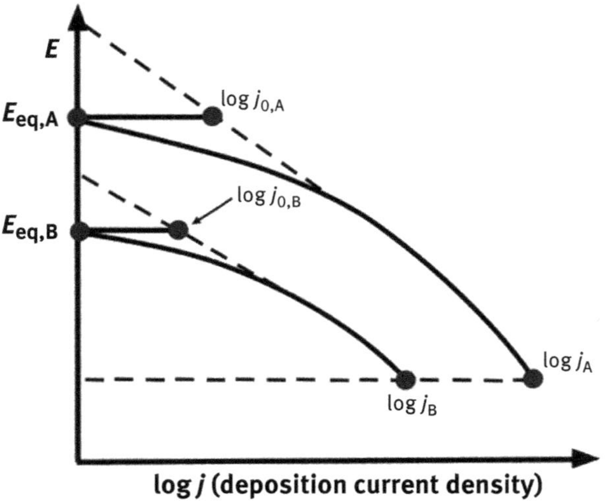

Fig. 5.8: Potential–current density plots presenting the influence of differences in the exchange current density ($j_{0,\mathrm{A}}$, $j_{0,\mathrm{B}}$) on alloy composition AB.

The ratio of current density is

$$\frac{j_\mathrm{A}}{j_\mathrm{B}} = \frac{j_{\mathrm{L,A}}}{j_{0,\mathrm{B}}} \exp\left(\frac{\alpha_\mathrm{B} F\left(E - E_{\mathrm{eq,B}}\right)}{nRT}\right) \tag{5.35}$$

The expression points out that when the current density j_B is higher than limiting current density $j_{L,A}$, the amount of metal B exceeds the amount of metal A in alloy even if the equilibrium potential of B/B^{n+} system is more negative than $E_{eq,A}$, Fig. 5.9.

Electrodeposition being widely applied for metals and metallic alloys, it is also developed for semiconductor (SC) thin film formation. In this way, binary III–V, II–VI (e.g., GaAs, CdS, ZnO), ternary ($CuInSe_2$), and even quaternary (Cu_2ZnSnS_4) compounds are formed [15, 16]. The electrodeposition of semiconductor compounds exhibits some specific features. The most of semiconductor compounds obtained electrochemically are binary compounds containing at least one metallic component such as Cd, Cu, Zn, In, and Ga and one nonmetallic component: S, Te, Se, As, P, and O. The equilibrium potentials and deposition potentials of nonmetals are very different from those of metals. For instance, the standard potentials of Cd/Cd^{2+} and Zn/Zn^{2+} are –0.4 V and –0.76 V versus NHE, respectively, while E^o of Te/Te^{4+} is +0.55 V versus NHE and E^o of Se/Se^{4+} is + 0.86 V versus NHE. Most of the semiconductor compounds have also the large values of negative Gibbs free energy of formation, for example, −154.7, −135.4, and −99.5 kJ/mol for CdS, CdSe, and CdTe, respectively. Similarly as in the case of alloy, this results in the shift of the cathodic deposition potentials of less noble component (E_{eq} potential of less noble component is more negative than E_{eq} of noble component) to the more positive value. The condition to be met to obtain the binary compounds of well-defined

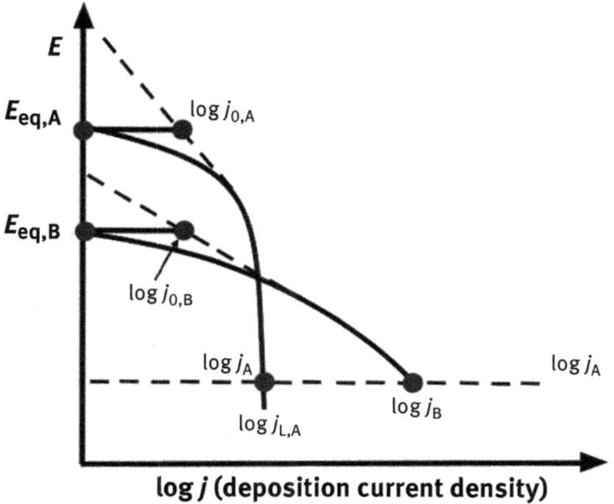

Fig. 5.9: Potential–current density plots presenting the influence of relation between limiting current density $(j_{L,A})$ and current density j_B on alloy composition AB.

stoichiometry has been discussed by Kröger [17]. This is behind the scope of this book. However, it is worth to mention that two component classes have been distinguished: class I with large difference in deposition potentials of individual components and class II with a smaller difference. In the first case, the deposition potential is determined by the less noble component over the whole range of metal–nonmetal composition. This is the case of CdX (X = S, Te, Se). For class II, the deposition potential may be determined by one component or the other, depending on the composition. Deposited SC films have much higher resistivity than films of metals or alloys. Therefore, after the deposition of the first layer, the distribution of charge and uniformity of current density may be changed resulting in composition changes. It was observed for CdTe. During the electrodeposition, the value of deposition potential and concentration of components are the crucial parameters that determine the type of conductivity and bandgap energy [18]. The richness of CdTe in Te causes changes in the type of CdTe conductivity from n-type to p-type. The same is observed in many cases for more complicated semiconductor materials I–III–VI_2 ($CuInGaSe_2$). The solubility of nonmetal salts used for the formation of semiconductor compounds is very low, and complicated equilibrium may be established in aqueous solutions, so very often the deposition kinetic is under mass-transfer control.

Bibliography

[1] Bagotsky VS. Fundamentals of electrochemistry. Hoboken, USA, John Willey & Sons Inc, 2000.

[2] Fuller TF, Harb JN. Electrochemical Engineering. Hoboken, USA, John Willey & Sons Inc, 2018.

[3] Pletcher D, Greff R, Peat L, Peter LM, Robinson J. (Southampton Group). Electrocrystallization. In Instrumental methods in electrochemistry. Amsterdam, The Netherlands, Elsevier, 2001, 286–316.

[4] Milchev A. Electrocrystallization: Fundamentals of nucleation and growth. Dordrecht, The Netherlands, Kluwer Academic Publisher, 2002.

[5] Budevski E, Staikov G, Lorentz WJ. Electrochemical phase formation and growth: an introduction to the initial stages of metal deposition. Weinheim, FRG, VCH, 1996.

[6] Bocris JO`M, Reddy AKN, Gamboa-Aldeco M. Modern electrochemistry. Fundamentals of electrodics. V 2A, New York, USA, Kluwer Academic Publisher, 2000.

[7] Harrison JA, Thirsk HR. The fundamentals of metal deposition. In Electroanalytical chemistry: a series of advance. Ed. Bard AJ. New York, USA, Marcel Dekker, 1971, 67–147.

[8] Milchev A. Electrocrystallization: nucleation and growth on nano-clusters on solid surfaces. Russ. J. Electrochem. 2008, 44, 619–645.

[9] Walsh FC, Herron ME. Electrocrystallization and electrochemical control of crystal growth: fundamental considerations and electrodeposition of metals. J. Phys. D: Appl. Phys 1991, 24, 217–225.

[10] Scharifiker B, Hills G. Theoretical and experimental studies of multiple nucleation. Electrochim. Acta. 1983, 28, 879–889.

[11] Scharificer B, Mostany J. Three dimensional nucleation with diffusion controlled growth. Number density of active sites and nucleation rate per site. J. Electroanal. Chem. 1984, 177,13–23.

[12] Gileadi E. Physical chemistry. Fundamentals, techniques, applications. Weinheim, FRG, Willey VCH Verlag GmbH, 2011.

[13] Plieth W. Electrochemistry for material science. Amsterdam, The Netherlands, Elsevier, 2008.

[14] Zangari G. Electrodeposition of alloys and compound in the era of microelectronics and energy conversion technology. Coatings 2015, 5, 195–218.

[15] De Mattel RC, Feigelson FS. Electrochemical deposition of semiconductors. In Mc Hardy J, Ludwig F. ed Electrochemistry of semiconductors and electronics. Processes and devices. NJ, USA, Noyes Publications 1992, 1–52.

[16] Bouroushian M. Electrochemistry of metal chalcogenides. Berlin, FRG, Springer Verlag, 2010.

[17] Kröger FA. Cathodic deposition and characterization of metallic or semiconducting binary alloys or compounds. J. Electrochem. Soc. 1978, 125, 2028–2034.

[18] Osial M, Widera J, Jackowska K. Influence of electrodeposition conditions on the properties of CdTe films. J. Solid State Electrochem. 2013, 17, 2477–2486.

6 Underpotential deposition (UPD)

6.1 General remarks

According to thermodynamics, the reduction of metal ions Me^{n+} on the corresponding Me substrate can take place at an equilibrium potential of Me/Me^{n+} couple (E_{eq}), but in practice, it occurs at a potential slightly negative than E_{eq} (e.g., deposition of Ag^{+} on Ag). Reduction of Me^{n+} ions on foreign metal substrate (S) and formation of bulk deposit usually require more energy and occur at much more negative potentials than E_{eq}. If the deposition potential is more negative than the equilibrium potential, the deposition process is named "overpotential deposition" (OPD). Plenty of experimental results pointed out that the reduction of metal ions on foreign metal substrate, resulting in the formation of monolayer, takes place at potentials more positive than the equilibrium potential of Me/Me^{n+}. In contrary to the OPD, such a process is referred to as underpotential deposition (UPD) Fig. 6.1. UPD is a self-limiting process and is terminated when a complete monolayer is formed.

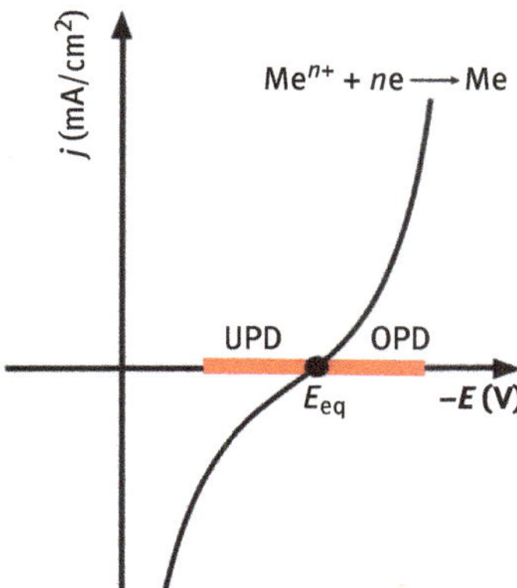

Fig. 6.1: Current density–potential plot showing the regions of underpotential deposition (UPD) and overpotential deposition (OPD).

What are the reasons of UPD?

UPD is the result of attractive interactions (Me–S) between the depositing metal ions Me^{n+} and the substrate S, being stronger than the atomic interaction Me–Me, S–S. As a result of these interactions, a metal is deposited in the UPD region as a monolayer (one atomic layer). Formation of such monolayer involves partial or complete discharging of Me^{n+} ions (partial or complete charge transfer) and potential-dependent adsorption referred to as Faradaic adsorption. If the UPD process is carried out on polycrystalline substrate or single crystal real surface, the nucleation and growth of monolayer occur preferentially at defect sites and expanded from there (see 2D nucleation and growth in previous chapter). Now and then, if the surface of the substrate is very rough, the formation of two atomic layers is observed.

https://doi.org/10.1515/9783111160986-006

At first look, it appears that the UPD process taking place at more positive potential than E_{eq} defines the laws of thermodynamics, but it does not, as we will see later [1, 2].

Let us consider the same two redox reactions of Me^{n+} ions, the first one on the corresponding Me and the other on a metal substrate S. In the first case:

$$Me^{n+} + ne(Me) \leftrightarrow Me\,(Me), \qquad E_{eq} = E^0 + \frac{RT}{nF}\ln\frac{a_{Me^{n+}}}{a_{Me}}$$

further

$$E_{eq} = E^0 + \frac{RT}{nF}\ln\frac{c_{Me^{n+}} f_{Me^{n+}}}{a_{Me}}$$

for the pure metal, we can assume that $a_{Me} = 1$,
then

$$E_{eq} = E^0 + \frac{RT}{nF}\ln c_{Me^{n+}} f_{Me^{n+}} \tag{6.1}$$

In the second case: $Me^{n+} + ne(S) \leftrightarrow Me\,(S)$, the surface of the substrate is only partially covered, so we can assume that $a_{Me} = f_{Me}\theta$, where θ is the degree of coverage of the substrate (S) surface. The product $f_{Me}\theta < 1$, since θ can be only ≤ 1 and $f_{Me} < 1$.

The potential of monolayer formation E_{ML} is

$$E_{ML} = E_{eq}(\theta) = E^0 + \frac{RT}{nF}\ln\frac{c_{Me^{n+}} f_{Me^{n+}}}{f_{Me}\theta} \tag{6.2}$$

and can be further written as

$$E_{ML} = E_{eq}(\theta) = E_{eq} - \frac{RT}{nF}\ln f_{Me}\,\theta \tag{6.3}$$

so $E_{ML} = E_{eq}(\theta) > E_{eq}$.

Conclusions from these equations are as follows:

i. the potential of the formation of metal monolayer, E_{ML}, is more positive than the equilibrium potential E_{eq};
ii. if θ is constant, the tenfold increase of Me^{n+} ions concentration shifts E_{ML} in the positive direction of 59/n mV;
iii. if θ is equal to 1, the formation of monolayer is completed.

6.2 Experimental examples – UPD features

In Fig. 6.2 we show the shape of cyclic voltammogram obtained during the deposition of Me^{n+} ions on S metal substrate (e.g., Pb^{2+} on polycrystalline silver). Two cathodic and two anodic peaks of current density are recorded. Peak 1 corresponds to a reduc-

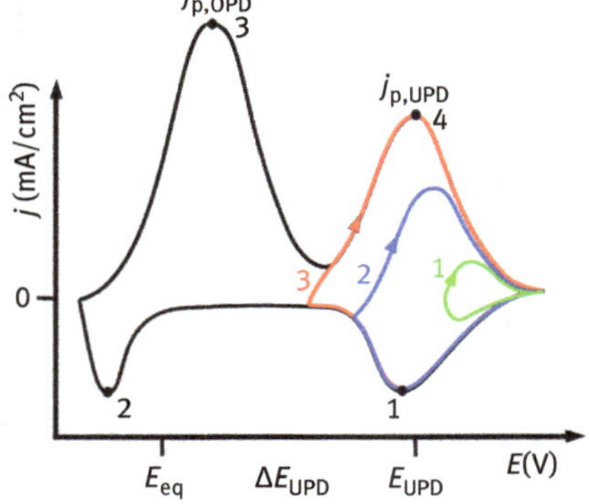

Fig. 6.2: Current density–potential curve recorded for the deposition of a metal. The end of the UPD region is marked by the red curve. At more negative potentials, the overpotential deposition takes place. The UPD shift is pointed on the potential axis. Adapted from [3].

tion of Me^{n+} ions at more positive potential (UPD) and formation of a monolayer, and peak 2 corresponds to bulk deposition of metal (OPD). The anodic current density peaks 3 and 4 correspond to striping the bulk deposit and monolayer, respectively. The difference between the potentials of the peak current density of formation of UPD monolayer (1) and the equilibrium potential of the same metal (2) is noted as UPD shift: $\Delta E_{UPD} = E_{p,UPD} - E_{eq}$. When $E_{p,UPD}$ is determined in solution with concentration (1 M) and UP deposition is reversible, this potential is chosen as the standard potential of the formation of UPD monolayer E_{ML} (under the assumption that $f_{Me}\theta$, $f_{Me^{n+}}$ are equal to 1). Since, $E_{p,UPD}$ and E_{eq} are described by the same Nernst equation, the ΔE_{UPD} shift does not depend on the concentration of metal ions in solution when the UPD process is reversible. In this case, the potentials of cathodic and anodic peaks on CV curves are not dependent on the scan rate up to some v value, characteristic for each system [3].

The UPD shift depends on the type of substrate and may range from 0.1 V even up to 0.5 V. For instance, when Ag is deposited on polycrystalline Pt, Pd, and Au, the UPD shift is equal to 0.44, 0.30, and 0.52 V, respectively. When different metals are deposited on the same substrate, the UPD shift varies too. In the case of the deposition of Tl, Cd, on polycrystalline Ag, it is equal to 0.28 and 0.16 V, respectively. The shift values depend also on the crystallographic structure of the single-crystal substrate.

Kolb [4] studying UPD phenomena on a large amount of M–S samples derived experimental correlation between the underpotential shift ΔE_{UPD} and work function Φ of metal and substrate:

$$\Delta E_{UPD} = 0.5\Delta\Phi \tag{6.4}$$

Let us go back to Fig. 6.2 to the part of the CV curve concerning UPD of the monolayer formation and removal. As potential is changed with time: $E(t) = E + vt$ (v is a sweep rate of potential), we can plot the current–time curve and easily calculate the charge:

$$q = \int_{E_{\mathrm{in}}}^{E} j_{\mathrm{F}} dt \tag{6.5}$$

where j_F is the current density associated with the UPD process.

The surface under the curve (red color) is now divided into several parts 1–3. These under lines 1 and 2 correspond to the formation and removal of partial monolayer while the surface under line 3 corresponds to the formation and striping of the complete monolayer. Let us denote the charge required for complete monolayer formation by q_i and fractional coverage of substrate (electrode) by θ, then the Faradaic charge consumed at any potential during formation of the partial monolayer is $q_F = q_i\theta$. As q_F and in consequence θ depends on the potential, we may now introduce the new notation – the adsorption pseudocapacitance, C_F:

$$C_{\mathrm{F}} = \left(\frac{dq_{\mathrm{F}}}{dE}\right)_{\mu}, \qquad C_{\mathrm{F}} = q_{\mathrm{i}}\left(\frac{d\theta}{dE}\right)_{\mu} \tag{6.6}$$

If we sweep the potential from its initial value E_{in} $(\theta = 0)$ to E $(\theta = 1)$ and back to E_{in} value, we record a Faradaic current j_F of the formation and removal of a monolayer:

$$j_{\mathrm{F}} = q_{\mathrm{i}}\left(\frac{d\theta}{dt}\right) = q_{\mathrm{i}}\left(\frac{d\theta}{dE}\right)\left(\frac{dE}{dt}\right) = C_{\mathrm{F}}\left(\frac{dE}{dt}\right) = C_{\mathrm{F}}\nu \tag{6.7}$$

The equation points out that the current density of formation and removal of UPD monolayer depend linearly on the sweep rate, ν.

The UPD was observed also for nonmetals such as hydrogen, oxygen, and halogens. Let us concentrate on a very important reaction from a scientific and technological point of view – on H_2 evolution (HER) [5]. In the first step, the hydrated ions H_3O^+ move to the electrode double layer and undergo discharge with the formation of electroadsorbed H on the electrode surface. Two types of electroadsorbed H are distinguished: the underpotential deposited H_{UPD} and overpotential deposited H_{OPD}. Experimentally, the electroadsorbed H_{UPD} is observed at potential E more positive than E^0 $(H^+|H_2)$, $[E^0$ $(H^+|H_2) = 0]$, while H_{OPD} is formed when $E < E^0$ $(H^+|H_2)$:

$$\mathrm{Me} + \mathrm{H_3O^+} + \mathrm{e} \rightarrow \mathrm{Me\text{-}H_{UPD}} + \mathrm{H_2O}, \qquad E > 0$$
$$\mathrm{Me} + \mathrm{H_3O^+} + \mathrm{e} \rightarrow \mathrm{Me\text{-}H_{OPD}} + \mathrm{H_2O}, \qquad E < 0$$

It is known that UPD of H takes place at Pt, Pd, Rh, and Ir, but not at Ag and Au electrodes, while OPD of H occurs on all metallic and other conducting substrates at which HER occurs. Both electroadsorbed H_{UPD} and H_{OPD} can undergo further change to the absorbate state, if the substrate (electrode) is capable of absorbing hydrogen atoms:

$$\mathrm{H_{UPD}} \rightarrow \mathrm{H_{abs}}, \quad E < 0; \qquad \mathrm{H_{OPD}} \rightarrow \mathrm{H_{abs}}, \quad E > 0$$

The formation of H_{UPD} at polycrystalline Pt is demonstrated in Fig. 6.3 In this figure, the $H_{UPD,1}$ and $H_{UPD,2}$ ascribe the peaks of desorption and adsorption of underpotential deposited hydrogen.

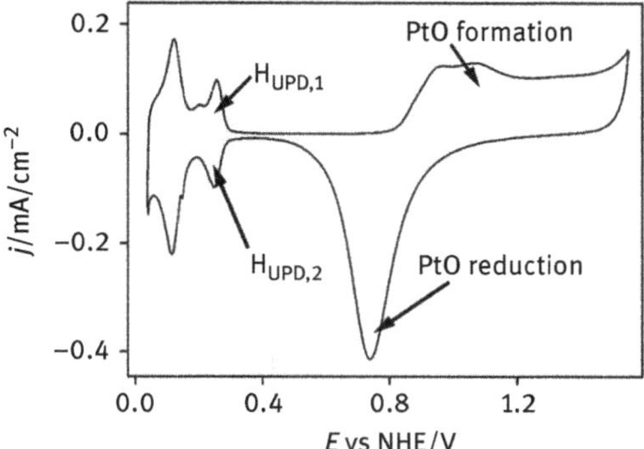

Fig. 6.3: Cyclic voltammogram recorded for Pt in 1 N H_2SO_4. It shows the regions of $H_{UPD,}$ electrosorption ($H_{UPD,2}$), electrodesorption ($H_{UPD,1}$), and PtO formation and reduction. Unpublished data, courtesy of B. Maranowski, M. Szklarczyk (Faculty of Chemistry, Warsaw University).

It is worth to mention that only H_{OPD} species participate in the reaction of H_2 evolution:

$$\text{Me-H}_{OPD} + H_3O^+ + e \rightarrow \text{Me} + H_2(g) + OH^-, \qquad \text{Volmer-Heyrowsky mechanism}$$

$$2\,\text{Me-H}_{OPD} \rightarrow 2\text{Me} + H_2(g), \qquad \text{Volmer mechanism}$$

The fact that UPD and OPD of H take place in different range of potentials (UPDH, $E > 0$; OPDH, $E < 0$) points to the differences in their Gibbs free energy of electrosorption (for UPDH $\Delta_{ec-ads}\, G < 0$, for OPDH $\Delta_{ec-ads}\, G > 0$). It implies further that these two species occupy two distinct adsorption sites on the electrode surface. Plenty of spectroscopic experiments were carried out but they are not fully conclusive. Interested readers will find much more about UPD of H in the review [5].

6.3 Underpotential deposits – catalytic properties

We can apply many electrochemical methods together with microscopic techniques for UPD, controlling in this way the amount, distribution, and structure of UP deposits. In fact, we can obtain the metal substrate decorated with nanostructural objects. It is worth to mention the two methods based on UPD: surface-limited replacement reaction (SERR) and electrochemical atomic layer epitaxy (ECALE) [6]. The SERR method has been developed by Brankovic et al. [7]. The process consists of the deposition of metal Me (2) on a metal substrate Me (1) under UPD condition and after that the replacement of M (2) monolayer by more noble metal ions, N, takes place, forming one atomic layer of N atoms. The driving force for such replacement is the difference in the redox potentials of Me/Me $(2)^{n+}$ and N/N^{n+} couple, so the process may take place at open-circuit potentials. An example is the replacement of Cu UPD monolayer on Pd single crystal by Au. The second method was introduced by Stickney [8]. This method enables the formation of semiconductor nanofilms "layer by atomic layer," each layer being deposited under the UPD control. This method can be used to obtain a wide range of semiconductors composed of metal and one of the nonmetal elements such as S, Se, Te, and As. The schemes of both methods SLRR and ECALE are presented in Fig. 6.4(a, b).

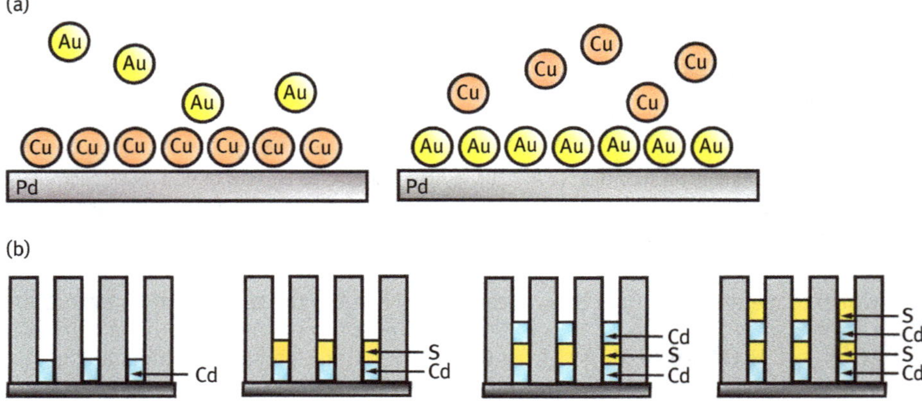

Fig. 6.4: (a) Scheme illustrating SERR (surface-limited replacement reaction) method. Adapted from [6]. (b) Scheme illustrating ECALE (electrochemical atomic layer epitaxy) method showing formation of thin semiconducting film in membrane. The formation may be carried out by using electrochemical methods or chemical vapor deposition.

UPD and derivative methods can be applied for the formation of nanostructures: nanoparticles and monolayers that are on the top of metal substrate changing its properties and playing an important role in electrocatalysis. The foreign ad-atoms may modify the electrocatalytic activity of substrate by (i) changing local electron density on the surface and work function of the substrate, (ii) preventing the poisoning of substrate surface by strong adsorption of intermediates, (iii) creating the energetically different active sites for adsorption, and so on. The enhancement of electrocatalytic properties manifests itself in the decrease of the overpotentials, in the improvement of reversibility of reaction, or in selectivity changes. There are many electrochemical reactions that are important from a practical point of view, such as oxygen reduction, hydrogen evolution, and oxidation of organics fuels. In all of them, the UP deposits were used as electrocatalysts to improve the efficiency and selectivity of the reactions. The reduction of oxygen was influenced by the UP deposit. This influence depended on the type of substrate [2]. The catalytic effects were observed in acidic solution when Pb_{UPD} and Bi_{UPD} were deposited on Au and Pb_{UPD} on glassy carbon, and also in alkaline solution for Pb_{UPD}, Tl_{UPD}, Bi_{UPD}, and Cu_{UPD} deposited on Pt. These effects mainly result from the decrease of overpotential of charge transfer and from changes in the mechanism of O_2 reduction from 2e to 4e (Chapter 4). However, in acidic solutions, Pb_{UPD}, Tl_{UPD}, Bi_{UPD}, and Cu_{UPD} deposited on Pt inhibit the reduction of O_2. Some experimental results also pointed out that the reaction of H_2 evolution is inhibited when Pt is covered with ad-atoms of heavy metals such as Pb_{UPD}, Sn_{UPD}, and Tl_{UPD} [2]. It was mentioned in Chapter 4 that the platinum has very good electrocatalytic properties, and they are strongly enhanced if Pt is modified with UP deposit. Underpotential deposited Pb_{UPD} and Bi_{UPD} on Pt catalyze the oxidation of HCOOH, HCHO, and CH_3OH, the potential fuel in the fuel cells (Chapter 8).

Interested readers can find much more about the theory and application of UPD in [9, 10].

Bibliography

[1] Sudha V, Sangaranarayanan MV. Underpotential of deposition of metals: structural and thermodynamic consideration. J. Phys. Chem. B 2002, 106, 2699–2707.

[2] Kokinidis G. Underpotential deposition and electrocatalysis. J. Electroanal. Chem. 1986, 201, 217–236.

[3] Gileadi E. Physical electrochemistry, fundamentals, techniques and applications. Weinheim, FRG, Wiley-VCH Verlag GmbH & Co.KGaA, 2011.

[4] Kolb DM. Advances in electrochemistry and electrochemical engineering. Eds. Gerischer H, Tobias CW. New York, USA, J. Wiley & Sons, 1978 (v.11).

[5] Jerkiewicz G. Electrochemical hydrogen adsorption and absorption. Part I. Underpotential deposition of hydrogen. Electrocatalysis 2010, 1, 179–199.

[6] Zangari G. Electrodeposition of alloys and compounds in the era of microelectronics and energy conversion technology. Coatings. 2015, 5, 195–218.

[7] Brankovic SR, Wang JX, Adzic RR. Metal monolayer deposition by replacement of metal adlayers on electrode surface. Surf. Sci. 2001,474, L.171–L.179.

[8] Stickney JL. Electrochemical atomic layer epitaxy (ECALE). Nanoscale control in the electrodeposition of compound semiconductor. In Advances in electrochemical science and technology. Eds. Alkiire RC, Kolb DM. New York, USA, Willey & Sons, 2009, 7, 1–105.

[9] Mayet N, Servat K, Kokoh B, Napporn TW. Probing the surface of noble metals electrochemically by underpotential deposition of transition metals. Surfaces. 2019, 2, 257–276.

[10] Oviedo OA, Reinaudi L, García SG, Leiva EPM. Underpotential deposition. From fundamentals and theory to applications at the nanoscale. Springer International Publishing AG, Switzerland, 2016.

7 Electrochemical methods in the formation of nanostructures

7.1 General remarks

The so-called dimensionality plays a crucial role in determining the properties of nanostructures obtained via the electrochemical deposition, such as quantum phenomena in the electron transport, photonic behavior of nanocrystals, and even magnetic properties of nanostructured transition metals [1–8]. In addition, the catalytic behavior of deposited nanoparticles, noble metals in particular, resulted in the tremendous resources being devoted to the development of new methods for preparing nanometer and micron-scale structures of various topologies.

As was introduced in the previous chapters, nanostructures can be formed by the two electrochemical methods: UPD and OPD. In both cases such synthesis starts at the surface of a conducting substrate (working electrode), *S*. This process is also referred to as *heterogeneous nucleation* and its mechanism has been described in Chapters 5 and 6. Because UPD is a self-limiting process and is terminated when a complete monolayer is formed, the deposition is limited to the surface of conducting substrate, *S*. Nevertheless, under controlled UPD potentiostatic or galvanostatic conditions, the deposition of nanostructures, sometimes called *0D nanostructures*, can be effected. The UPD, offering a "bottom-up" approach, has been widely used to control the growth of nanostructures of different types and shapes. This includes Au octahedral nanocrystals, nanorods, bimetallic Au–Pd hexoctahedra, Ag "flakes," Pt–Cu hierarchical structures, and Pd concave nanostructures. For more detailed library of different surface grown structures electrodeposited for catalytical purposes, the interested reader is referred to excellent resource books [1, 2].

These nanostructures can subsequently act as "seeds" under the OPD conditions for further growth of 2D structures: nanowires, nanotubes, or nanorods. For these types of depositions, the nucleation and growth can be controlled separately by varying the potential or current values, their duration, and profile. In order to obtain well-ordered and uniform in size and shape nanostructures, it is necessary to use templates at the electrode surface or the structural characteristics of the electrode surface itself. The following sections will describe briefly such an approach yielding 0D and 2D nanostructures (Fig. 7.1).

https://doi.org/10.1515/9783111160986-007

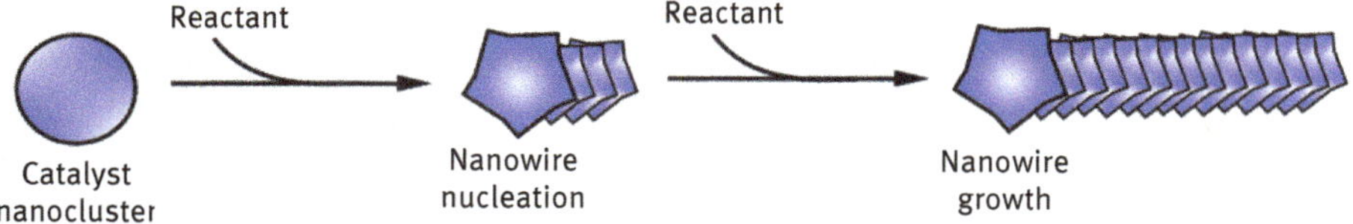

Fig. 7.1: 0D nanocluster acting as a "seed" for further nucleation and growth of 1D nanowire.

7.2 Template-assisted electrodeposition of nanostructures

Generally, the template-assisted electrodeposition can be divided into two groups [9]:
- active template-assisted electrodeposition and
- restrictive template-assisted electrodeposition.

7.2.1 Active template-assisted electrodeposition

Surface template-assisted electrodeposition is widely used to synthesize various kinds of nanostructures of tailored morphology and topology. The key issue in using such electrodeposition is the preparation of proper templates because the surface characteristics of the electrode surface affects – "templates" – the obtained nanostructures. The nucleation and subsequent growth of nanostructures is influenced by Gibbs' surface free energy of the substrate and crystallographic lattice mismatch at the boundary between the surface and nucleus of reduced metal. For instance, electrodeposited metal nanostructures, such as Ag, Au, or Pd, tend to form small nanoclusters on surface defects, such as step edges at the surface (see schematic drawing, Fig. 7.2).

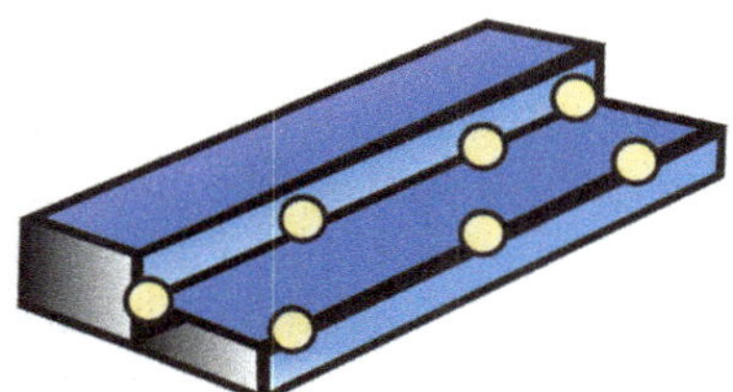

Fig. 7.2: Nanostructures deposited on the energetically favored step edges in the crystalline surface of a substrate electrode.

Since for this type of electrodeposition it is required to have a detailed control over the electrode surface, numerous works were devoted to the highly oriented pyrolytic graphite (HOPG) and graphene surfaces, as well as single crystal metal surfaces as templates. For instance, it was found that in the case of HOPG, palladium deposition reaction at early stages proceeds through progressive nucleation under diffusion control. Palladium islands and the development of their dense distribution indicated the key role of step edge energy barrier in determining the growth pattern of Pd electrodeposition [10]. Similar results were obtained in the case of atomically flat terraces of HOPG and Ag deposits. Preferential deposition of silver nanostructures was observed at step edges and other surface defects [11, 12]. Other metals, like Pt nanocrystals, were deposited on the basal plane of HOPG with the help of pulsed potentiostatic method with controlled pulse amplitude, duration, and temporal resolution. Similar approach was used for one-step, pulsed potentiostatic electrodeposition of Ag on Au(111) single crystal via the UPD. The analysis of scanning tunneling microscopy images revealed that stable Ag nanoclusters are formed all over the surface of Au(111) single crystal, without preferential tendency of deposition at the step edges, nor diffusional agglomeration on the defect sites in time. Moreover, by controlling the potential step it was possible to obtain Ag clusters of monoatomic height [13, 14].

7.2.2 Restrictive template-assisted deposition

As we have shown above, the template-based electrochemical synthesis of various conductive nanostructures is a very general and relatively simple, low-cost method. The subsequent growth of nanostructures on "seed" nanoclusters to form nanowires or nanotubes will continue as long as the electric field is applied. Current density on the tips of growing 2D structures will increase with their elongation, because of decreasing distance between these tips and counter electrode. Therefore, the new deposits will be more likely formed on the tips, resulting in the propagation of growth. However, the use of surface topology of a working electrode as a template for such synthesis has its limitations, since it is almost impossible to control the surface distribution, organization, and length of the deposited structures. Therefore, this method is hardly used in practice for the synthesis of nanowires or nanotubes. Instead, porous 3D matrices with desired channel distribution and length are used to control the growth of nanowires and nanotubes via the electrochemical deposition. Figure 7.3 illustrates the scheme of the typical setup used for the electrochemical deposition of 2D nanostructures with the help of an inert, nonconducting porous membrane. Examples of such matrices include, but are not limited to, track-etched polycarbonate filtration membranes, porous alumina, porous titanium dioxide (TiO_2, anatase), mesoporous silica, conducting polymers, and so on. Some of them can be prepared electrochemically, such as porous alumina and porous titanium dioxide, finally yielding well-organized 2D nanotubes, and therefore we will outline their formation in more detail. They can serve as electrochemically fabricated arrays of nanotubes – nanostructures themselves, but, of course, they can be used also as inert matrices (porous alumina in particular) for electrochemical deposition of conducting nanorods and nanowires, as shown in Fig. 7.3.

The most facile electrochemical synthesis of vertically aligned metal oxide nanotubular structures is the so-called self-ordering electrochemical anodization (SOA). It can be carried on a suitable metal, for example, Al, Ti, W, Zn, and Zr. Exposing such

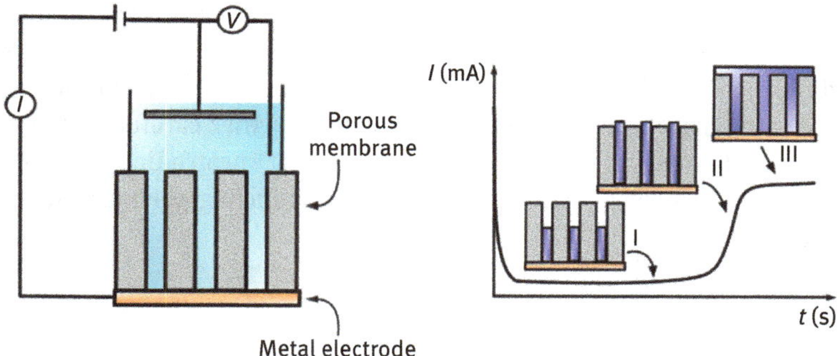

Fig. 7.3: Typical experimental setup for the restricted template-based electrochemical growth of 2D nanostructures. Right-hand side panel shows the current–time profile for the deposition of metallic nanowires growing inside the nanometer-sized pores in a nonconducting, templating membrane.

metal to a sufficiently high anodic voltage, an oxidation reaction will be triggered. The reaction path can proceed through different mechanisms, depending mainly on the electrolyte content and anodization parameters, that is, voltage and/or current densities. The first path is obvious: the metal ions are formed, and the metal is dissolved continuously (corrosion, electropolishing). The second path, for example, in the presence of oxygen-saturated solution, involves the formation of insoluble oxide, and a passivation layer is formed. However, under specific conditions of competition between oxide formation and dissolution, a self-organized nanotube or nanorod arrays can be formed. The best of investigated systems, in which almost perfect organization was achieved, was the self-ordered nanotubular arrays of aluminum oxide structures. It is formed in acidic electrolytes and has tunable porous morphology and mechanical and chemical strength. Its formation triggered hundreds of research work and numerous technological applications, including templating and nanostructuring of metal, semiconductor, or polymer materials [15, 16]. Many of the principles and mechanisms described for porous alumina SOA can be adapted for the formation of self-organized 1D structures on other metals, taking into account some specifics of electrochemical behavior of their oxides.

7.2.2.1 Porous anodic alumina (PAA) matrices

The porous structure of aluminum oxide was first reported in 1953 [17]. Since that time a whole gallery of nanotubular arrangements of different pore geometries has been proposed, such as hexagonal, square, and triangular openings [18]. When PAA matrix is used for the formation of metal or carbon nanowires, the pattern of their growth inside the pores of PAA under potentiostatic conditions can be followed by measuring the current versus time curve. Three zones can be distinguished, as shown in the right panel of Fig. 7.3: zone I corresponds to the initial deposition of metal inside the pores, and zone II reflects the complete filling of some pores which gives rise to the formation of an array of hemispherical nanoelectrodes. They finally coalesce into a continuous metal layer, completely covering the PAA surface (zone III). The current saturates due to semi-infinite diffusion control. Before zone III is reached, the cathodic deposition of metal can be stopped by disconnecting the applied voltage. Further careful chemical dissolution of PAA template leaves the metal nanowire "forest" protruding from the substrate surface. Similar nanowires can be formed from conducting polymers, semiconductors, and oxides [19].

7.2.2.2 Porous anodic titanium dioxide matrices

Hollow titanium dioxide (TiO_2) nanostructures, such as nanotubes, have attracted a broad range of scientific attention, not only as templates but also as nanostructured components for solar energy conversion and as possible nanocarriers for drug delivery systems [20]. As templates for growing well-aligned 1D nanostructures, titania nanotubes were successfully used for coating or filling the active materials on/into TiO_2, for fabrication of electrodes in supercapacitors and batteries with an enhanced capacitance due to the enlargement of electrode area [21–24].

Formation of compact tubular layers of TiO_2 is usually carried out in aqueous electrolytes, as in the case of porous alumina, at a constant potential in the range of 10–30 V, applied to pristine Ti electrode. In the absence of competing factors, such as fluoride ions, this will result in a compact TiO_2 layer. However, in the presence of F^- ions that form water-soluble complexes TiF_6^{2-}, complexation of Ti^{4+} ions formed during the anodic process as well as dissolution of TiO_2 under the influence of fluorides occurs, both processes competing with the formation of compact TiO_2 layer. Under the appropriate electrochemical conditions balance [23], well-organized tubular array of TiO_2 is obtained (Fig. 7.4c).

7.2.3 Concluding remarks

As described earlier, electrochemical methods are extremely suitable for the preparation of various nanostructured materials. Both cathodic and anodic polarization can be applied to obtain such structures of controlled shape, distribution, dimensions, and properties. While the reduction procedures are predominantly used for the electrodeposition of metals, alloys, bimetallic structures, polymers, semiconductors, and oxides with precise control over their shapes, oxidation processes are mainly used to prepare oxides; porous alumina and titania being their most frequent representatives.

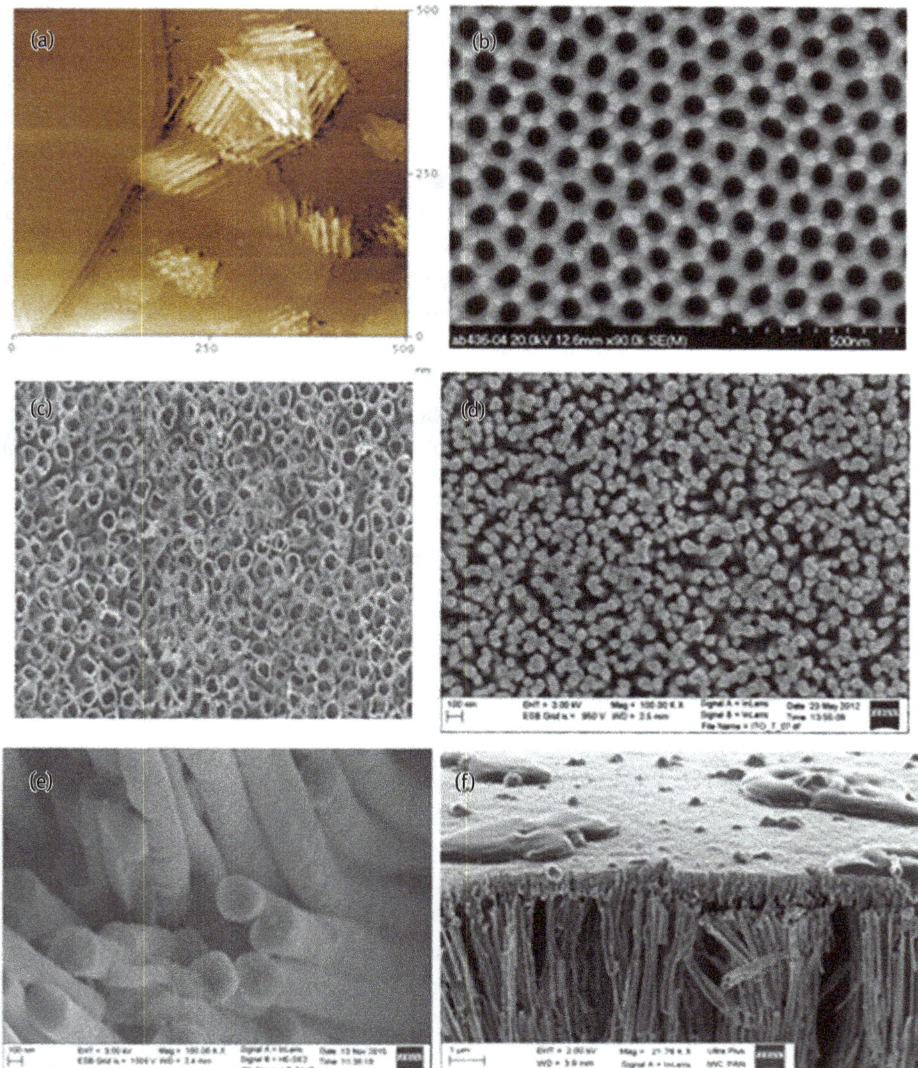

Fig. 7.4: Panel a: Silver nanoneedles electrodeposited on HOPG electrode (SEM image, 500 nm frame). These nanoneedles were formed randomly at the flat surfaces of HOPG. Their needlelike growth was controlled by mercaptopropionic acid (MPA) preferential adsorption. Courtesy of Prof. M. Mazur, Faculty of Chemistry, University of Warsaw. **Panel b**: SOA of porous alumina matrix deposited electrochemically on Al electrode (top view). Courtesy of Dr A. Brzózka, Faculty of Chemistry, Jagiellonian University, Kraków. **Panel c**: P.K. own SEM image of TiO_2 nanotube array, ca. 50 nm in diameter. Somehow tilted bundles resulted from nonoptimized anodization process. **Panel d**: ZnO nanorods grown electrochemically onto FTO electrode from $Zn(NO_3)_2$ aqueous electrolyte. Courtesy of Dr K. Zarębska, Faculty of Chemistry, University of Warsaw. **Panels e, f:** Template-assisted electrochemical growth of conducting polymer (polyindole) nanowires from its monomer solution. Courtesy of Dr M. Osial, Faculty of Chemistry, University of Warsaw.

7.3 Applications of nanostructured materials

Every time, when the high surface area versus small dimensions for efficient catalysis, energy conversion, and storage is required, present-day engineering turns to nanomaterials. This is because conducting and semiconducting materials at the nanometer scale exhibit unusual and frequently advantageous behavior compared to their bulk counterparts. The origin for these differences stems from the fact that a significant fraction of the total number of atoms and molecules in the nanostructures reside on the surface and therefore the surface energy plays the dominant role in their properties. Nanostructured materials offer an extremely broad possibilities of applications, ranging from electronics, photonics, energy storage devices, catalysis, medical applications, and so on. It is unrealistic to summarize all possible applications or foresee the new ones within this section. Therefore, here we will focus on several, most common uses of electrochemically synthesized nanostructures. More broad information can be found in [1–8], as well as in [25, 26].

7.3.1 Catalysis

Electrocatalysis was described in detail in Chapter 4. Here, we will only outline the nanoparticle-based electrocatalysis. Noble-metal-based nanomaterials have attracted much interest in their promising potentials in fields of energy-related and environmental catalysis. Tailoring and controlling the surface/interface structure of such metal-based nanomaterials at the atomic scale have great significance for optimizing the performances in practical catalytic applications. Nanostructures are used to control electrochemical reactivity, such as catalytical oxygen electroreduction on several reactions, exemplified here as O_2 electroreduction (ORR), CO_2 electroreduction, and ethanol electro-oxidation. The presence of low-coordinated sites or facets, particle size, shape, and composition in nanostructures can be used to tune reactivity and selectivity. Since the price of noble metals, platinum, in particular, is skyrocketing, while their availability is decreasing, other nanostructured catalysts began to emerge, with low or no Pt content. And so, in recent years, many alloy systems and core–shell systems with lower Pt content was reported and used, particularly for ORR in fuel cells and for decreasing the emission of CO_2 and CO. Systems such as PtPd, PtAu, PtCu, PtNi, and PtW have been studied of various morphologies and compositions. Much research has been devoted to Pt-free nanocatalysts, from which palladium-, ruthenium-, and iridium-based nanostructures were expected to at least replace Pt-based electrocatalysts. However, even though Pd is ca. three times more expensive than Pd, the catalytic efficiency of palladium, the best of other metals studied, was at least five times lower than that of Pt [27–29]. Therefore, the efforts to design nanostructures with better activity than Pt in catalysis are still required and in progress.

7.3.2 Batteries: supercapacitors

As will be discussed in Chapter 8, nanostructuring of electrodes plays a crucial role in lithium-ion batteries and supercapacitors. Moreover, commercialization requirements direct the researchers toward environmentally safer and cheaper electrode materials with good capacitive behavior, such as oxides and hydroxides of Ni, Co, Mn, W, Ru, and V, all in a form of nanostructures. In addition, carbon-derived materials, such as carbon nanotubes, have attracted much attention because of their advantageous electronic and mechanical properties. As an example, a simple method of direct electrodeposition of hexagonal arrays of $Ni(OH)_2$ on titanium substrate has been proposed for high specific capacitance material for supercapacitors [25]. In the case of Li-ion secondary batteries, also discussed later in this book, the improvement of storage lifetime and the number of charging/discharging cycles can be achieved by nanostructuring of the electrode material. This approach takes an advantage from the experimental and theoretical observations that the smaller the dimensions of the electrode materials that react with lithium (nanostructuring), the better the electrochemical performance of the cell due to the improved Li intercalation densities [26].

7.3.3 Photoelectrochemical devices: solar cells

Similar considerations as in the case of batteries and supercapacitors can be applied for nanostructures used for solar energy conversion, discussed in Chapter 11. Here, we will mention titanium-derived nanostructures (nanoparticles, nanotubes, and nanorods) as their use in the photoelectrochemical devices seem to be most explored. These nearly translucent structures can be easily dye-sensitized, tailoring their spectral/electrochemical behavior for more efficient light energy conversion. More detailed information can be found in excellent review [23] and references therein. TiO_2 nanotubes show a much higher electron diffusion length compared to nanoparticles. It was proved experimentally that 15–20 μm long nanotubes show the maximum solar cell efficiency. Along the same line, the organized anatase (thermally annealed form of rutile TiO_2) nanotube layers are more efficient than the rutile phase, as the electron transport is faster, leading to high-efficiency solar cells [18].

7.3.4 Biological applications: nanozymes

In recent years a broad range of electrosynthesized nanostructures have been considered as possible theranostic tools. In particular, due to the luminescent and superparamagnetic behavior of certain metal oxides (e.g., Gd_2O_3, Fe_3O_4), they were tested as tissue imaging tools. Biocompatibility of oxides allowed for their promising use as a drug delivery platform for multimodal treatment of numerous diseases, including cancer [30, 31]. However, it should be stressed that for this type of applications, techniques other than electrochemical synthesis are employed, more suitable for obtaining large amounts of nanostructures suspended in liquids, such as coprecipitation, thermal decomposition of precursors, bulk redox reactions, and so on. So, here we will outline the applications of those nanostructures that can be obtained electrochemically on the electrode surface. Along this line, it was reported lately that anodic titanium dioxide nanotubular structures can be used for intracellular drug delivery. When compared to solid-core nanostructures, such hollow nanotubes offer substantially higher capacity for drug loading and can be optimized for the appropriate drug release kinetics. Additionally, they can be loaded with magnetic nanoparticles for site-specific guidance in an external magnetic field [32, 33]. Metallic nanostructures of different shapes can also be used in biomedical applications. Among them, gold nanoparticles received broad interest, because they are known to be stable, easily electrosynthesized (even though they are obtained mostly in a form of suspension). Their shape can be controlled experimentally (e.g., gold nanorods); they possess tunable optical properties and tailored surface chemistry. For more detailed biological applications of metallic, carbon-derived, oxide, and semiconductor nanoparticles, interested reader is directed to Ref. [34].

Nanozymes: Although we already mentioned the electrocatalytic activity of various nanostructures, noble-metal nanoparticles in particular, their broad applications were generally focused on energy-related and environmental catalysis areas. However, some nanomaterials have been found to exhibit unexpected enzymelike activities, and great advances have been made in this area due to the unique characteristics of such nanomaterial-based artificial enzymes (nanozymes). From the electrochemical point of view, four types of redox enzymes have been reported to be mimicked by either metallic (e.g., Au) or oxide nanoparticles (e.g., Fe_3O_4, CeO_2). These redox enzymes include superoxide dismutase, peroxidase, oxidase, and catalase. In all these cases, the enzymelike activities are due to the intrinsic behavior of nanoparticle cores instead of the functional groups present on their surfaces. For more comprehensive information regarding nanozymes, their structure, and properties, we suggest two successive review papers showing the research progress and references therein [35, 36].

Bibliogaphy

[1] Parak WJ, Manna L, Simmel FC, Gerion D, Alivisatos P. Quantum dots. In Nanoparticles. From theory to application. Schmid G. Wiley-VCH, Germany, 2004, 4–362.

[2] Nasirpouri F. Electrodeposition of nanostructured materials. Springer AG, Switzerland, 2017.

[3] Petri OO, Electrosynthesis of nanostructures and nanomaterials. Russ. Chem. Rev. 2015, 84, 159–193.

[4] Zhang YW. Ed. Bimetallic nanostructures: Shape-controlled synthesis for catalysis. VCH Wiley, Germany, 2018.

[5] Efekhtari A. Ed. Nanostructured materials in Electrochemistry. Wiley-VCH, Germany, 2008.
[6] Losic D, Santos A. eds. Electrochemically engineered nanoporous materials. Springer, Switzerland, 2015.
[7] Cao G. Nanostructures and nanomaterials. Synthesis, properties and applications. London, Imperial College Press, 2006.
[8] Rauh A, Carl N, Schweins R, Karg M. Role of absorbing nanocrystal cores in soft photonic crystals: A spectroscopy and SANS study. Langmuir 2018, 34, 54–867.
[9] Bera D, Kuiry SC, Seal S. Synthesis of nanostructured materials using template-assisted electrodeposition. JOM 2004, 56, 49–53.
[10] Gimeno Y, Hernandez Creus A, Carro P, Gonzalez S. Salvarezza RC, Arvia AJ. Electrochemical formation of palladium islands on HOPG: Kinetics, morphology, and growth mechanisms. J. Phys. Chem. B 2002, 106, 4232–4244
[11] Potzschke RT, Gervasi CA, Vinzelberg S, Staikov G, Lorenz WJ. Nanoscale studies of electrodeposition on HOPG (0001). Electrochim. Acta 1995, 40, 1469–1474.
[12] Zoval JV. Electrochemical deposition of silver nanocrystallites on the atomically smooth graphite basal plane. J. Phys. Chem. 1996, 100, 837–844.
[13] Mazur M. Electrochemically prepared silver nanoflakes and nanowires, Electrochem. Comm. 2004, 6, 400–403.
[14] Borissov D, Tsekov R, Freyland W. Pulsed electrodeposition of two-dimensional Ag nanostructures on Au(111). J. Phys. Chem. B 2006, 110, 15905–15911.
[15] Sulka GD. Highly ordered anodic porous alumina formation by self-organized anodizing. In: Eftekhari A. ed. Nanostructured materials in electrochemistry. Wiley VCH, Germany, 2008, pp.1–97.
[16] Brzózka A, Budzisz A, Hnida K, Sulka GD, Losic D., Santos A. eds. Electrochemically engineered nanoporous materials. Springer, Switzerland, 2015, 219–272.
[17] Keller F, Hunter MS, Robinson DL. Structural features of oxide coatings on aluminium. J. Electrochem. Soc. 1953, 100, 411.
[18] Masuda H, Fukuda K. Ordered metal nanohole arrays made by a 2-step replication of honeycomb structures of anodic alumina. Science 1995, 268, 1466–1468; Masuda H, Asoh H, Watanabe M, Nishio K, Nakao M, Tamamura T. Square and triangular nanohole array architectures in anodic alumina. Adv. Mat. 2001, 13, 189–192.
[19] Choi J, Sauer G, Nielsch K, Wehrspohn RB, Gosele U. Hexagonally arranged monodisperse silver nanowires with adjustable diameter and high aspect ratio. Chem. Mater. 2003, 15, 776–779.
[20] Kafshgari MH, Mazare D, Distaso M, Goldmann WH, Peukert W, Fabry B, Schmuki P. Intracellular drug delivery with anodic titanium dioxide nanotubes and nanocylinders. ACS Appl. Mater. Interfaces 2019, 11, 14980–14985.
[21] Global A, Faraji M. Fabrication of nanoporous nickel oxide by de-zincification of Zn–Ni/(TiO_2-nanotubes) for use in electrochemical supercapacitors. Electrochim Acta 100, 2013, 133–139.
[22] Wang Y-G, Wang Z-D, Xia Y-Y. An asymmetric supercapacitor using RuO_2/TiO_2 nanotube composite and activated carbon electrodes. Electrochim. Acta 2005, 50, 5641–5646.
[23] Roy P, Berger S, Schmuki P. TiO_2 nanotubes: Synthesis and applications, angew. Chem. Int. Ed. 2011, 50, 2904–2939.
[24] Lee K, Mazare A, Schmuki P. One-dimensional titanium dioxide nanomaterials: Nanotubes. Chem. Rev. 2014, 114, 9385–9454.
[25] Zhao DD, Bao, SJ, Zhou WH, Li HL. Preparation of hexagonal nanoporous nickel hydroxide film and its application for electrochemical capacitor. Electrochem. Comm. 2007, 9, 869–874.
[26] Beaulieu LY, Eberman KW, Turner RL, Krause LJ, Dahn JR. Colossal reversible volume changes in lithium alloys. Electrochem. Solid-State Lett. 2001, 4, A137–A140.

[27] Nie Y, Li L, Wei Z. Recent advancements in Pt and Pt-free catalysts for oxygen reduction reaction. Chem. Soc. Rev. 2015, 44, 2168–2201.

[28] Duan S, Du Z, Fan H, Wang R. Nanostructure optimization of platinum-based nanomaterials for catalytic applications. Nonomaterials, 2018, 8, 949.

[29] Mistry, H.; Varela, A.S.; Strasser, P.; Cuenya, BR. Nanostructured electrocatalysts with tunable activity and selectivity. Nature Rev. Mater. 2016, 1, 1–14.

[30] Kafshgari M, Harding F, Voelcker N. Insights into cellular uptake of nanoparticles. Curr. Drug Deliv. 2015, 12, 63–77.

[31] Steinhauser I, Spaenkuch B, Strebhardt K, Langer K. Trastuzumab-modified nanoparticles: optimization of preparation and uptake in cancer cells. Biomaterials 2006, 27, 4957–4983.

[32] Kafshgari MH, Mazare A, Distaso M, Goldmann WH, Peukert W, Fabry B, Schmuki P. Intracellular drug delivery with anodic titanium dioxide nanotubes and nanocylinders. ACS Appl. Mater. Interfaces, 2019, 11, 14980–14985.

[33] Kafshgari MH, Kah D, Mazare A, Nguyen NT, Distaso M, Peukert W, Goldmann WH, Schmuki P, Fabry B. Anodic titanium dioxide nanotubes for magnetically guided therapeutic delivery. Nature Research Scientific Reports 2019, 9, 13439.

[34] Ashikbayeva Z, Tosi D, Balmassov D, Schena E, Saccomandi P, Inglezakis V. Application of nanoparticles and nanomaterials in thermal ablation therapy of cancer, Nanomaterials, 2019, 9, 1195. doi:103390/nano9091195.

[35] Wei H, Wang E. Nanomaterials with enzyme-like characteristics (nanozymes): next-generation artificial enzymes. Chem. Soc. Rev. 2013, 14, 6060–6093.

[36] Wu J, Wang X, Wang Q, Li S, Zhu Y, Qin L, Lou Z, Wei H. Nanomaterials with enzyme-like characteristics (nanozymes): next generation of artificial enzymes (II). Chem. Soc. Rev. 2019, 48, 1004–1076.

8 Electrochemistry in energy conversion and storage

8.1 Batteries

8.1.1 Electrochemical cell – fundamentals

The simplest electrochemical cell contains two electronic conductors (electrodes, mostly metals) and ionic liquid conductor containing ionic species (electrolyte) between them. The electronic conductor and its interface with electrolyte serve as the place where the electrochemical reactions occur. To prevent any unwanted reaction with electrolyte species, it is often necessary to apply diaphragm dividing the cell into two parts (half cells). In this case, the additional potential is formed, called the *liquid junction potential*, which can be limited by means of a salt bridge or binary electrode. The schemes of exemplary cells without (A) and with junction (B) are as follows:

$$\mathrm{Ag}|\mathrm{Ag_2O},\ \mathrm{KOH_{aq}},\ \mathrm{HgO}|\mathrm{Hg}, \qquad \mathrm{Zn}|\mathrm{ZnSO_{4,\,aq}}||\mathrm{CuSO_{4,\,aq}}|\mathrm{Cu} \quad (\mathrm{A})$$

$$\mathrm{Zn}|\mathrm{ZnSO_{4,\,aq}}(a_1)|\mathrm{CuSO_{4,\,aq}}(a_2)|\mathrm{Cu}, \qquad \mathrm{Zn}|\mathrm{ZnCl_{2,\,aq}}(a_1)|\mathrm{AgNO_{3,\,aq}}(a_1)|\mathrm{Ag} \quad (\mathrm{B})$$

The other kind of electrochemical cells are the concentration cells, in which the two half cells differ only in the concentration of the same electroactive species taking part in electrochemical reaction:

$$\mathrm{Ag}|\mathrm{AgCl},\ \mathrm{HCl_{aq}}(a_1),\ \mathrm{H_2}(p_1)|\mathrm{Pt}-\mathrm{Pt}|\mathrm{H_2}(p_2),\ \mathrm{HCl_{aq}}(a_2),\ \mathrm{AgCl}|\mathrm{Ag} \quad (\mathrm{A})$$

$$\mathrm{Ag}|\mathrm{AgCl},\ \mathrm{HCl_{aq}}(a_1)|\mathrm{HCl_{aq}}(a_2),\ \mathrm{AgCl}|\mathrm{Ag} \quad (\mathrm{B})$$

As the voltage of concentration cells is low, they are not used practically in energy production and will not be considered further.

Note that a slash | represents a phase boundary, a double slash || represents the phase boundary whose potential is negligible as a result of application of separator between the two electrolytes, and coma separates two components in the same phase.

The overall chemical reaction in an electrochemical cell is formed by two independent half-reactions taking part in half cells, each of them characterized by the interfacial potential difference $\Delta\varphi$ (Galvani potential, see Chapter 1). It is accepted that, in the notation of cells, the electrode with a lower value of potential ($\Delta\varphi_1$, E_1) is placed on the left side of such notation. At this electrode, being negatively charged, named *anode*, the oxidation reaction takes place. At another electrode, which is positively charged, cathode, ($\Delta\varphi_2$, E_2), the reduction reaction occurs:

https://doi.org/10.1515/9783111160986-008

$$\text{anode (oxidation reaction):} \qquad \mathrm{Red_1 \longrightarrow Ox_1 + ne}$$

$$\text{cathode (reduction reaction):} \qquad \mathrm{Ox_2 + ne \longrightarrow Red_2}$$

The overall reaction taking place in the cell is the sum of these two reactions:

$$\mathrm{Red_1 + Ox_2 \longrightarrow Ox_1 + Red_2}$$

Electrochemical cells can operate in the three modes: equilibrium mode, galvanic mode, and electrolytic mode.

Let us consider the behavior of the cell being in equilibrium mode (at equilibrium). The most convenient for laboratory purpose is the Daniell cell (Fig. 8.1(a)):

$$\mathrm{Zn}\left|\mathrm{ZnSO_{4,\,aq}(a_1)}\right|\left|\mathrm{CuSO_{4,\,aq}(a_1)}\right|\mathrm{Cu}$$

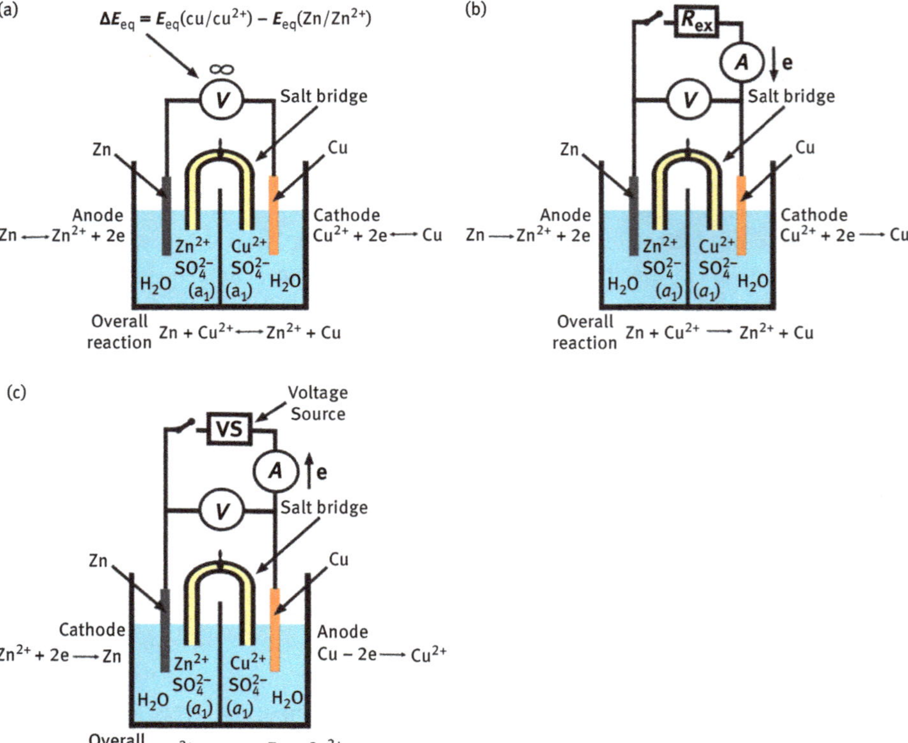

Fig. 8.1: (a) Scheme illustrating Daniell cell in the state of equilibrium. (b) Scheme illustrating Daniell cell in galvanic mode. Discharge of the cell takes place. The potential of the cell decreases in time. (c) Scheme illustrating Daniell cell in electrolytic mode. Note that comparing with galvanic mode, in the electrolytic mode, the anode becomes the cathode.

If we put metallic Zn in one part of the cell containing Zn^{2+} ions and Cu in the other part containing Cu^{2+} ions, some kind of equilibrium is established at the interfaces: $Zn|Zn^{2+}$, $Cu|Cu^{2+}$, resulting of the following reactions:

$$\mathrm{Zn} \leftrightarrow \mathrm{Zn}^{2+} + 2\,\mathrm{e}$$

$$\mathrm{Cu}^{2+} + 2\,\mathrm{e} \leftrightarrow \mathrm{Cu}$$

The equilibrium of the whole system is characterized by the equilibrium potentials of anode and cathode: $E_{\mathrm{eq},1}(\mathrm{Zn}^{2+}|\mathrm{Zn})$ and $E_{\mathrm{eq},2}\,(\mathrm{Cu}^{2+}|\mathrm{Cu})$. Putting the electronic voltmeter of very high resistance between the electrodes or compensating the equilibrium potentials of electrodes by an external potential source of the same absolute value, but of opposite sign, we can measure the equilibrium potential of the cell:

$$\Delta E_{\mathrm{eq}} = E_{\mathrm{eq},2} - E_{\mathrm{eq},1} \tag{8.1}$$

This difference was called previously "electromotive force (emf)" but now the term "zero-current cell potential" is often used.

Assuming that the conditions of thermodynamic reversibility at equilibrium are fulfilled and that the volume work ($-pdV$) is zero, we can apply the equation relating the changes in Gibbs free energy with nonvolume work, w_{rev}, at constant pressure and temperature:

$$dG = dw_{\mathrm{rev}} \tag{8.2a}$$

As the processes are reversible, the work done by such a system has its maximum value ($w_{\mathrm{rev}} = w_{\max}$), so further we can write:

$$\Delta G = w_{\max} \tag{8.2b}$$

In the electrochemical cell T, p are constant, and the only source of energy (work) is electrode reactions, energetically characterized by changes in Gibbs free energy ($\Delta_r G$). These reactions can produce electrical work pushing the electrons through an external circuit. We can calculate the maximum electrical energy ($-nF\Delta E_{\mathrm{eq}}$), which can theoretically be produced by the cell, using the following equations:

$$w_{\max} = \Delta_r G = -nF\Delta E_{\mathrm{eq}} \tag{8.3}$$

$$\Delta_r G = \Delta_r G^0 + RT\ln Q, \qquad Q = \Pi_i a_i^{\nu_i} \tag{8.4}$$

$$-nF\Delta E_{\mathrm{eq}} = nF\Delta E^0 + RT\ln Q \tag{8.5}$$

$$\Delta E_{\mathrm{eq}} = \Delta E^0 - \frac{RT}{nF}\ln Q \tag{8.6}$$

$\Delta_r G^0$ is standard Gibbs free energy of reaction, ΔE^0 is standard cell potential, and Q is the reaction quotient.

The reaction quotient Q means the product of activities a_i of all substrates and reaction products raised to the power equal to their stoichiometric numbers v_i (they are positive for reaction products and negative for substrates).

The value of $\Delta_r G$ gives us the value of chemical energy, which we can obtain from the reactions; this value decreases to zero in time when the reaction is terminated and the activities of substrates and products attain the equilibrium values. Then

$$\Delta_r G = 0 \text{ and}$$

$$\Delta_r G^0 = -RT\ln K \tag{8.7}$$

In this equation, K is equilibrium constant, $\Delta E_{eq} = 0$, and, according to eq. (8.5)

$$\Delta E^0 = \frac{RT}{nF}\ln K \tag{8.8}$$

The cell, in which the overall reaction: $Red_1 + Ox_2 \rightarrow Ox_1 + Red_2$ has not reached equilibrium state, and $\Delta_r G$ of reaction is lower than 0 ($\Delta_r G < 0$), operates in galvanic mode, producing electrical energy during the spontaneous discharging of the cell. The chemical reaction drives the electron through the external circuit (load, device) connecting the two electrodes (Fig. 8.1(b)).

Maximum amount of energy that can be produced depends on the thermodynamic driving force, meaning the equilibrium potential of the cell ΔE_{eq}. In practice, for the cell working in galvanic mode, the terms open circuit potential (OCP) or discharging cell potential U_d are often used. Note that they show the potential difference between the two electrodes (half-cells) and do not have thermodynamic meaning. As was described earlier, the cell working in the galvanic mode converts the chemical energy only into electrical energy.

The third mode of the electrochemical cell is the electrolytic mode. In this mode, the flow of current is forced through the cell by an external source of potential (voltage). In this case, the anode in galvanic mode becomes cathode in the electrolytic cell (reduction reaction) and cathode becomes anode (oxidation reaction; see Fig. 8.1(c)). In the electrolytic mode, the electrical energy from an external source is converted into the chemical energy and is stored in the reaction products at the electrode – the cell is being charged and works as an electrochemical energy storage. It should be mentioned that the cell operates in electrolytic mode only when the applied voltage is larger than that of the equilibrium potential of this cell.

Understanding how the electrochemical cells work gives us the necessary background for understanding how the electrochemical sources of energy (EPS – electrochemical power sources) function, how to characterize them, and what parameters are important for comparison and for the decision of further application and improvements. Similar to the simple electrochemical cell, the EPS can provide the electrical energy while working in galvanic mode discharging themselves (primary batteries, fuel cells), and subsequently, in the case of secondary batteries [1, 2], the EPS can be recharged by an external voltage source, to store the electrical energy in the form of chemical energy for future use.

8.1.2 Characteristics of batteries

The main parameter that should be considered before construction and practical application of a battery is its intrinsically available work (energy) of a chemical reaction. The maximum energy available for conversion to electrical work is equal to $\Delta_r G$ and is given by eq. (8.3). Based on this equation, one can conclude that the intrinsic maximum efficiency ε_{max} of the electrochemical converter working reversibly is equal to 100%. However, the better expressions determining the intrinsic maximum efficiency of electrochemical converter are:

$$\varepsilon_{\text{max}} = \frac{\Delta_r G}{\Delta_r H} = \frac{\Delta_r G}{\Delta_r G + T\Delta_r S} \tag{8.9}$$

or

$$\varepsilon_{\text{max}} = \frac{\Delta E_{\text{eq}}}{\Delta E_{\text{eq}} - T\left(\frac{\partial \Delta E_{\text{eq}}}{\partial T}\right)} \tag{8.10}$$

$\left(\frac{\partial \Delta E_{eq}}{\partial T}\right)$ is the temperature coefficient of cell.

Equation (8.10) points out that the theoretical intrinsic efficiency can depend on the sign of the temperature coefficient of a cell and may be lower or higher than 1. When this coefficient is positive, the working cell takes some amount of heat from the surroundings (equal $T\Delta S$ when working reversibly) and changes it to work.

When the cell undergoes discharge, the decrease of the starting potential (ΔE_{eq}, OCP, U_d) is observed. The degree of potential changes is described by the potential efficiency, $\varepsilon_p = \frac{U_d}{\Delta E_{\text{eq}}}$. For the characterization of the degree of energy conversion, the current efficiency is also considered. In many reactions the current efficiency is equal to 1; it means that all substrates are fully converted to products, and so there is no alternative reaction consuming the electrons.

During work in galvanic mode, the cell is characterized by the dependence of the discharge potential U_d upon flowing current (discharge current), I_d:

$$U_d = I_d R_{\text{ex}} = \Delta E_{\text{eq}} - \sum|\eta| - I_d R_{\text{in}} \tag{8.11a}$$

$$U_d = I_d R_{\text{ex}} = \Delta E_{\text{eq}} - (\eta_{aC,} + \eta_{cC} + \eta_{aA} + \eta_{cA}) - IR_{\text{in}} \tag{8.11b}$$

R_{ex} is external resistance, R_{in} is internal resistance of a cell, usually the electrolyte resistance and other resistances resulting from the cell design, $\sum|\eta|$ – means the sum of concentration ($\eta_{c,C}$, $\eta_{c,A}$) and activation overpotentials ($\eta_{a,C}$, $\eta_{a,A}$) at both electrodes: cathode (C) and anode (A) (see Chapter 1).

Based on such dependences, the power delivered by a cell during its discharge can be calculated using the equation: $P = U_d I_d$. The common plots U_d versus I_d and P versus I_d are presented in Fig. 8.2. It is clearly seen from the equation and the plots that the power output of a battery is low when discharging current I_d is low, but the power is also low at the highest value of current, as a result of U_d drop. Convenient parameters that can be determined from such plots are admissible discharge current I_{adm}, maximum admissible power P_{adm}, and cut-off potential – the minimum available voltage. The battery cannot operate above I_{adm} or below cut-off potential.

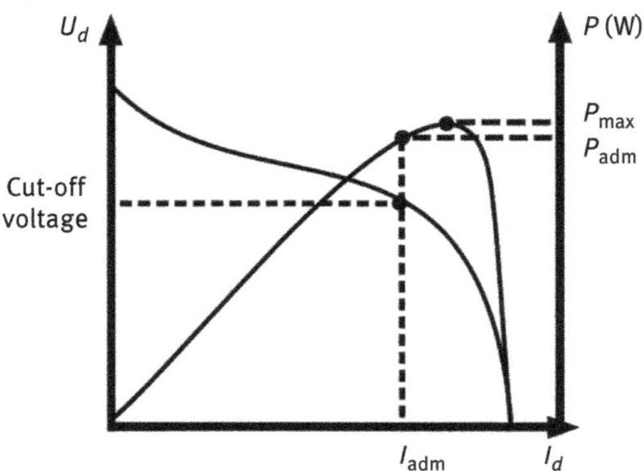

Fig. 8.2: Voltage (U_d)–current (I_d) curve during discharging and power (P) dependency on discharging current.

Considering Equation (8.11b) and taking into account the expressions determining the activation and concentration overpotentials of cathode and anode (see Chapter 1), this equation can be written as follows:

$$U_d = I_d R_{ex} = \Delta E_{eq} - \left[\frac{RT}{\alpha nF} In \frac{I_d/A_C}{j_{0,C}} + \frac{RT}{nF} \ln \left(1 - \frac{I_d/A_C}{j_{L,C}} \right) + \frac{RT}{\beta nF} \ln \frac{I_d/A_A}{j_{0,A}} + \frac{RT}{nF} \ln \left(1 - \frac{I_d/A_A}{j_{L,A}} \right) \right] - I_d R_{in} \tag{8.12}$$

j_0 is the exchange current density, j_L is the limiting current density, A is the electrode surface, indices A, C mean anode and cathode, and α and β are the transfer coefficients, different for anode and cathode, respectively.

This equation points out how the hindrance in electrode reactions and the internal resistance of the cell affect the shape of potential – discharging current plot (Fig. 8.3(a, b, c)).

The curve (1) in Fig. 8.3(a, b, c) presents the idealized U_d Versus I_d dependence. The shape of this curve is changed by the large activation overpotentials, meaning the low value of exchange currents (curve 2, Fig. 8.3(a)) The changes of the curve (1) in Fig. 8.3(b)

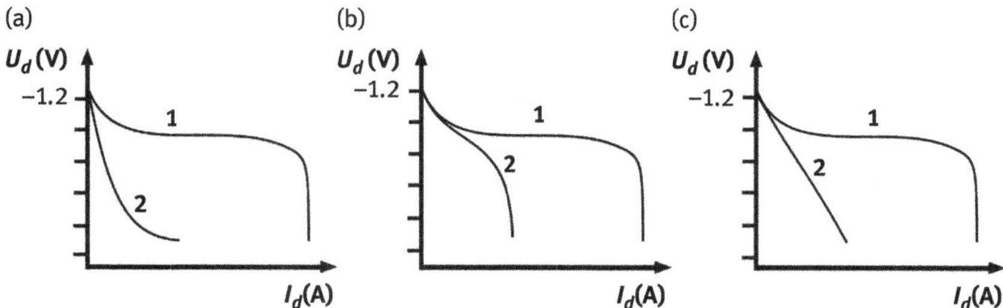

Fig. 8.3: Influence of different factors on idealized $U_d - I_d$ dependence.
Idealized behavior – curve 1 (a,b,c), the influence of large activation overpotential – curve 2 (scheme, a), influence of large concentration overpotential – curve 2 (scheme, b), and the influence of large internal resistance – curve 2 (scheme c).

result from the large values of concentration overpotentials and low values of limiting currents. Figure 8.3(c) presents the influence of the internal resistance of the cell on the shape of the curve (1). The results of calculations and resulting plots can be found in [1]. During discharging the dependence of discharge voltage, U_d, upon discharging time t_d is also recorded. The typical discharging curve for a primary battery is shown in Fig. 8.4. The usefulness of battery can be judged by the length of plateau being long for good battery (changes of U_d are very small for many hours under given load), while for worse battery the time of reaching cut-off potential is short.

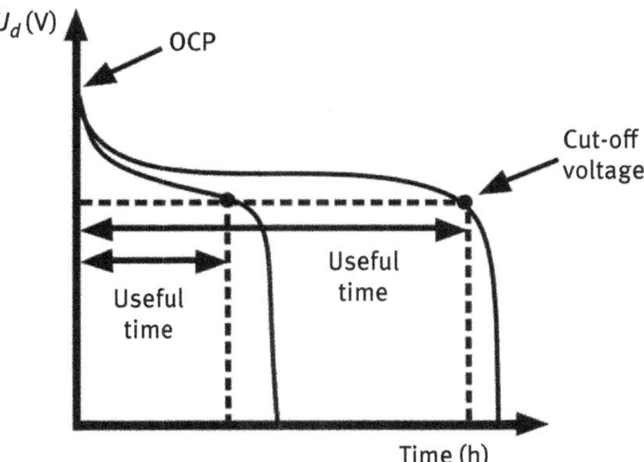

Fig. 8.4: Comparison of discharging curves for good and worse batteries.

As was mentioned earlier, during the process of charging the secondary batteries, the electrical energy is converted into chemical energy and stored in the reaction products. In analogy to U_d, the charging potential U_{ch} is given by eq. (8.13), where I_{ch} means the current flowing during the charging process:

$$U_{ch} = I_{ch}R_{ex} = \Delta E_{eq} + \sum|\eta| - I_{ch}R_{in} \tag{8.13}$$

Comparing Eqs. (8.13) with (8.11a) it is clearly seen that the charging of battery demands more electrical energy than one can obtain during discharging. This additional energy is needed to overcome the overpotentials of backward electrode reactions. The example of discharging and charging curves for a secondary battery are shown in Fig. 8.5.

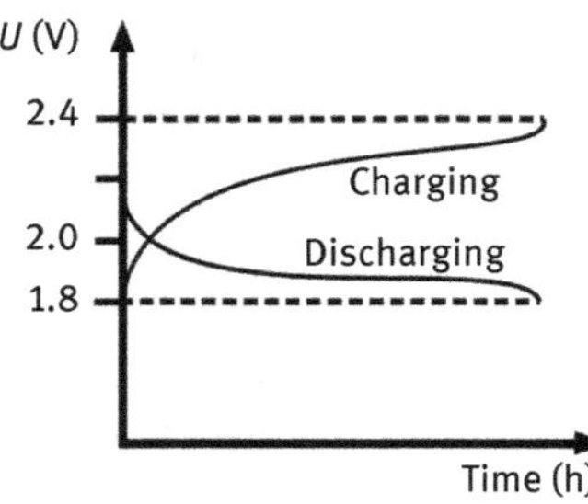

Fig 8.5: An example of discharging and charging curves of secondary battery. Note that the charging of battery demands more electrical energy than we can obtain during discharging.

Another parameter that can be determined from I_d versus t_d curves is the total charge $Q_d = I_d t_d$ delivered by the battery during full discharge. It is called the *capacity* (Ah, 1 Ah = 3,600 C) of a battery. The product of capacity and potential (voltage) determines the total energy (J, Wh) delivered by a battery. Since the capacity of the battery depends upon a magnitude of external resistance, cut-off voltage, and temperature, the total energy is not a characteristic quantity. The practical quantities that give us the possibility to compare batteries are as follows: specific energy – the energy delivered by unit mass of battery (Wh/kg) and specific power (W/kg) – the power delivered by a unit mass of battery in a brief period of time. The Ragone diagram, the dependence of specific power upon specific energy, is presented in Fig. 8.6. From this diagram plots, it is clearly seen that some batteries that can provide high power do not provide a high amount of energy per unit weight and vice versa.

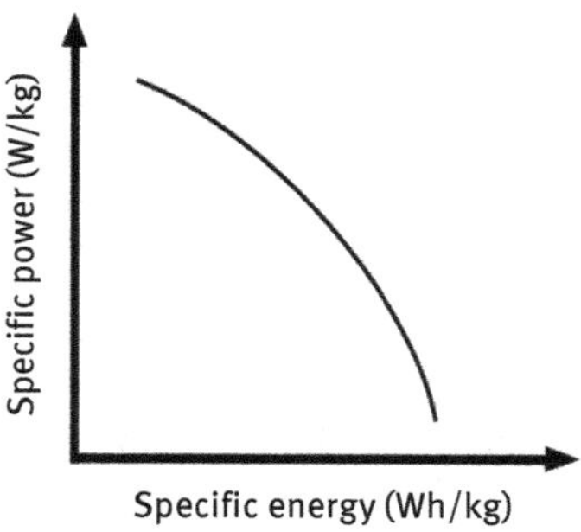

Fig. 8.6: Dependence of specific power of battery on its specific energy. The Ragone diagram.

The other important parameters ("life" parameters) for primary and secondary batteries are as follows: the rate of self-discharge, shelf life, and time of useful work (time of full discharging and the number of charge-discharge cycles). Self-discharge of the battery is a result of additional reactions that are going in a battery, like corrosion, redox reactions of electrolyte impurities, and so on. The rate of self-discharge of battery influences the shelf life – the maximum time between production and utilization of battery. The number of charge–discharge cycles determines how many times the battery can be charged/discharged before losing its usefulness for practical applications.

8.1.3 Classification of batteries and examples

Batteries, as was shown above, operate like single electrochemical cells. However, to meet the operational requirements and conditions, their constructions are much more complicated. When the voltage of the single battery is not high enough to run the devices, the number of batteries is connected in series and works as a unit. Below we briefly have described some examples of different kind of batteries: primary batteries, secondary (rechargeable) batteries, and fuel cells, focusing mainly on secondary batteries and fuel cells. For much more detailed information about fundamentals, constructions, and applications of different types of batteries, interested readers will find in books, reviews, and references therein [1–11].

8.1.3.1 Primary batteries

Primary batteries often called *single-discharge batteries* contain a finite amount of active reagents. When these reactants are consumed to some level during the battery discharge, such a battery cannot be used again. They produced the electrical energy when a load (external resistance, a device) is connected between the two electrodes of a cell, and discharging reaction takes place. For quite a long time in the past, the zinc–carbon battery has been used as a major primary battery, first patented in 1868 by G.L. Leclanché (Leclanché battery). A more modern version of this battery contains a metallic Zn body working as an anode. The cathode is made from a carbon rod (collector of electrons) in direct contact with paste containing MnO_2, mixed with carbon powder. The electrolyte is NH_4Cl or $ZnCl_2$ or a mixture of both, with a small amount of water, in the form of a paste.

The electrode processes occurring in zinc–carbon batteries $Zn\,|\,NH_4Cl,\ MnO_2\,|\,C$ are quite complicated, and in a rough approximation, the discharging reaction, which is the source of electrical energy, can be written as

$$Zn + 2\,MnO_2 + 2\,H_2O \rightarrow Zn(OH)_2 + 2\,MnO(OH)$$

In some books, the other overall reaction is proposed:

$$Zn + 2\,MnO_2 + 2\,NH_4Cl \rightarrow Zn(NH_3)_2Cl_2 + 2\,MnOOH$$

Freshly manufactured batteries have the OCP of about 1.55 V. These batteries have some disadvantages: the OCP strongly depends on the composition of electrode and electrolyte, which decreases during discharging at the shelf storage. The specific energy (~70–80 Wh/kg) is relatively low. Taking advantage of the construction and of the electrode materials utilized in the Leclanché cell, much more advanced battery was made, the so-called alkaline manganese dioxide battery: Zn | 27%–40% KOH aq | MnO_2 | C. In this battery, Zn is in the form of a paste with a high surface area, and so the specific energy is much higher than in the case of Leclanche battery. The manganese dioxide can be replaced by other oxides (HgO, Ag_2O). All these batteries called "alkaline" are successors of the original Leclanche battery. The main advantage of modern Leclanche batteries is their low costs, and so they are widely used for simple applications as power devices.

8.1.3.2 Secondary batteries

Secondary batteries (rechargeable) after their discharge can be recharged by forcing a flow of electric current in the opposite direction. This should regenerate the discharging products to the original substrates. Secondary batteries do not only produce the electrical energy during discharging, but also can store this energy in a form of chemical energy. (products of the electrochemical reaction that proceeds during charging). The efficiency of energy storage in rechargeable batteries is high (about 60–90%) and properly manufactured batteries can be charged and discharged thousands of times (charging/discharging cycles) without any substantial loss of their properties.

Currently, the most widely used secondary battery is the lead–acid battery, first invented by G. Planté in 1859. The original battery contained two lead plates separated by special materials and immersed in sulfuric acid. The layer of lead dioxide was electrodeposited on Pb during the first charging cycle. Further development of this battery led to the changes in electrode construction and composition, resulting in substantial mass decrease and much better performance [5, 6]. The present lead–acid battery contains Pb anode in the form of small Pb particles deposited onto the Pb–Sb grid and lead dioxide cathode (active material) deposited onto Pb (collector of electrons) screen. As the electrolyte, sulfuric acid of various concentrations is used. The acid anions are involved in electrode reactions, and so the OCP of batteries depends on acid concentration and is equal to 2.15 V at acid concentration 5.2 M (25° C) and decreases during discharging (anions are consumed in the reactions).

The scheme of lead–acid battery can be written:

$$Pb|PbSO_4,\ H_2SO_4(25-40\%),\ PbO_2|Pb$$

at such concentration, sulfuric acid practically exists in form of H^+ and HSO_4^- ions.

The reactions during discharging at the anode and cathode are as follows:

anode (negative electrode): $Pb + HSO_4^- \rightarrow PbSO_4 + H^+ + 2e$

cathode (positive electrode): $PbO_2 + HSO_4^- + 3H^+ + 2e \rightarrow PbSO_4 + 2H_2O$

The reactions during charging of lead–acid battery from an external source of potential are as follows:

cathode (negative electrode): $PbSO_4 + H^+ + 2e \rightarrow Pb + HSO_4^-$

anode (positive electrode): $PbSO_4 + 2H_2O \rightarrow PbO_2 + HSO_4^- + 3H^+ + 2e$

Overall reaction in this battery is

$$Pb + PbO_2 + 2HSO_4^- + 2H^+ \leftrightarrow 2PbSO_4 + 2H_2O$$

During discharge, the dense layer of lead sulfate is deposited on electrodes (Pb, PbO_2), resulting in the passivation of electrode active materials and in a decrease of discharge voltage. The other effect influencing the battery performance is a decrease of sulfuric acid concentration to about 12–24% during a discharge. This is due to water formation and causes a decrease in electrolyte density, compromising the specific energy of acid–lead battery. The theoretical specific energy is 171 Wh/kg, but in practice, the specific energy is much lower and does not exceed 40 Wh/kg. The lifetime of acid–lead battery allows for ca. 100–500 discharging–charging cycles. The specific energy is really low, but for application as a car battery, it is sufficient. The main advantage of lead–acid batteries lay in its low cost compared to other available secondary batteries.

Other rechargeable batteries include nickel–cadmium batteries or nickel–iron batteries. They contain Cd (Fe) anode and NiO(OH) cathode, its active material being pressed in the pores of a sintered nickel plate (metallic Ni is a collector of electrons). Concentrated KOH solution is used as an electrolyte. The scheme of Ni–Cd battery is Cd| 9M KOH_{aq}| NiO(OH). The cell reactions are as follows:

anode: $Cd + 2OH^- \rightarrow Cd(OH)_2 + 2e$, discharging

cathode: $2NiOOH + 2H_2O + 2e \rightarrow 2Ni(OH)_2 + 2OH^-$, discharging

cathode: $Cd(OH)_2 + 2e \rightarrow Cd + 2OH^-$, charging

anode: $2Ni(OH)_2 + 2OH^- \rightarrow 2NiOOH + 2H_2O + 2e$, charging

The overall reaction in this battery is

$$Cd + 2NiOOH + 2H_2O \leftrightarrow Cd(OH)_2 + 2Ni(OH)_2$$

The average OCP (cell voltage) is 1.29 V, and the theoretical energy density is 208 Wh/kg. The main advantage of Ni–Cd as compared to lead–acid batteries is their longer cycle life – about 3,500 cycles, in comparison with 500 cycles of lead–acid battery. The Ni–Cd batteries can last 5–10 years depending on their configuration and applications, whereas the lead–acid batteries do not last more than 5 years. As the cost of Ni–Cd batteries decreased over the last years, they replaced the alkaline primary batteries. However, because of cadmium toxicity, the Ni–Cd battery will be probably replaced by nickel–metal hydride (NiMeH) batteries, which have the same potential characteristics but higher capacity. The overall reaction occurring in NiMeH cell is

$$\mathrm{MeH(s) + NiOOH(s) \leftrightarrow Me(s) + Ni(OH)_2(s)}$$

As a cathode, like in Cd–Ni batteries, the nickel oxide hydroxide NiOOH is used. The anode (instead of Cd) consists of hydrogen-absorbing alloy – an intermetallic compound. The compositions of the alloys are patented but typically two components are used. The one which possesses a high affinity to hydrogen incorporation in its structure is a mixture of La with rare earth elements or Zr and Ti, whereas the second component is the metal that forms weak hydrides (Ni, Co, Al). NiMH battery contains a 30% solution of potassium hydroxide as the electrolyte, like in Cd–Ni cell. A fully charged battery has the cell voltage of ca. 1.25 V, the energy density in the range 40–110 Wh/kg, and the capacity of such battery is much higher than that of Cd–Ni battery.

8.1.3.3 Lithium batteries

The huge development of portable electronics results in the need for a new type of batteries that have higher voltage, larger capacity, and can be conveniently formed in various sizes and shapes. Numerous experiments were carried out on applications of alkali metal in cells because of their high negative potential in the electrochemical scale. Lithium has the most negative value of electrode potential, so the application of Li /Li^+ as an anode in the batteries automatically increases the OCP of lithium cell to 4.5 V. This value is two times higher than for batteries described above. However, Li is very active chemically and decomposes water; therefore, in lithium cell, only aprotic organic solvents can be used. The special cleaning of solvent and lithium salt used as a source of lithium ions increases the cost of lithium cells. The lithium salts and aprotic organic solvents widely applied in lithium cells are: LiBr, $LiAsF_6$, $LiPF_6$, AC (acetonitrile), DMF (dimethylformamide), EC (ethylene carbonate), and PC (propylene carbonate). To choose proper salt and solvent for lithium cells, one should consider the conductivity of the solution and electrochemical stability of the solvent. The electrochemical stability of the solvent is defined by the potential range in which the solvent does not decompose. Li is aggressive metal and after the first contact with an aprotic solvent, the passivation of Li occurs resulting from the formation of thin film on the electrode surface. This film, on one hand, protects the Li anode against corrosion and, on the other hand, it causes a decrease of potential in the first phase of battery discharging. The lithium batteries can be designed and work as primary or secondary batteries. The examples of primary lithium batteries include batteries with liquid cathode such as Li-$SOCl_2$ battery, Li-SO_2 battery, or batteries with solid cathode (metal oxide, sulfide).

The scheme of Li-thionyl chloride cell is: Li | Li $AlCl_4$, $SOCl_2$ | C, chloride thionyl works in cell as liquid cathode and also as a solvent for Li salt. In such types of cells instead of thionyl chloride, sulfuryl chloride or phosphoryl chloride can also be used as a liquid cathode. The reactions that occur during discharging of the cell are as follows:

$$\text{anode (negative electrode):} \quad Li \rightarrow Li^+ + e$$

$$\text{cathode (positive electrode):} \quad 2\,SOCl_2 + 4\,e \rightarrow SO_2 + S + 4\,Cl^-$$

$$\text{and the overall reactionis:} \quad 4\,Li + 2\,SOCl_2 \rightarrow SO_2 + S + 4\,LiCl$$

Theoretical reversible potential and specific energy of the cell were estimated as 3.5 V and 1.48×10^3 Wh/kg, respectively; however, in the commercial cells, the value of specific energy of ca. 700 Wh/kg was reached. Similar batteries containing SO_2 were manufactured. The schemes of these cells are: Li | SO_2, LiBr, AN | C, and the reactions of discharging of these cells can be written: $2Li + 2SO_2 \rightarrow Li_2S_2O_4$. These batteries are characterized by good value of energy density – 333 Wh/kg, stable discharging potential (2.8–2.9 V) even at high discharging currents. Many of lithium primary cells have solid cathode of metal oxides, metal sulfides, and oxy-salts. Compounds such as CuO, Bi_2O_3, MnO_2, CuS, FeS, Ag, Cu oxy-salts were tested in lithium batteries. Very popular is the cell with Ag_2CrO_4 cathode:

$$Li|LiClO_4, PC|Ag_2CrO_4, C$$

The discharging reaction is

$$2\,Li + Ag_2CrO_4 \rightarrow 2\,Ag + Li_2CrO_4$$

The OCP of this battery is about 3 V and its specific energy is ca. 200 Wh/kg or even higher. The battery is manufactured in a button shape and is widely used to power the pacemakers.

In 2019 year, John B. Goodenough, M. Stanley Whittigham, and Akira Yosino received the Nobel prize for their contribution to the development of rechargeable lithium-ion battery. Contrary to the primary Li batteries, in the secondary (rechargeable) batteries no metallic Li is used. Graphite intercalated with Li ions serves as anode in these batteries, whereas cobalt oxide, manganese oxide, or other oxides intercalated with Li ions are used as cathodes. The electrolyte is the solution of lithium perchlorate or lithium hexafluorarsenate, $LiAlF_6$, in mixed organic solvents. The scheme of an exemplary cell is as follows:

$$C|LiClO_4, EC|Mn_2O_4, \quad EC - \text{ethylene carbonate,}$$

overall reaction in the cell is:

$$LiC_6 + Mn_2O_4 \leftrightarrow 6\,C + LiMn_2O_4$$

As may be seen, lithium ions do not participate in overall reaction but take place in partial electrode reactions:

$$\text{anode:} \quad LiC_6 \rightarrow Li^+{}_{(solv)} + 6\,C + e$$

$$\text{cathode:} \quad Li^+{}_{(solv)} + Mn_2O_4 + e \rightarrow LiMn_2O_4$$

The Li-ion battery operates in a voltage range from about 4.0 to 2.5 V, and its specific energy is about 150 Wh/kg. Some disadvantages are connected with relatively unattractive shelf life and safety conditions. The information about the last developments and new challenges in the lithium batteries field can be found in the bibliography [7–9].

8.1.3.4 Fuel cells

A fuel cell is an electrochemical device, which, unlike to other devices, can be continuously fed with substrates, and the reaction products are also removed continuously. A fuel cell consists of an electrolyte sandwiched in some matrix (membrane) between thin porous anode and cathode.

Fuel cells emerged as promising green substitutes for fossil fuels. They are unique in terms of the variety of fuels and feedstocks for energy conversion into electricity. The potential to reduce overall energy consumption, carbon emission, and high energy density makes these cells suitable for a wide range of applications, across multiple sectors, including transportation, industrial, commercial, residential buildings, and long-term energy storage for the grid in reversible systems. In the last few years, research interest in fuel cells has been boosted by the growth in green energy requirements.

In the most known hydrogen–oxygen fuel cell, hydrogen is delivered to the anode where the oxidation reaction takes place, and the oxygen is reduced at the cathode. In the fuel cell, hydrogen gas (or hydrogen-rich fuels) and oxygen (pure or from the air) are directly converted into electrical energy and heat through the electrochemical reaction of water formation. The reactions that take place in the cell are as follows:

$$\text{anode:} \quad H_2 \rightarrow 2\,H^+ + 2\,e,$$

$$\text{cathode:} \quad 1/2\,O_2 + H_2O + 2\,e \rightarrow 2\,OH^-$$

The overall reaction is

$$H_2 + 1/2\,O_2 \rightarrow H_2O, \ \Delta_r G^0 = -237\ \text{kJ/mol}, \ E^0 = 1.229\text{V}$$

Depending on the electrolyte used in hydrogen–oxygen cells, we can distinguish:

1. Alkaline fuel cells (AFC), containing 30–70% solution of KOH. The development of alkali fuel cells started in the thirties of the twentieth century when the British scientist Francis T. Bacon demonstrated the first fuel cell, now the so-called Bacon

cell. About 20 years later, the stack of Bacon cells was demonstrated, the technology was patented, and stacks were applied in Apollo spacecraft. The electrical power was 5 kW.

In the modern hydrogen–oxygen fuel cell, gas-diffusion electrodes are used, consisting of a porous layer of carbon black and of catalytic layer. Different catalysts, such as carbon, manganese, nickel mesh, and so on, are tested and applied. The working temperature of cells is in the range from 60 to 240° C. One of the main problems with AFC is the degradation of electrodes by carbonate, formed in the reaction: $CO_2 + 2\ OH^- \leftrightarrow CO_3^{2-} + H_2O$ (carbon dioxide is the contaminator of gas reactants used). The crystals of carbonate (K_2CO_3) block the pores of gas-diffusion electrodes, affecting the performance of the cell. To avoid this problem, stable solid alkaline polymers are applied – mainly polymers containing methyl-ammonium or methyl-pyridine groups. Such solid-state cell can work at temperatures up to 120 °C.

2. Phosphoric acid fuel cells (PAFC) contain 85–95% solution of phosphoric acid as the electrolyte. The acid fills the silicon carbide matrix. The reactions at the anode, cathode, and the overall reaction are the same as in other hydrogen–oxygen cells. The reactions take place on highly dispersed catalyst Pt or Pt alloys with Cr, Ti on porous carbon. The operating temperature of the cell is in the range 150–200 °C.
3. Membrane fuel cell (MFC) – in these cells, the polymeric ion-exchange membrane as an electrolyte is used. These fuel cells are also named the *solid polymer fuel cell* or *proton-exchange membrane fuel cell* (PEMFC). Modern PEMFCs are built of membrane electrode assemblies (MEA), which include the electrodes, solid electrolyte, catalyst, and gas diffusion layers pressed together. The Nafion polymer is mainly used as a solid electrolyte. This polymer contains sulfonic acid groups – SO_3H, and serves as an acid electrolyte, a separator, and a cation-exchange membrane. As sulfonic acid is a strong acid, the sulfonic group dissociates, and therefore the proton conductivity of the membrane is high (0.1 S/cm at 80 °C). For the same reason, the membrane allows for the passage of protons but stops the passage of other ions. The MEA is made as follows: an ink of catalyst suspension (mainly Pt or Pt–Ru), carbon, and electrode material are sprayed or painted onto the solid electrolyte, and then the carbon paper is hot pressed on either side to protect the inside of the cell, acting also as electrodes. Carbon paper can also play the role of a gas diffusion layer. Their operating range of temperatures is 60–100 °C. Splitting of the hydrogen molecules is relatively easy by using a Pt catalyst. Unfortunately, splitting the oxygen molecule is more difficult and this causes significant electric losses, about 20% in the efficiency of energy conversion. Difficulties in the reduction process can lead to a two-electron reduction of oxygen to hydrogen peroxide, instead of four-electron reduction to water or OH^- (Chapter 4). Up till now, there is a huge effort for finding the appropriate catalysts for this process. The scheme of MFC is shown in Fig. 8.7.

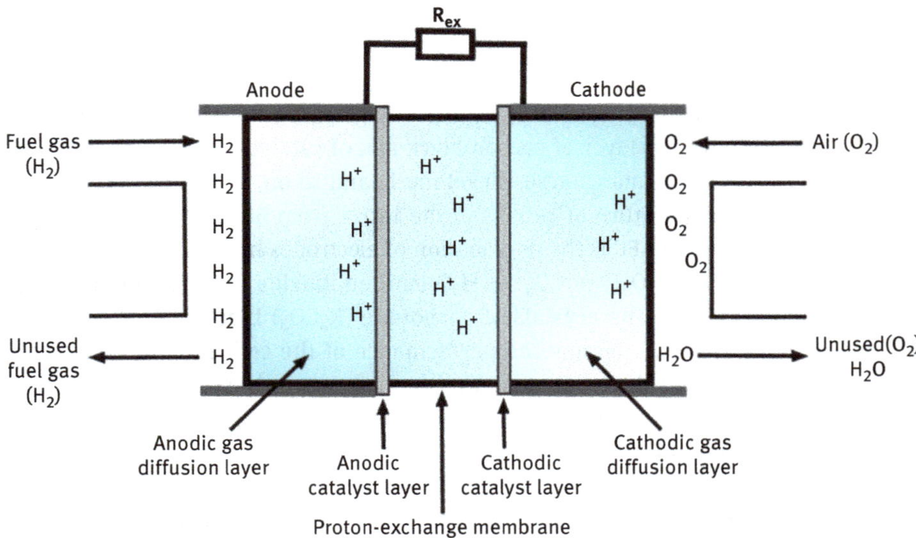

Fig. 8.7: Scheme of membrane fuel cell.

As we mentioned at the beginning of this chapter, a double slash ||, appearing in the schemes of the cells, represents the phase boundary whose potential is negligible as a result of the application of a separator between the two electrolytes. This is theory; in real cells and fuel cells operating under galvanic or electrolytic conditions, the presence of a separator, being necessary to separate physically (but not electrically) the contents of the anodic and cathodic compartments (as it is shown in Fig. 8.7 in the case of proton exchange membrane (PEM)), introduces an additional component to the U_d, (eq. 8.12), diminishing it (or increasing the U_{ch}). Therefore, the design of such a separator is a key factor in the construction of high-performance fuel cells. The most desired properties of membraneous separators are:
- high permeability for counterions, impermeable to coions;
- low electrical resistance – the permeability of an ion-exchange membrane for the counterions under the driving force of an electrical potential gradient should be as high as possible;
- good mechanical stability – mechanically strong, should have a low degree of swelling, shrinking or embrittlement, and cracking during the long-term operation;
- high chemical stability – the membrane should be stable over the entire operation.

These properties determine to a large extent the technical feasibility and the economics of the MFCs.

The basic processes and mechanisms involved in the functioning of the ion-exchange membranes were briefly described in Chapter 1.4 of this book.

Ion-exchange membranes can be classified by their function as separators:

- cation-exchange membranes (CEM) that contain fixed negatively charged ions and have a selective permeability for cations. A special case of such membranes utilized widely in the MFC is the proton exchange membrane (PEM). With their development, PEM fuel cells emerged as promising substitutes for fossil fuels with great potential to reduce overall energy consumption, zero carbon emission, and high energy density. In the last few years, research interest in PEM fuel cells has been boosted by the growth of green energy requirements for the transportation sector and mobile energy frameworks. However, the commercialization of PEM fuel cells suffers from two major hurdles, that is, high cost and low durability,
- anion-exchange membranes (AEM) that contain fixed positively charged ions and which have a selective permeability for anions. Certain reports prove the suitability of AEM compared to CEM, where high performance can be related to reduced proton accumulation and pH splitting.

More complex structures involve amphoteric ion-exchange membranes in which both negatively and positively fixed ionic groups are randomly distributed, bipolar membranes consisting of a cation- and an anion-exchange membrane laminated together, and mosaic ion-exchange membranes which are composed of macroscopic domains of negatively fixed ions and those of positively fixed ions, randomly distributed in a polymer matrix.

Another classification is based on the structure and basic material with the representative example being a solid oxide fuel cell, operating at high temperatures and mentioned briefly below.

All these fuel cells operate in moderate temperatures range from 60 to 240 °C. At higher temperatures, molten-carbonate (MCFC) and solid oxide fuel cells (SOFC) are used. MCFC contains a molten mixture of carbonates (sodium, potassium, lithium) as an electrolyte; the operating temperature of such a cell is about 650 °C. Solid oxide fuel cell with the solid electrolyte – zirconium oxide doped with other metal oxides can work at temperatures up to 1,000 °C.

For the last few decades, a huge development in science and technology in designing fuel cell is observed, as a result of wide applications in power plants, automotive transportation, space vessels, and so on. There is a strong interest in the development of fuel cells working at temperatures < 100 °C, like the PEMFC described above, or direct methanol fuel cell (DMFC), where hydrogen is replaced by liquid methanol. They are very promising cells and leading candidates for replacing the alkaline batteries. DMFCs, like PEMFCs, are built of MEAs that include the electrodes, proton-exchange membrane, and the porous carbon layers. The fuels are methanol solution in water (concentration between 2 and 5 M/L) and oxygen (from air). The electrodes in DMFC consist of noble metal catalysts, Pt–Ru alloy is used as the anode, and Pt as the cathode. Reactions proceeding during discharging of DMFC are as follows:

$$\text{anode:} \quad CH_3OH + H_2O \rightarrow CO_2 + 6\,H^+ + 6\,e$$

$$\text{cathode:} \quad 3/2\,O_2 + 6\,H^+ + 6\,e \rightarrow 3\,H_2O$$

The overall reaction:

$$CH_3OH + 3/2\,O_2 \rightarrow CO_2 + 2\,H_2O$$

$$\Delta_r G^0 = -706.2\,\text{kJ/mol}, \quad E_{eq}{}^0 = 1.22\text{V}$$

The main disadvantages of DMF cells are the processes called "methanol crossover" and slow electro-oxidation of methanol at the anode. The crossover process means the penetration of methanol across membrane separator to the cathode compartment, resulting in losses of methanol and in decrease of the working potential of the cell. The decrease in the rate of methanol electro-oxidation is caused by the formation of various intermediates (CO, HCO, HCOO), poisoning the catalyst surface. However, the cells possess a number of advantages, such as easy transport, storage, low cost of methanol, and the possibility of compact design, small size, and weight that make them suitable for future potential applications. At present, the challenge for scientists is to find a new type of catalysts that will yield higher values of exchange currents not only for electro-oxidation of methanol but also for electro-oxidation of other organic compounds like hydrazine, formic acid, and ethanol. There is a huge literature concerning fuel cells; a few you will find in the bibliography [10, 11].

8.2 Supercapacitors

8.2.1 General remarks

Nowadays, it is difficult to imagine the everyday life without mobile phones, tablets, or laptops – in fact any kind of small, portable electronic devices. All of them require frequent charging, boosting the development of novel, sustainable, and efficient energy storage systems, fulfilling sometimes also demands of high power and energy of personal and commercial transportation systems. It is well known that all these requirements can be met by designing devices with high density of stored energy and power, large number of charging/discharging cycles, capable of miniaturization and compatible with modern electronics. Supercapacitors perfectly fulfill the above requirements [12–15]. Therefore, they find a wide range of applications of high-power sources in short time intervals, for example, for start/stop systems in electrical and hybrid vehicles, but also as energy sources for medical devices like defibrillators (e.g., AED) or pacemakers [16–18]. The latter, however, does not require high power, but a backup is needed for short-time powering, before battery replacement. Recently, due to a strong dependence of solar cell operation on weather conditions, as well as fluctuation in solar radiation, a new approach, based on supercapacitors, was proposed to provide uninterrupted use of solar energy-generated electricity. Usually, such uninterrupted supply was realized by secondary batteries connected in parallel to the solar panel. Replacing them with supercapacitors provides the advantage of long-term stability and fast dynamics upon repeated charging/discharging cycles, relatively low energy loss, and effective buffering of voltage fluctuations in solar radiations [19–22].

8.2.2 Capacitor versus electrochemical cell/battery

A capacitor (condenser) is a passive electronic device, with two conducting terminals, separated by a dielectric, that stores electrical energy. The relation between the charge Q stored between the conducting terminals (electrodes) and the potential developed across the dielectric, U, defines the capacitance C of the capacitor:

$$C = \frac{Q}{U} \tag{8.14}$$

The unit determining the capacitance is Farad [F; 1 F = 1 C/1 V] relating the charge accumulated within the capacitor and potential developed between the conducting terminals.

The physical form and construction of capacitors vary widely, depending upon their applications as many types of capacitor are in common use. Most capacitors contain metallic surfaces as conducting terminals separated by a dielectric medium. The conducting terminal can be in a form of foil or thin film or an electrolyte in the case of electrolytic capacitors. The nonconducting materials commonly used as dielectrics include ceramic, polymer film, paper, mica, air, and oxide layers. The relationship between the electric field intensity, E, developed within the dielectric, and the potential across this dielectric, U, is given by

$$U = \int_0^d E(x)dx \tag{8.15}$$

The energy stored within the capacitor is equivalent to the electric work of charge separation within the dielectric, W_{el}, and therefore is given by

$$W_{el} = \int UdQ = \int \frac{Q}{C}dQ = \frac{Q^2}{2C} = \frac{CU^2}{2} = \frac{QU}{2} \tag{8.16}$$

As we discussed earlier, the energy stored in the electrochemical systems (cells, batteries) is due to the faradaic reactions that are characterized by changes in Gibbs free energy, $\Delta_r G$. The maximum electrical energy that can be produced due to these reactions is (eq. 8.3)

$$w_{\max} = \Delta_r G = -nFE_{eq} \quad (8.17)$$

Please note that the units of both U and E_{eq} are Volts, but while E_{eq} is defined by the equilibrium potentials of redox electrodes, U is the potential difference between the polarized conducting terminals and not the reversible redox reactions. Now, keeping in mind that the product nF is the charge that can flow through the cell as a result of faradaic reaction, it is easily seen that the energy stored in the electrochemical cell is twice that of the capacitor charged with the same amount of $Q = n\,F$, with potential U equal to the zero-current cell potential E_{eq}:

$$w_{\max} = 2W_{el} \quad (8.18)$$

So, if the energy stored in the conventional capacitor is half that of the energy stored in the electrochemical cell, then what are the advantages of practical use of capacitors or supercapacitors over batteries or cells as power supplies in nowadays electronics? First of all, batteries and cells are characterized by totally different mechanism of charge storage. This mechanism is based on the conversion of chemical into electrical energy as a result of electrochemical reactions on the electrodes (anode and cathode). Sometimes, it is also accompanied by the processes of insertion/disinsertion of electroactive ions within the whole electrode material, as in the case of lithium batteries. All these mechanisms result in relatively slow charging/discharging processes. Limited kinetics of the electrode reactions additionally determines the relatively low power of batteries and electrochemical cells. In the case of capacitors, however, the mechanism of charge storage is different and depends upon the dielectric and electrode materials used. The charge is electrostatically stored between the conducting material (electrodes) upon the external polarization. The overall capacitance C of, for example, parallel plate capacitor, scales proportionally to the electrode area, A, and dielectric permittivity, $\boldsymbol{\varepsilon}$, of the material between the electrodes, and inversely to the distance between the conducting plates, d. Consequently, the dielectric capacitance of such capacitor is small. However, if we use for the "plates" of conducting materials of large specific surface area, for example, porous or nanostructure conductive electrodes separated by a few nanometers of electrolyte, then we will design the so-called ***supercapacitor***, where the charge is stored within the electrical double layers at the interfaces of the electrolyte and the electrodes [23]. The performance difference between a dielectric and supercapacitor is considerable. Typical specific capacitance of large capacitors is up to 15–50 μF/cm^2, but a supercapacitor has been shown to have capacitance as high as 1,500 F/g. Figure 8.8) summarizes graphically the differences between the three types of energy storage devices: dielectric capacitor, supercapacitor, and battery.

The overall construction scheme of a supercapacitor relies on the two nanostructured or porous electrodes in contact with current collectors. The electrodes are separated with a thin membrane saturated with concentrated electrolyte solution. The membrane enables free movement of ions, but protects against the elec-

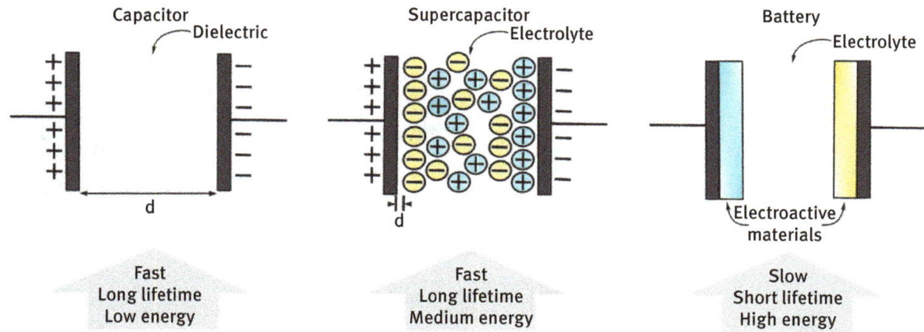

Fig. 8.8: Graphical representation of differences between the three types of energy storage devices: dielectric capacitor, supercapacitor, and battery.

trode short-circuit. Depending on the mode of electrical energy storage, there are different types of supercapacitors:

- *EDLC supercapacitors* (electrical double layer capacitors) where the energy is stored via the reversible adsorption of ions. This process is purely electrostatic.
- *Pseudocapacitors*, in which the electrical energy is additionally stored with help of fast redox reactions localized on the surfaces of the electrodes.

If both electrodes are of the same material, then we have the symmetric system, but, of course, they can be made of different conducting materials, yielding the asymmetric supercapacitors, where each electrode has different specific capacitance.

In the case of EDLC supercapacitors, apart from the electrode area, their performance depends on the surface concentration of adsorbing ions, and therefore the charging/discharging cycle of this device depends on ionic mobilities and solvent reorientation/reorganization. Moreover, they are simple to load and their lifetime is very long with extremely high number of charging/discharging cycles. Since there are no electrochemical reactions, the potential and charge accumulated within the supercapacitor are not limited by these reactions (Fig. 8.9).

For pseudocapacitors, their charge storage capability is additionally enhanced by nanostructured transition metal mixed oxide materials, such as RuO_2 or MnO_2. Other oxides include TiO_2, VO_2, MoO_2, NiO_2, CoO_2, SnO_2, and LiO_2. In addition, thin layers of electroactive polymers or polymers with metallic centers embedded within (such as metallocenes) can be included in the design of pseudocapacitors. Interesting structures include also polyoxometallates of W, Nb, and Ta.

Pseudocapacitance (or capacitance redox) appears when the adsorbed ions of mixed valency can undergo fast and reversible surface redox reactions with no diffusion involved. In acidic media, such reactions can be written as follows:

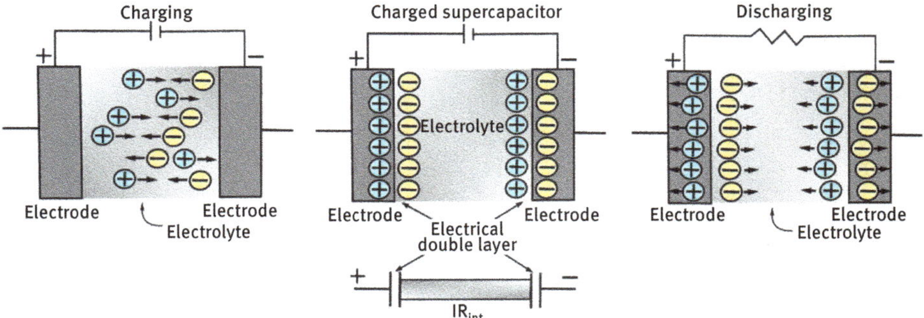

Fig. 8.9: EDLC supercapacitor: charging, storage, and discharging processes. Below the middle capacitor is its electrical equivalent circuit.

$$RuO_x(OH)_y + \delta H^+ + \delta e^- \leftrightarrow RuO_{x-\delta}(OH)_{y+\delta}$$

$$Mn^{IV}O_2 + xe^- + xH^+ \leftrightarrow H_xMn_x^{III}Mn_{1-x}^{IV}O_2$$

The stoichiometric coefficients describe only the surface layer, because the overall reactions of the mixed oxide layers are more complex. In the case of nanostructured, ruthenium-based oxides, the specific capacitance can reach up to 900 F/g; however, these materials are quite expensive and toxic. Less costly are manganese-based mixed oxides, but they give lower specific capacitance of ca. 200 F/g.

In certain cases, these processes can be accompanied by intercalation of cations within the mixed oxide matrix. For instance,

$$Mn^{IV}O_2 + e^- + H^+ \leftrightarrow Mn^{III}OOH \qquad \text{intercalation/reduction}$$

Or, more generally (C^+ stands for a cation):

$$Mn^{IV}O_2 + e^- + C^+ \leftrightarrow Mn^{III}OOC \qquad \text{intercalation/reduction}$$

$$Mn^{IV}O_2 + e^- + C^+ \leftrightarrow Mn^{III}OO^-C^+ \qquad \text{surface adsorption/reduction}$$

Supercapacitors have a very long life cycle of almost one million charging/discharging cycles with cycle efficiency of about 95%, and are charged and discharged to the maximum limit in seconds. They have a high specific power of up to 15 kW/Kg as compared to, for example, lithium ion battery (up to 2 kW/Kg). As in the case of cells and batteries, specific energy and power of supercapacitors are most frequently defined for unit mass. However, contrary to batteries and cells, this is typically the mass of electrodes, because they are the major contribution to the total mass. Due to their design, supercapacitors are also characterized by low internal resistance, R_{int}, which allows for high currents to be supplied by such device. This resistance depends on the ionic conductivity of electrolyte, ion diffusion within the porous structure of electrodes and electrical contact between the grains of electrode material.

Electrode materials used for the construction of supercapacitors should be of large surface (even though of small external dimensions), of high conductance, thermally, mechanically, and chemically stable and of low costs and ease of manufacturing. Nanostructured, conducting materials are of primary use, such as carbon materials (graphene, nanotubes, activated carbon, carbon fiber-cloth, carbide-derived carbon, carbon aerogel, graphite, nanofibers), conducting polymers, transition metal oxides, polyoxometallates, their mixtures, and the like. Similar requirements should be fulfilled by the electrolyte filling the space within the porous structure of supercapacitors. For this purpose aqueous electrolytes are of lesser use because of their relatively narrow potential window of up to ca. 1.2 V in the neutral pH. Other types of electrolytes include nonaqueous, organic electrolytes, ionic liquids, solid gel, and polymer electrolytes [24–29].

Table 8.1 compares the major performance parameters for ECDL capacitors, pseudocapacitors, lithium ion battery, and hybrid systems. It is easily seen that although the specific power supply of ECDL and pseudocapacitors is high, their energy density is ca. 10% of a lithium ion battery. On the other hand, a battery, even though it stores more specific energy, has limited number of charging/discharging cycles, shorter lifetime, limited power delivery, and significant impact on the environment. It is obvious then that there should exist a device combining features of rechargeable batteries and supercapacitors. These are the hybrid structures in which a supercapacitor is coupled with a battery, forming a kind of "super battery" [27–29]. This hybrid structure provides more effective energy supply in connection with faster charging. Their performance is presented in the last column (Hybrid) of Table 8.1.

Table 8.1: Comparison of general performance parameters of different energy storage systems.

Parameter	ECDL	Pseudocapacitor	Li/ion battery	Hybrid
Potential /V	1.2–3.3	2.2–5	2.5–4.5	2.2–5
Charging/Discharging cycles	10^5–10^6	10^5–10^6	500–10^4	10^4–10^5
Capacitance /F	0.1–470	100–12000	-	300–3300
Specific energy Wh/kg	1.5–3.9	4–9	100–300	10–15
Specific power kW/kg	2–10	3–10	0.3–2	3–15
Lifetime /years	5–10	5–10	3–5	5–10
Self-discharge	weeks	weeks	months	months

Even better comparison of various energy storage systems currently available in the market is shown in a form of Ragone diagram discussed earlier (Fig. 8.10).

It is clearly seen from this plot that charge storage devices that could fulfill all needs of current electronics, having high specific energy as well as high specific power output, are yet to be designed.

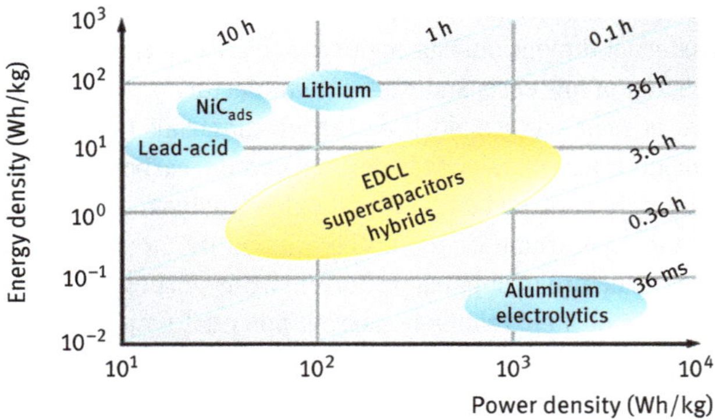

Fig. 8.10: The Ragone diagram of specific energy density versus specific power density of energy storage and conversion devices. It is worth to note that the electrolytic capacitor was developed in 1896 by Polish engineer, Karol Pollak.

8.2.3 Working characteristics of supercapacitors

The relationship between the capacitance of, for example, a parallel plate capacitor and its geometry can be described by

$$C = \frac{\varepsilon\varepsilon_0}{d} A \tag{8.19}$$

where ε_0 and ε are the dielectric permittivities of vacuum and dielectric (inside the capacitor), respectively, A is the area of current collector, and d is the distance between the two plate collectors. This distance is typically in a nanometer scale and is limited by the thickness of the adsorbed ions in the e.d.l. at the electrode/electrolyte interface. The area A should be very large due to the nanoporous or nanostructured surface of the electrodes of the supercapacitor. As it is evident from the above equation, a decrease of the distance d and an increase of the area A results in an increase of capacitance C. These two parameters determine the conducting materials that can be used for the design of supercapacitors. The third parameter that affects the charge storage capability of supercapacitor (and capacitors in general) is the dielectric permittivity, ε, of electrolyte between the electrodes/collectors.

Even in the case of symmetric supercapacitors, due to the differences in size of cations and anions in the electrolyte, both electrodes can have different capacitances. This is shown in the middle panel in Fig. 8.9. Then the overall capacitance C is related to capacitances of both electrodes according to the equation:

$$\frac{1}{C} = \frac{1}{C_+} + \frac{1}{C_-} \tag{8.20}$$

where C is the total capacitance of supercapacitor [F], and C_+, C_- are the capacitances of + and – electrodes, respectively [F]. According to the above equation, the electrode of smaller capacitance will determine the overall system capacitance.

As we discussed above (eq. 8.16) the total electrical energy, W_{el}, accumulated within the supercapacitor is directly related to its capacitance, C, and working potential, U:

$$W_{el} = \frac{CU^2}{2} \tag{8.21}$$

If we express the value of W_{el} per unit mass (or volume) of the electrode material, we will get the specific energy.

Maximum specific power depends also on the working potential U:

$$P_{max} = \frac{U^2}{4R_{int}} \tag{8.22}$$

where R_{int} is the internal resistance described above. As in the case of specific energy it can be expressed per unit mass or unit volume.

Another important parameter determining the performance of supercapacitors as charge storage and energy supply systems is their rate of charging and discharging. This is described by the charging/discharging time constant, τ [s]:

$$\tau = R_{int} \cdot C \tag{8.23}$$

Its values, shown in the Ragone diagram as skewed, blue lines, describe the dynamics of charging/discharging of electrical power sources and for the case of supercapacitors are very short, in the range of 0.1 to 20 seconds. It is worth to note that although the fast charging rate is very advantageous for a charge storage device, the fast discharge characteristics is disadvantageous from the point of view of supporting relatively constant potential over an extended period of time. Therefore, supercapacitors find their primary use where the large charge has to be delivered over a short period of time. The potential-time discharge characteristics of battery and supercapacitor are shown in Fig. 8.11. Please keep in mind that these curves are intended solely to show the different shapes of these two characteristics, but neither the potential nor time axes are comparable for both systems.

Nevertheless, since supercapacitors can be constructed at low cost practically in any size, they find extensive use as short-time power support or for applications where the buffering of fluctuating loads are required, such as hand-held devices, laptops, photographic flashes, to name a few. On a larger scale, they can stabilize voltage and buffer unwanted power fluctuations on the grids. They are also suitable as temporary energy harvesting and storage systems [17, 18, 22]. Low internal resistance of supercapacitors triggered the applications requiring short-term high currents, such as in the case of engine start-up [30]. It is worth to note that the first

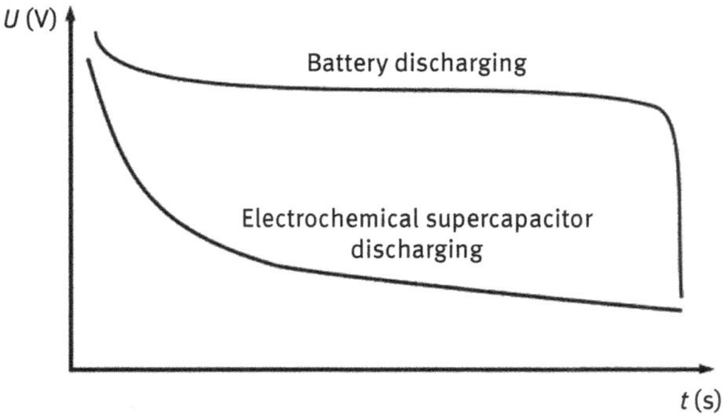

Fig. 8.11: Discharging curves for a battery and electrochemical supercapacitor. See text for details.

hybrid bus in Europe, partially utilizing supercapacitors, was constructed and put into operation in 2012 under the nickname Ultracapbus in Nuremberg, Germany.

Bibliography

[1] Bocris J O`M, Reddy AKN. Modern electrochemistry. Fundamentals of electrodics. V 2B. New York, USA, Kluwer Academic Publisher, 2000.

[2] Bagotsky VS. Fundamentals of electrochemistry. Hoboken. USA. J. Willey & Sons Inc, 2006.

[3] Linden D, Reedy TB. Handbook of batteries (3 ed). New York. USA. Mc. Graw-Hill, 2002.

[4] Fuller TF, Harb JN. Battery fundamentals. In Electrochemical Engineering. Hoboken. USA. J. Willey& Sons Inc, 2018.

[5] Davidson A, Monahov B. Lead batteries for utility energy storage: a review. J. Energy Storage 2018, 15, 145–157.

[6] Lach J, Wróbel K, Wróbel J, Podsadni P, Czerwinski A. Application of carbon in lead-acid batteries a review. J. Solid State Electrochemistry 2019, 23, 23693–705.

[7] Etacheri V, Marom R, Elezari R, Salitra G, Aurbach D. Challenges in the development of advanced Li-ion batteries: a review. Energy & Environmental Science 2011, 4, 3243.

[8] Schipper F, Ericson EM, Erk Ch, Chesneau FF, Aurbach D. Review – recent advances and remaining challenges for Lithium ion battery. Nickel rich, $LiNi_xCo_yMn_zO_2$. J Electrochem Soc 2017, 164, A 6620–A 6628.

[9] Heidari EK, Kamyabi-Gol A, Sohi HM, Ataie A. Electrode materials for lithium ion batteries: a review. Journal of Ultrafine Grained and Nanostructured Materials 2018, 1, 1–12.

[10] Carrette L, Friedrich K, Stimming U. Fuel cells-fundamentals and applications. Fuel Cells 2001, 1, 5–39.

[11] O`Hayre R, Suk-Wan Cha, Collela W, Prinz FB. Fuel cells. Fundamentals. Hoboken. USA. J. Willey& Sons Inc, 2016.

[12] Zhang Y, Feng H, Wu X, Zhang A, Xia T, Dong X, Li X, Zhang L. Progress of electrochemical capacitor electrode materials: A review. Int. J. Hydrogen Energy 2009, 34 4889–4899.

[13] Gonzalez A, Goikolea E, Barrena JA, Mysyk R. Review on supercapacitors: Technologies and materials. Renewable & Sustainable Energy Reviews. 2016, 58, 1189–1206

[14] Raza W, Ali FZ, Raza N, Luo YW, Kim KH, Yang JH, Kumar S, Mehmood A, Kwon EE. Recent advancements in supercapacitor technology. NanoEnergy 2018, 52, 441–473.

[15] Poonam SK, Arora A, Tripathi SK. Review on supercapacitors: Materials and devices. J. Energy Storage 2019, 21, 801–825.

[16] Slesinski A, Fic K, Frackowiak E. New trends in electrochemical supercapacitors. In: VanEldik R., Macyk W. eds. Materials for sustainable energy. Book Series: Advances in Inorganic Chemistry. Elsevier Acad. Press Inc., San Diego CA, USA . . . 2018, 72, 247–286

[17] Huang Y, Zhu MS, Huang Y, Pei ZH, Li HF, Wang ZF, Xue Q, Zhi CY. Multifunctional energy storage and conversion devices. Adv. Mat. 2016, 28, 8344–8364

[18] Baptista JM, Sagu JS, Wijayantha UKG, Lobato K. State-of-the-art materials for high power and high energy supercapacitors. Performance metrics and obstacles for the transition from lab to industrial scale – A critical approach.

[19] Meng H, Pang S, Cui G. Photo-supercapacitors based on third-generation solar cells. ChemSusChem 2019, 12, 3431–3447

[20] Xu X, Li S, Zhang H, Shen Y, Zakeeruddin SM, Graetzel M, Cheng YB, Wang M. A Power Pack Based on Organometallic Perovskite Solar Cell and Supercapacitor. ACS Nano 2015, 9, 1782–1787.

[21] Yang PH, Xiao X, Li YZ, Ding Y, Qiang PF, Tan XH, Mai WJ, Lin ZY, Wu WZ, Li TQ, Jin HY, Liu PY, Zhou J, Wong CP, Wang ZL. Hydrogenated ZnO core-shell nanocables for flexible supercapacitors and self-powered systems. ACS Nano 2013, 7, 2617–2626.

[22] Lau D, Song N, Hall C, Jiang Y, Lim S, Perez-Wurfl I, Ouyang Z, Lennon A. Hybrid solar energy harvesting and storage devices: The promises and challenges. Materials Today Energy 2019, 25, UNSP 100852.

[23] Fuller TF, Harb JN. Electrochemical double-layer capacitors. In: Electrochemical Engineering, Wiley & Sons, Inc. 2018, pp.251–273.

[24] Pal B, Yang S, Ramesh S, Thangadurai V, Jose R. Electrolyte selection for supercapacitive devices: a critical review. Nanoscale Advances. 2019, 1, 3807–3835.

[25] Choi H, Yoon H. Nanostructured electrode materials for electrochemical capacitor applications. Nanomaterials 2015, 5, 906–936.

[26] Wang G, Zhang L, Zhang J. A review of electrode materials for electrochemical supercapacitors. Chem. Soc. Rev. 2012, 41, 797–826.

[27] Muzaffar A, Ahamed MB, Deshmukh K, Thirumalai J. A review on recent advances in hybrid supercapacitors: Design, fabrication and applications. Renewable & Sustainable Energy Reviews 2019, 101, 123–145.

[28] Dubal DP, Ayyad O, Ruiz V, Gomez-Romero P, Hybrid energy storage: the merging of battery and supercapacitor chemistries. Chem. Soc. Rev., 2015, 44, 1777–1790.

[29] Xie J, Yang PP, Wang Y, Qi T, Lei Y, Li CM. Puzzles and confusions in supercapacitor and battery: Theory and solutions. J. Power Sources. 2018, 401, 213–223.

[30] Kouchachvili L, Yaici W, Entchev E. Hybrid batter/supercapacitor Energy storage system for the electric vehicles. J. Power Sources. 2018, 237–248.

9 Interfacing applied electrochemistry and biology

9.1 General remarks

The tight and direct relation between biology and current flow has been known since the works of Luigi Galvani in his villa at the suburban hills of Bologna at ca. 1780. Then he discovered that static electricity generator connected to the severed frog's leg stimulated its twitch. This discovery gave a new thrust to the understanding of the workings of nervous system. The acknowledgment of such connection culminated in the famous phrase by Albert Szent-Györgyi: "Life is nothing but an electron looking for a place to rest" (Nobel Prize, Physiology or Medicine, 1937). And so, the direct transformation of chemical energy into electrical energy via the electrochemical reactions involving transfer of electrons through the biochemical pathways become the focus of bioelectrochemistry. In the sections that follow we will focus on applied electrochemistry that involves enzymatic catalysis for anodic and cathodic processes defined previously in the text.

The main stream of research and applications of enzymatic catalysis focuses on the electroactive enzymes – oxidoreductases, in search for the best, stable, and reproducible constructs for an extended and low-cost, disposable use. Among the oxidoreductases, glucose oxidase (GOx), laccases, bilirubin oxidase (BOx), peroxidase, tyrosinase, dehydrogenases, for example, cellobiose dehydrogenase, and catalases play key roles. Their substrates can be oxygen, hydrogen peroxide, phenols, quinones, sugars, lignins, and so on. Enzymes as biological electrocatalysts are unique as reaction accelerators, because for some of them the number of elementary charge transfer reactions occurring at the so-called active center of the enzyme reaches up to 10^6/s. This shows their advantage over the inorganic catalysts currently used in industry. Another advantage of enzymes as bioelectrocatalysts is their specificity in the reactions of a single target compound or a very small group of compounds. Moreover, enzymes work at ambient temperatures and pH from slightly acidic (e.g., 4.5) to slightly alkaline (e.g., 8.5), whereas commercially used inorganic catalysts require elevated temperature, above 200 °C and, very often highly acidic media, posing potential threat to the user and environment. The mechanism of catalytic activity of the enzymes also differs them from that of inorganic catalysts. Even though the first step is always the substrate binding, for the case of enzymes this step is already very selective and specific. This is achieved by structural binding pockets, complementary to the particular substrates with respect to shape, charge, and hydrophilic/hydrophobic balance. These features allow the biocatalysts to distinguish between very similar molecules. The specificity of enzymes was initially explained by the so-called *lock and key* model that assumed that both the enzyme and its substrate must have specific complementary shape, so that the substrate can fit exactly into the binding pocket. This model can explain the specificity of enzyme–substrate interactions, but does not account for the energetic stabilization of the transition state of the resulting complex. A modification of this rigid model – the *induced*

https://doi.org/10.1515/9783111160986-009

fit model – acknowledges that both the enzyme and substrate are flexible and during mutual interactions can reshape, and adapt to each other to the final precise positions that enable the enzyme to catalytically convert its substrate into product. At this precise position, the final shape and charge distribution of the enzyme–substrate complex is attained with lower standard Gibbs free energy of activation, $\Delta G^{0\#}$ (see the Eyring–Polanyi theory, eq. (1.92)) when compared with noncatalyzed reaction).

The sequence of the amino acids specifies the structure of the enzyme, thus determining its catalytic activity; however, only a small part of enzyme structure (ca. 2–6 amino acids) is directly involved in catalysis, forming the catalytic site. The catalytic site in turn is located next to one or more binding sites with functionalities orienting properly the substrates with respect to the catalytic sites. The catalytic and binding sites together form the active site of the enzyme. The remaining majority of the enzyme structure protects and maintains the precise orientation and dynamics of the active site. In some enzymes, no amino acids are directly involved in catalysis; instead, the enzyme contains sites to bind and orient catalytic cofactors. There are a large variety of natural oxidoreductase inorganic and organic cofactors, such as metal ion centers, iron–sulfur clusters, hemes, pyrolloquinoline quinone, nicotinamide adenine dinucleotifde (NAD^+), nicotinamide dinucleotide phosphate ($NADP^+$), and flavin adenine dinucleotide (FAD). Their role is to change the oxidation state during the substrate catalysis. "Wiring" these cofactors or enzymes to the electrode surface, either directly or with the help of linker molecules or nanostructures, is a challenge that underlines the effectiveness and performance of the biocatalysts either as biosensors or as electrodes in biofuel cells (BFCs). Since the more detailed description and classification of enzymes is well beyond the scope of this chapter, interested reader is directed instead to several excellent literatures [1–5].

9.2 Bioelectrocatalysis

Bioelectrocatalysis, similarly to electrocatalysis, is a type of heterogeneous catalysis that results in the acceleration of electrochemical reaction rate at the electrode/electrolyte interface due to the biological catalysts: enzymes, or even whole cells [5]. The applications of bioelectrocatalysis range from the preparations of electrodes for BFCs to biosensors and bioelectronic devices [6].

Until recently, the enzymes have not been used in electrocatalysis for a number of reasons, such as inherent instability of enzymes in artificial environment, on the electrode surface in particular, and the absence of pure enzyme preparations on a larger scale. However, due to the improvements in methods of immobilization of enzymes in their active forms on electrodes, as well as in the methods of enzyme isolation and purification, the above-mentioned problems are largely reduced. Therefore, the following advantages of enzymatic bioelectrocatalysis play a crucial role in its development and applications:

- Possibility of electrochemical reactions at ambient temperatures and neutral pH, where no effective inorganic catalysts are currently available
- Possibility of electrochemistry of complex organic reactions with high efficiency at moderate overpotentials and ambient conditions
- Possibility of utilization in BFCs
- Possibility of the development of highly selective bioelectrochemical sensors

Last, but not least, with the progressive depletion of natural sources of noble metals as catalysts and rapid increase of their costs, the relatively cheap production of enzymes for modification of the electrode surface becomes more and more attractive, particularly when the usage of enzymes for the above purposes is sufficient to merit the large-scale production, decreasing further their costs.

Some examples of redox reactions that can be catalyzed by enzymes immobilized on the electrode surface, largely exploited from the point of view of BFCs and biosensors, are:

$$O_2 + 4H^+ \xrightarrow{\text{Laccase}} 2H_2O$$

This cathodic reaction can be catalyzed by laccases (as in the above reaction) or, for example, BOx. The direct, four-electron reduction of oxygen to water is most favorable from the point of view of BFCs and is discussed in Section 8.1.

Peroxidases (e.g., horseradish peroxidase) catalyze the two-electron reduction of oxygen, energetically less favorable cathodic reaction:

$$H_2O_2 + 2H^+ + 2e \xrightarrow{\text{Peroxidase}} 2H_2O$$

Hydrogenases catalyze the reduction of protons:

$$2H^+ + 2e \xleftrightarrow{\text{Hydrogenase}} H_2$$

There exist several approaches to bind the biocatalysts with the electrodes for applications in bioelectrocatalysis:

- Mechanical/physical entrapment into the nonconducting or conducting polymeric network on the electrode
- Covalent binding of the enzyme to the modified surface, for example, via the carbodiimide procedure
- Enzyme cross-linking
- Noncovalent (e.g., electrostatic, affinity) binding to the electrode modified with the appropriate functional groups
- Direct adsorption on the electrode.

The last approach is less frequently used, because the enzymes tend to undergo denaturation in contact with metals, losing their activity. There are several conditions to be fulfilled for successful immobilization of biocatalysts: they should be stable, enzymatic activity unaffected, and with proper orientation of the enzyme's active center toward the electrode – the latter feature indispensable for the case of the direct electron transfer (DET) between the enzyme and the electrode. The last condition is less important for the case of mediated electron transfer (MET), where a redox couple – a mediator – transfers the electrons between the active center and the electrode in concert with enzymatic conversion of a substrate *S* into product *P*. Moreover, the access of the substrate to the active center of the immobilized enzyme should not be compromised [7–10].

The differences between DET and MET is shown in Fig. 9.1.

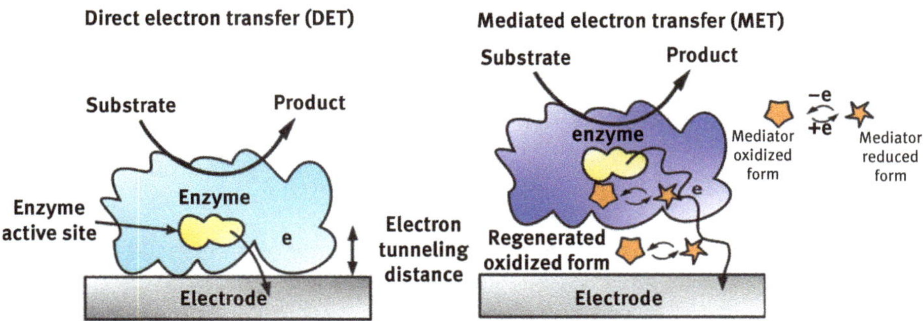

Fig. 9.1: The differences between the direct electron transport (DET) and mediated electron transport (MET) of an enzyme at, or in the vicinity of, the electrode surface.

In the direct bioelectrocatalysis (DET), the rate of catalyzed reaction is determined by the activity of an enzyme and/or by the thermodynamics of the reaction; however, in mediated bioelectrocatalysis (MET), apart from the enzymatic activity, it is the reaction of mediator molecules that may screen the substrate–enzyme reaction of interest.

In order to assess the bioelectrocatalytic properties of enzyme-modified electrodes, we will use the Michaelis–Menten model of enzymatic reaction. We have to keep in mind, however, that this model was developed for free enzyme and substrate in the solution, whereas here we have a surface-confined enzyme and such confinement affects its accessibility by a substrate molecule. Moreover, it may affect the enzyme's activity, too. Therefore, the obtained kinetic parameters of such immobilized system cannot be compared with the very same enzyme in the solution. Thus, now the scheme of reaction can be written as follows:

$$E + S \underset{k-1}{\overset{k_1}{\rightleftharpoons}} ES \overset{k_{\text{cat}}}{\rightarrow} E + P$$

where E and S are the enzyme and substrate, respectively, ES is the enzyme–substrate complex, P is the product, k_1 and k_{-1} are the rate constants of complex formation and dissociation, respectively, and k_{cat} is the catalytic rate constant of product P formation. Assuming a steady-state condition, that is, $d[ES]/dt = 0$, and using kinetic equations:

$$\frac{d[ES]}{dt} = k_1[E][S] - [ES](k_{-1} + k_{cat}) = 0 \tag{9.1}$$

Defining the rate of the enzymatic reaction as $v = (d[P]/dt) = -k_{cat}[ES]$, we can get $v = k_{cat}[E_0]\frac{[S]}{K_M + [S]}$, where $k_{cat}[E_0]$ defines the maximum rate of catalytic reaction, v_{max}. Here, square brackets are used to denote concentrations and $[E_0]$ is the total concentration of an enzyme ($[E_0] = [E] + [ES]$). To find the maximum value of a biocatalytic reaction, the substrate concentration is increased until a constant rate of product formation is achieved. At the maximum reaction rate, all the enzyme active sites are occupied by the substrate, and the amount of ES complex is the same as the total concentration of enzyme. The catalytic rate constant, k_{cat}, is also known as the turnover number, describing the number of substrate molecules converted by one active site of an enzyme per second. K_M is the Michaelis–Menten constant:

$$K_M = \frac{k_{-1} + k_{cat}}{k_1} \tag{9.2}$$

The Michaelis–Menten constant characterizes the affinity of an enzyme to the substrate; the lower the value of K_M, the greater the enzyme affinity to the substrate. Each enzyme has a characteristic K_M for a given substrate. The efficiency of a biocatalyst is frequently expressed by a ratio of k_{cat}/K_M, called the specificity constant, reflecting the affinity and catalytic capability of an enzyme toward a given substrate. Therefore, it is useful for comparing catalytic properties of the same enzyme for different substrates or different enzymes with the same substrate. The theoretical maximum value for the specificity constant is limited by the diffusion rate of a substrate toward the enzyme and can reach up to 10^9 $[M^{-1}s^{-1}]$ for a free enzyme in the solution.

Since the electrocatalytic reaction of the immobilized enzyme requires the transfer of electrons between the enzyme active center and the electrode (even in MET), then, calling upon the Faraday's law relating the rate of the reaction with charge being transferred (eqs. (1.31) and (1.91)), we can finally get the catalytic current due to the enzymatic reaction on the electrode:

$$i_{cat} = nFA\Gamma k_{cat}\frac{[S]}{[S] + K_M} \tag{9.3}$$

where Γ is the surface concentration of the enzyme and A is the electrode area. This is the electrochemical version of Michaelis–Menten equation relating the current measured on the enzyme-modified electrode with the enzyme electrocatalytic properties: k_{cat} and K_M. The maximum current, i_{max}, at the plateau of i_{cat} versus $[S]$ is equal to $nFA\Gamma k_{cat}$. This is a key equation used for amperometric biosensors. In a form of Lineweaver–Burk plot (double reciprocal plot), it predicts the linearity of $1/i_{cat}$ versus $1/[S]$:

$$\frac{1}{i_{cat}} = \frac{1}{nFA\Gamma k_{cat}} + \frac{K_M}{nFA\Gamma k_{cat}} \times \frac{1}{[S]} \tag{9.4}$$

By measuring current as a response to the controlled addition of substrate and plotting its reciprocal $1/i_{cat}$ versus $1/[S]$, we can get all characteristics of biocatalyst-modified electrode: i_{max}, k_{cat}, and K_M. This is shown in Fig. 9.2.

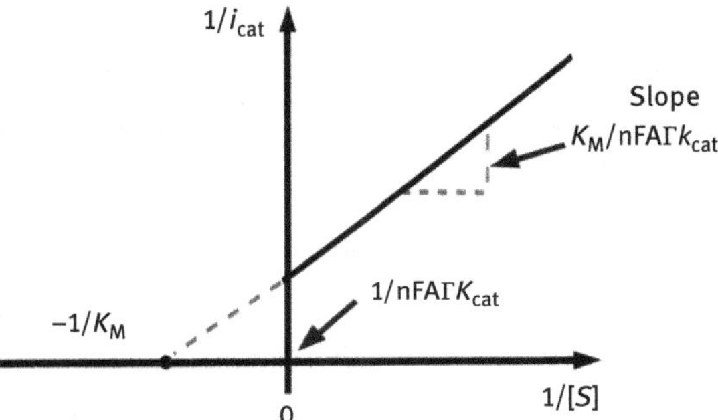

Fig. 9.2: Lineweaver–Burk plot, showing characteristic features for retrieving the electrocatalytic properties of enzyme-modified electrode.

Then, if needed, we can apply the Eyring–Polanyi theory (transition-state theory) for the bioelectrocatalytic process, to account for the dependence of rate constants on the energetics of the electron transfer rate (eq. (1.92)).

Above, we have described several methods of enzyme immobilization on the electrodes for further use as anodes and cathodes for BFCs and/or biosensors. Now, we shall briefly focus on materials that can be utilized as electrode support for biocatalyst immobilization. The proper selection is essential for the biocatalyst applications [4–10]. These materials should fulfil several requirements:

- They should be inexpensive, reproducible, and with highly developed active surface for immobilization and electron transfer.
- The immobilization procedure for such support should be simple and without significant compromise of biocatalyst activity and specificity.
- Should be chemically stable in the analytical environment and inert to the enzyme, detected analytes, and mediator (for MET).
- If they contain functionalities mediating in the electron transfer, these functionalities should be reversible electrochemically.
- The conductivity of such materials should be high.

Many efforts have been made to improve the properties of conducting supports for enzyme deposition. The use of conducting porous and mesoporous materials as well as nanoscale materials has been widely explored. Examples of such electrode materials are metal or oxide nanoparticles, carbon nanowires, carbon nanotubes, modified carbon nanotubes, carbon felt, carbon paper, carbon paste, graphene, graphite in form of felt, woven felt, graphite cloth platinized, and/or polyaniline/modified polyaniline conducting polymers [12, 13]. Table 9.1 explores some of the most interesting solutions:

Table 9.1: Materials utilized for enzyme immobilization.

Material(s) for catalytic bioelectrode formation	Immobilized biocatalyst	Reference
Citric acid-coated superparamagnetic iron oxide nanoparticles on gold electrode	Fructose dehydrogenase	[14]
Carbon paper disk coated with carbon nanotubes	Laccase, bilirubin oxidase	[15]
Carbon nanomaterials, nanofibers, nanotubes	Laccase, bilirubin oxidase, cellobiose oxidase, glucose oxidase	[16]
Metal nanoparticles (Au, Ag) Au/Pt/Pd/TiO_2NT/Ti	tyrosinase, sensors 13	[17]
Hybrid, osmium redox copolymer/carbon	Glucose oxidase, bilirubin oxidase	[18]
Chitosan, TiO_2/CeO_2/chitosan/glassy carbon electrode	Tyrosinase	[19]

The catalytic properties of oxidoreductases immobilized on conducting supports can be used either in biosensors or BFCs.

9.3 Biosensors

Monitoring of phenolic compounds in the food industry and for environmental (e.g., to detect polyphenols) and medical applications (monitoring the levels of glucose or neurotransmitters) has become increasingly relevant in recent years. Enzymatic biosensors based on oxidoreductases represent a fast method for online and in situ monitoring of these compounds. They rely mostly on the amperometric detection of:
- the product(s) of enzymatic reaction, or
- direct enzyme-MET between the electrode and the analyte.

The first type of detection is used, for example, in the case of re-reduction of phenols, polyphenols, and neurotransmitters, oxidized by the enzyme–tyrosinase immobilized on the electrode. The presence of enzyme improves the selectivity versus the interfering molecules (ascorbic acid, uric acid, etc.). The direct detection is used when the enzyme transfers the electrons between the conducting support and the analyte, causing its reduction or oxidation. This can be seen in the laccase biosensors "wired" to the electrode surface and monitoring the oxygen level by its direct reduction to water. For both cases, of course, the measured catalytic current should be proportional to the analyte concentration.

Figure 9.3 presents a relatively simple biosensor layout consisting of a detection layer of enzyme(s) immobilized onto a working transducer electrode. As discussed earlier, enzymes are optimal catalysts/recognition elements because they provide excellent selectivity and specificity for their substrates and have high catalytic activity. However, at the same time, enzymes are the shortest lived components of biosensors, BFCs, because they gradually lose activity in time, even if unused, determining therefore the lifespan of the enzyme-based electronic devices.

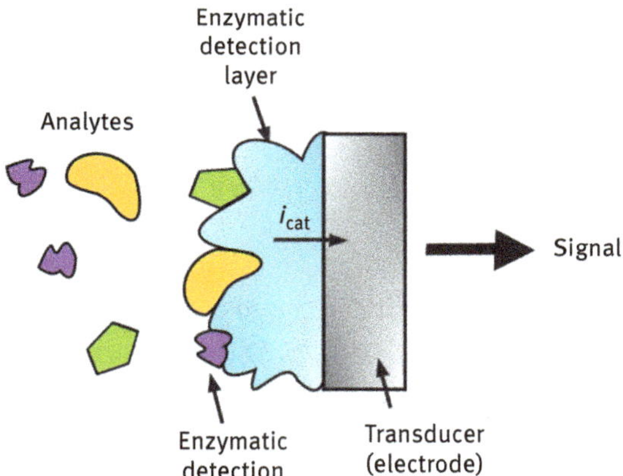

Fig. 9.3: A typical design of the enzyme-coupled electrochemical biosensors.

During the sensing event, a controlled voltage is applied to the electrode inducing redox reaction of the analyte, generating a signal that scales with the analyte's concentration. In the potentiometric mode, a change in the electrode potential as a result of redox reaction can also be used as the measurable response. Finally, an appropriate electronic device (e.g., computerized potentiostat) collects and displays the resultant signal. This type of electrochemical detection also offers additional selectivity of biosensor by controlling the potential applied to the electrode, because different electroactive molecules can be oxidized/reduced at different potentials. The overall system is compatible with modern miniaturization trends and can be also made biocompatible for implanted bioelectronics. Biocompatibility is often an important issue, since biological fluids, such as blood, are the most common sample matrices for enzyme electrodes in clinical applications. The components of such matrices, mostly proteins, can affect or even deteriorate the sensor performance by fouling its surface.

Apart from the above-mentioned food industry and environmental monitoring, the electrochemical biosensors are most commonly used in health care, by monitoring the glucose level in blood being the dominant application. Using a glucose biosensor as an example, we will present the evolution of biosensors, based on the sensing event and bioelectrochemistry of signal transduction between the electroactive enzyme and the electrode. This evolution is shown in Fig. 9.4.

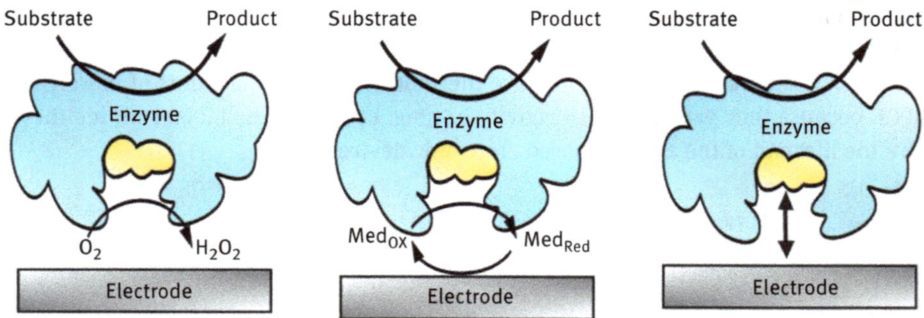

Fig. 9.4: Schematic representation of differences between the first through the third generation of biosensor design and characteristics.

Figure 9.4(a) shows schematically the main characteristics of the so-called first-generation glucose biosensor: the trapped GOx, oxidizes β-D-glucose (S) to β-D-gluconolactone (P), with a simultaneous reduction of GOx cofactor FAD to $FADH_2$. $FADH_2$ is subsequently re-oxidized to FAD by its natural substrate – dissolved O_2 producing H_2O_2. Appropriate voltage applied to the electrode oxidizes H_2O_2 at the electrode surface, yielding an electric signal as a result of biosensing event. Unfortunately, the first generation of sensor has several disadvantages and shortcomings. The active site and FAD are buried deeply within the protein shell, hindering drastically the diffusion of reagents [11]. Additionally, as we discussed earlier, the Marcus theory states that the electron transfer requires reorganization energy, λ, and also decays exponentially with increasing distance. The active sites of many oxidoreductases are hidden within the proteins. Moreover, O_2 has a limited solubility in aqueous media and is the reagent of unstable concentration (due to the fluctuations in partial pressure in the air), leading to oxygen deficiency at higher glucose concentrations affecting the sensor response. The oxidation of hydrogen peroxide as a sensing event also proceeds at relatively high potential, leading to the simultaneous electro-oxidation of interfering species typically present in biomatrices, for example, ascorbic acid, uric acid, and even possible drugs such as paracetamol.

Second-generation biosensor is shown schematically in Fig. 9.4(b). Its design solved many of the first-generation biosensor problems by the incorporation of a redox mediator molecule that shuttles the electrons between the enzyme and electrode. As stated earlier, DET is not possible, so the role of mediator molecules is to facilitate the transfer between deeply buried FAD redox center of GOx. In the mediator redox cycle, its oxidized form Med_{Ox} regenerates FAD by oxidation of $FADH_2$. The resultant reduced form of mediator, Med_{Red}, is re-oxidized at the electrode generating the electric signal. Ideal mediator molecules should show fast and reversible redox kinetics with both reduced and oxidized forms being nontoxic, chemically stable, and ideally should maintain constant concentration within the biosensor, that is, they should not diffuse out of the sensor construct. Mediators such as poly(vinylimidazole) and poly(vinylpyridine) linked with osmium complexes, or $Os^{+/2+}$ redox couples co-immobilized in a hydrogel [18] serving as electron relays provided close proximity for the redox center of the polymers and the FAD redox center of the enzymes resulting in fast sensor response and high current output [20]. This approach was reported also for immobilized dehydrogenases with azine/polyazine relays between the electrode and NAD^+/NADH [21, 22]. The use of synthetic mediators in second-generation biosensors eliminated also the problems related to oxygen and allowed also for lower applied potentials. However, the observed gradual decrease of sensors performance due to the steady leaching of mediators to the surrounding solution stimulated the design of the third generation of biosensors.

The design of third-generation sensors involves “wiring” of the enzyme to the modified transducer surface in such a way to provide a DET from the redox site to the electrode or through the immobilized polymeric relays (Fig. 9.4(c)). This approach is typically achieved by direct co-immobilization of the enzyme and mediator on the electrode surface or within the conductive polymer matrix [23] (in the case of some peroxidases having peripheral redox centers, a DET can be achieved by depositing the enzymes directly on the electrode). The mediators cannot escape from the sensor and their concentration effectively facilitates the transport of electrons between the enzyme active and the electrode. Of particular interest is, of course, direct electrical contact between the enzyme and electrode by connecting the active center through the tailored molecular communication path. Such configuration should greatly increase the charge transfer kinetics, resulting in larger sensing currents. This design is also advantageous from the point of view of BFCs, where the efficient electron transfer at the bioelectrodes is required. Other enzyme and mediator immobilization methods include the layer-by-layer deposition of polyelectrolytes, entrapment in hydrogels, and electropolymerization in the presence of the enzyme and the mediator, such as carbon or osmium copolymer “wires” [24–26].

9.4 Biofuel cells

Even though the concept of BFC dates back to the beginning of the past century, the real progress in this area began in the 1960s with NASA space program and the demand for alternative power sources. However, the initial results were poor and the BFC concept was dormant until the energy crisis of the late 1970s. And currently, due to the permanently increasing demand for cheap, renewable, and sustainable fuel, the enzymatic BFCs represent an emerging advantageous technology that can generate electrical energy from biologically renewable catalysts and fuels, addressing the above issues. In principle, a BFC can be considered as a special type of an electrochemical fuel cell, where inorganic catalysts (mainly metals, such as Pt, Pd, Ru, and Rh) are replaced with biocatalysts in the form of microorganisms, organelles, or enzymes. In a conventional fuel cell, an oxidation reaction occurs at the anode/solution interface and the reduction reaction takes place at the cathode/solution interface. Both reactions use relatively simple inorganic chemistries (Chapter 8), releasing the electrons at the anode that subsequently travel through the external circuit producing electrical energy that can be used for different engineering purposes. However, those fuel cells that operate in the so-called low-temperature regime (ca. 80–100 °C) typically require expensive metallic p-group catalysts. Contrary to the above, the BFCs use enzymes or whole cells as catalysts and therefore they can operate under mild, physiological conditions: temperature in the range of 20–40 °C and near-neutral pH. However, these conditions are not their limitations, though advantageous from the point of view of BFC applications, because extremophilic organisms and enzymes are capable of operation under a wide range of external conditions. In the sections that follow, we will briefly describe the principles and basic concepts of enzymatic BFCs only, directing the reader interested in microbial fuel cells to the review papers [27, 28]. We will consider recent advances in mediated and DET-based BFCs capable of producing relatively significant power while being quite stable for a long period of time. A wide variety of redox enzymes have been tested and employed to create electrodes (both cathodes and anodes) for use in BFCs as energy sources powering small electronic devices, self-powered sensors, and implantable power sources. Here we will limit our discussion to the oxidoreductases – a group of enzymes that catalyze oxidoreduction reactions by transferring electrons from an oxidant to the reductant. Some of these enzymes were presented in the preceding sections.

From the point of view of their applications and competition with fuel cells, the requirements for successful BFCs can be summarized as follows:

- BFC should convert chemical energy into electrical with the help of low-cost, easy-accessible, stable redox enzymes as catalysts.
- They should work under mild conditions, neutral pH, and ambient temperatures.
- Should use natural sources for fuel, such as ethanol, methanol, glucose, and oxygen.
- BFC should be renewable and cheap (biomass is cheap).
- The only products of their reaction should be water and carbon dioxide.

Further comparison of both types of currently used fuel and BFCs is given in Table 9.2:

Table 9.2: Comparison of fuel cells and biofuel cells focusing on catalyst used and operating conditions.

	Fuel cell	Biofuel cell
Catalyst	Noble metals	Microorganisms or enzymes
pH	Highly acidic or basic	Between 5.0 and 8.0
Temperature	Above 80 °C	25–40
Efficiency (%)	40–60	40
Fuel	H_2, methanol, O_2	Carbohydrates, methanol, O_2, etc.

As any of the electrochemical cell, the BFC contains cathode and anode made out of well-conducting material, most frequently carbonaceous, placed in separate compartments – cathodic and anodic – typically separated by a semipermeable membrane to avoid mixing of the electroactive species (fuel and biocatalysts) in the electrolyte. If the biocatalysts are immobilized on the electrodes separately, there is no strict demand for the use of separating membranes. There are some exemptions, of course, and the most common is when GOx is used on the anode for oxidation of glucose and oxygen used as fuel on cathode interferes with anodic processes or both cathodic and anodic processes. For example, reactive oxygen species (ROS, such as $O_2^{\cdot-}$), produced by possible incomplete reduction of O_2, can seriously affect the activity of enzymes. However, it has been shown that suitably immobilized GOx gives an anodic current (rate of the reaction) very close to the theoretical maximum [29], indicating that the presence of oxygen in the anodic compartment does not affect the biocatalytic activity of this enzyme. This BFC employed also cytochrome C oxidase immobilized on cathode. The selectivity of the immobilized enzymes separated the reduction and oxidation reactions. Such an approach eliminates issues of crossover. By removing the membrane separator, such design allows both electrodes to be as close to each other as possible, decreasing the internal resistance of the cell without short-circuiting, and allowing for further miniaturizations. This leads to subsequent research on improvement of the system performance [open circuit potential (OCP), power output] to make it more economically effective as portable energy devices and compatible with current electronics and also directing the technology toward the implantable systems [30].

Finally, the generalized schemes of both fuel and BFCs is shown in Fig. 9.5, exemplifying the latter in the case of glucose/oxygen BFC – the *Holy Grail* of current research:

This scheme reveals advantageous feature of BFC discussed above: the fuel cell requires the proton exchange membrane (PEM), with H^+ as connecting ion, whereas BFC

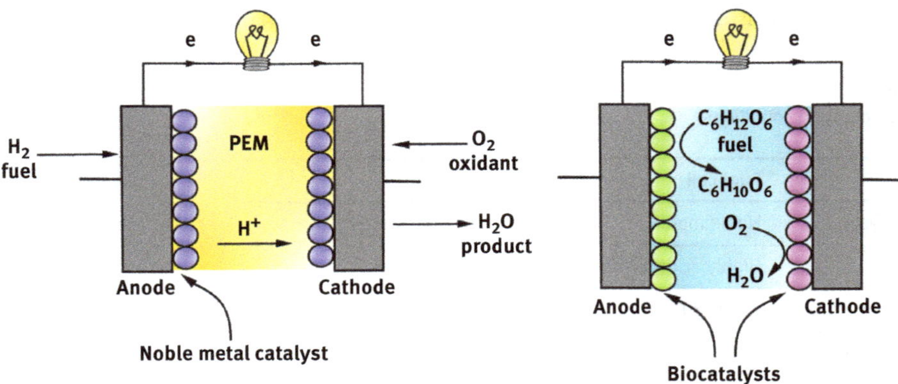

Fig. 9.5: Generalized schemes of both fuel and biofuel cells. Biofuel cell is shown as a glucose/oxygen biofuel cell.

does not necessarily require any membrane separating the anodic compartment from the cathodic one.

9.4.1 Design and enzymes used in biofuel cells

Obviously, in light of the above discussion, the enzymes that show high catalytic activity, the ability to DET, and low overvoltage of catalytic reactions are typically used in BFC studies. Although DET is highly desirable for enzymatic BFCs, many of the currently investigated systems rely on the MET, because of higher current densities that can be achieved. The incorporation of organometallic redox mediators into the conducting redox polymers pioneered by the Heller group [31 and all subsequent related works] provided a means of "indirect wiring" of the enzymes. These redox mediators – osmium complexes – mentioned earlier in the text, characterized by high stability, facile, and reversible kinetics of the electron transfer, were connected with polymer chains, typically polyvinylpyrrolidone, polyacrylamide, or polyvinylimidazole. Osmium redox centers in such molecules by tuning the ligands can mediate cathodic or anodic reactions and therefore were utilized for biocathodes and bioanodes. In summary, the engineering approaches to create well-performing bioelectrodes are mainly focused on tailoring the properties of scaffolds for enzyme immobilization and improving the electron transfer rates between enzymes and electrodes.

9.4.1.1 Biocatalysts and design of biocathodes

Multicopper oxidases (MCO) such as laccase (Lac) or BOx have been intensively investigated as bioelectrocatalysts for cathodic reduction of oxygen to water. Even though

BOx shows better tolerance against chloride inhibition, the availability of laccases results in more frequent utilization of this enzyme for the design of cathode in BFC.

Laccases have oxidative activity toward *ortho*- and *para*-diphenols and therefore are also used for biosensor design. However, for the use as a part of biocathode in BFC, it is imperative that the enzyme takes four electrons necessary for the reduction of oxygen to water only from the electrode material. It is instructive to follow the electron path inside this multicopper enzyme, from the electrode material to oxygen. The catalytic mechanism of the laccase starts with the entry of an electron at the T1 copper site, followed by an internal electron transfer from the reduced T1 to the T2 and T3 copper sites. The T3 copper in the presence of the T2 site functions as an acceptor in the aerobic oxidation process. The reduction of oxygen to water takes place at the T2 and T3 clusters [32]. Now, we should recall the term "internal resistance" and all its components that were discussed earlier in the case of power outputs and performance of cells in general (Chapter 8). In the case of BFCs, the overpotentials of the electron pathways within the biocatalyst and its immobilizing layer should be considered as part of this inner potential drop, diminishing the BFC performance, and as such should be minimized by proper design.

Laccases have defined Nernst (equilibrium) redox potentials. However, some of them are branded as "low" potential, because for T1 and T3 copper sites, this potential falls within the range of approximately 0.4–0.5 V (vs NHE), whereas for the so-called high potential laccases E_N is approximately equal to 0.8 V. Regardless of these values, for the case of T2 Cu, the equilibrium potential, E_{eq}, is approximately 0.4 V for both types of laccases [32]. Of course, the higher the overall potential established on enzymatic cathode, the better from the point of view of the BFC performance, in particular from the point of view of the OCP, discussed in Chapter 8.

The design and charge transfer mechanism of MCO biocathodes is directly related to the technology of enzyme immobilization. The use of redox mediators (MET) improves the current output and stability of BFCs; however, the redox mediators commonly used are of limited choice, due to the fact that they need to have similar redox potential as the active site of immobilized enzymes, have small overpotential (good reversibility), should be nontoxic, biocompatible, and stable. Moreover, they should not leak from the cathodic space. We addressed these issues earlier in the case of biosensors. Due to the above limitations, other constructs are being developed, such as conductive polymers that can serve as both the immobilization scaffold for the enzyme and conducting wires between the copper centers and electrode. Lately, an extensive effort of researchers has been directed toward the nanomaterials, such as carbon nanomaterials and metallic (gold) and metal-oxide nanoparticles. These materials greatly improve the catalytic performance of biocathodes, exhibiting high conductivity, good biocompatibility, and enhancement of DET by addressing/approaching directly the electron entry catalytic site of MCOs, that is, the T1 site of laccases. Among these nanomaterials, multiwalled carbon nanotubes, either surface-modified or pristine, play a crucial role. High surface area provides also a statistically improved orientation of MCOs at such modified cathodes, enhancing also this way the electron transfer from the active site to the electrode.

The generalized scheme of oxygen enzymatic (laccase) electroreduction can be written in the following steps:

1. The formation of "active complex" laccase – oxygen in the presence of water molecules

$$\text{laccase} + O_2 + H_2O \rightarrow \text{laccase} - HO_2OH$$

2. Transfer of two electrons from the electrode to the complex

$$\text{laccase} - HO_2OH + 2e + 2H^+ \rightarrow \text{laccase}(OH)_2 + H_2O$$

3. Transfer of the next two electrons followed by dissociation

$$\text{laccase}(OH)_2 + 2e + 2H^+ \rightarrow \text{laccase} + 2H_2O$$

This scheme also provides an explanation of the pH dependence of laccases and BOx: they require a slightly acidic medium with the highest activity at approximately pH 5.5 (laccase).

9.4.1.2 Biocatalysts and design of bioanodes

On the other part of the BFC shown in Fig. 9.5 (right panel), the anode is designed to oxidize glucose. Most of the oxidoreductase enzymes used for the design of anodes contain redox cofactors changing their oxidation state during substrate catalysis. One of such cofactors is nicotinamide adenine dinucleotide phosphate $NAD(P)^+$, in NAD(P)-dependent dehydrogenases. This is a diffusional mediator being reduced to NAD(P)H upon oxidation of fuel (sugar). Subsequently, it has to be re-oxidized/regenerated at the electrode. However, NAD(P)H tends to quickly passivate the electrode surface. Additionally, it has large overpotential, between 0.5 and 0.9 V, depending on the electrode material, so the NAD(P)-dependent dehydrogenases are not the best choice for anode design; nevertheless, they are still commonly used because they are oxygen insensitive comparing to oxidase-based bioanodes [30, 33]. Other bioanodes used for the construction of BFCs rely on flavoprotein oxidases, such as GOx or glucose dehydrogenase (GDH), carrying a flavin cofactor, (FAD), tethered within the protein shell. The role of FAD, as in the case of $NAD(P)^+$, is to change its oxidation state during the anodic reaction of sugars. Now, the proposed electrochemical oxidation of glucose can be written as follows:

In the case of GOx:

$$\text{glucose} + \text{GOx(FAD)} \rightarrow \text{gluconolactone} + \text{GOx}(FADH_2);$$

$$\text{Regeneration at the electrode:} \quad \text{GOx}(FADH_2) \rightarrow \text{GOx(FAD)} + 2H^+ + 2e$$

In the case of GDH:

$$\text{glucose} + \text{GDH(FAD)} \rightarrow \text{gluconic acid} + \text{GDH(FADH}_2\text{)}$$

$$\text{Regeneration at the electrode:} \quad \text{GDH(FADH}_2\text{)} \rightarrow \text{GDH(FAD)} + 2\text{H}^+ + 2\text{e}$$

The electrons then flow through the external circuit to the cathode, generating the power output of BFC. It should be mentioned here that FAD utilizes O_2 as an electron acceptor, producing another ROS species – H_2O_2 – which can impart a negative effect on both enzymes in the system. However, FAD-dependent GDH, *GDH(FAD)*, appears to be oxygen insensitive, unlike FAD-dependent GOx, *GOx(FAD)* [33, 34].

The engineering strategies to create efficient bioanodes, as in the case of biocathodes, are mainly focused on tailoring the properties of scaffolds for enzyme immobilization and improving the electron transfer rates between enzymes and electrodes. Among such strategies, nanomaterials and carbon nanomaterials in particular that improve direct electron communication between the enzyme and electrode emerge as crucial materials [30, 33–38].

9.4.2 Energy and power of the biofuel cells

Critical factor and challenge of BFCs in their applications as "green," sustainable, and renewable energy sources are their output voltages and power that are generally incompatible with the operational requirements of commercially available microelectronic devices. Such devices are typically built to operate at a minimum of 1–3 V voltage supply; therefore, BFC will need a step-up power system [39–41]. It is easy to evaluate the magnitude of the equilibrium potential, ΔE_{eq}, also called the OCP, at standard conditions of the electrochemical reaction, using its relation with the standard Gibbs free energy, $\Delta_r G^0$, derived previously for the general case (eq. (4.1)):

$$\Delta_r G^0 = -nF\Delta E^0_{\text{eq}} \tag{9.5}$$

All symbols used here have their meaning defined previously.

In the example of glucose as an anodic fuel and oxygen as a cathodic one, we can write:

$$C_6H_{12}O_6(\text{glucose}) + \frac{1}{2}O_2 \rightarrow C_6H_{10}O_6(\text{gluconolactone}) + H_2O \tag{9.6}$$

At 25 °C and p = 1 bar, 0.1 M glucose concentration at pH 7, the modulus of $\Delta_r G^0$ is approximately 227.2 kJ/mol. Then, with n = 2 the value of ΔE_{eq} (OCP) is equal to approximately 1.2 V. This is also the thermodynamically maximal value of OCP in an aqueous media; otherwise, we begin to split water at higher potential values.

As in the case of batteries and fuel cells, the typical polarization curve in galvanic mode is a superposition of contributions due to bioanode and biocathode kinetic overpotential regions, mass transport limitations, and ohmic losses (compare Fig. 8.2):

$$U_d = I_d R_{ex} = \Delta E_{eq} - \left(\sum \eta_{a,\,c,\,Cathode} + \sum \eta_{a,\,c,\,Anode}\right) - I_d R_{in} \tag{9.7}$$

where U_d is the potential of BFC under a given load (R_{ex}) biasing the cell from its equilibrium potential value, ΔE_{eq}, and inducing a discharge current flow I_d (Fig. 9.6). $\sum \eta_{a,\,c,\,Cathode}$ is the sum of activation and concentration overpotentials at the biocathode, while $(\sum \eta_{a,\,c,\,Anode})$ describes the sum of activation and concentration overpotentials at the bioanode. R_{in} is the internal BFC resistance of ionic transport within the electrolyte, ohmic drop in the polymer scaffold of the enzyme on the electrode, if present, but also the resistance of molecular "wires" transferring electrons between the electrode and the enzyme.

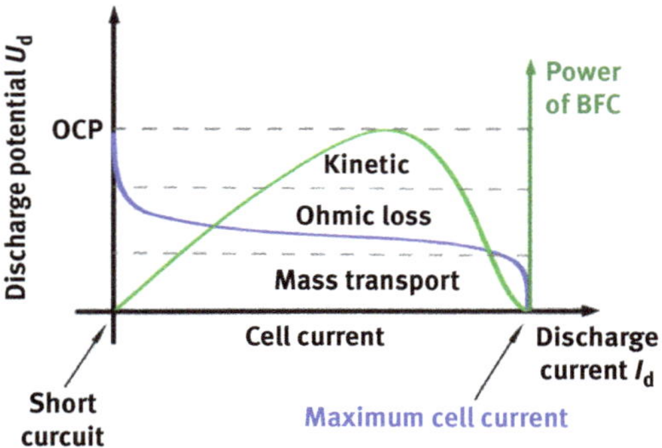

Fig. 9.6: Schematic graph of the BFC discharge curve at decreasing load, R_{ex}, and increasing the discharge current, I_d, identifying different mechanisms responsible. Each point of this discharge curve, $U_d = f(I_d)$, where U_d and I_d are the discharge potential and current, respectively, can be described as in the case of battery (compare Fig. 8.2).

It is important to understand that in the case of BFCs, these overpotentials are directly related to the electrochemical mechanisms of redox reactions of enzymes deposited on the electrodes. For biocathodes, MCO-based bioelectrodes undergoing DET with low overpotentials are widely adopted, whereas for bioanodes operating through the cofactors, their low overpotentials determine the choice of the enzymes. Of course, these should also correlate with possibly large ΔE_{eq}. For a better understanding of basic issues related to the performance of the BFC, let us split the BFC into anodic and cathodic parts, and examine each of the electrodes separately. This is shown in Fig. 9.7.

As we have shown before, in an ideal case, the OCP is determined by the difference between the thermodynamic equilibrium potentials between the cathodic reduction and anodic oxidation reactions, $E_{eq,c}$ and $E_{eq,a}$, respectively (for our BFC in Fig. 9.5,

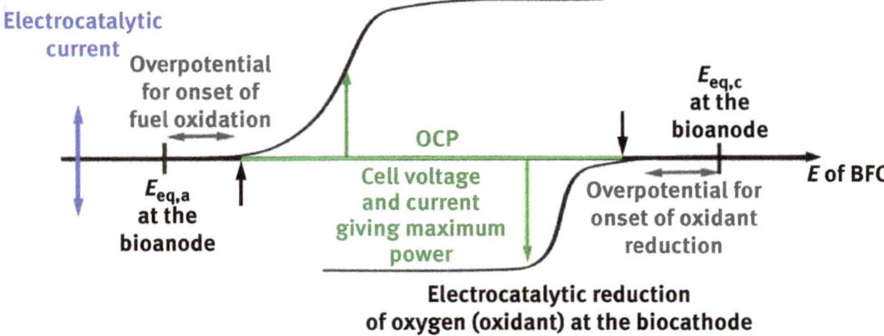

Fig. 9.7: Schematic graph of cathodic and anodic electrocatalytic current/voltage characteristics plotted along the overall BFC voltage (cathodic current is negative). Green arrows show the value of current I_d at maximum power of BFC, P_{max}, under a given load R_{ex}.

these are the reduction of oxygen and oxidation of glucose), taking into account the nonstandard conditions of the BFC. At very low discharge currents, the measured U_d is determined by the difference between the onset potentials for catalysis on cathode and anode, respectively (red arrow). Then, the U_d decreases with decreasing load, until the maximum value of the product $U_d{\cdot}I_d$ is achieved, which is equal to the maximum power of a given BFC, according to the definition of power: $P_{BFC} = U_d{\cdot}I_d$. Finally, with still decreasing load R_{ex}, we reach the so-called short-circuit conditions (Fig. 9.6), where $R_{ex} = 0$, and anode and cathode are directly connected electrically. Then, again as in the case of the equilibrium state, no useful electrical work is done. Useful power can be delivered by a cell, also by a BFC cell, at voltages between these two limiting cases. The magnitude of current is determined by the electrode with the lowest electrocatalytic rate (in Fig. 9.6, this is the reduction of oxygen at the cathode, as indicated by the lower catalytic "saturation" current, limited by the slow diffusion and low concentration of oxygen).

In contrast to the conventional fuel cells with noble metal catalysts, BFCs are of very low power outputs. Typical power outputs of BFCs are in the range of up to milliwatts per square centimeter of their geometric area (see Table 9.3), compared to the kilowatt outputs from fuel cells. Therefore, their applications are limited to the wearable or implantable short-term devices in medicine or self-powered sensors, such as glucose sensors, harvesting energy from biofuel inside the body, or from the environment that is monitored. These applications include self-powered portable electronic devices for point-of-care measurements [41].

In order to even better visualize the challenges that have to be overcome for future commercial utilization of BFC, let us present the graphical representation of power outputs for various electrochemical power sources [46]. Similarly to the Ragone diagram, it shows the current status of research on performance of BFCs (Fig. 9.8).

Table 9.3: Power outputs of several enzymatic BFCs.

Anodic enzyme	Cathodic enzyme	Maximum power, P_{max} (mW/cm^2)	Ref.
GOx	BOD	0.97	[42]
GDH	Laccase	0.60	[43]
GOx	Laccase	1.12	[44]
GOx	Laccase	1.2	[35]
GOx	BOD	1.54	[45]

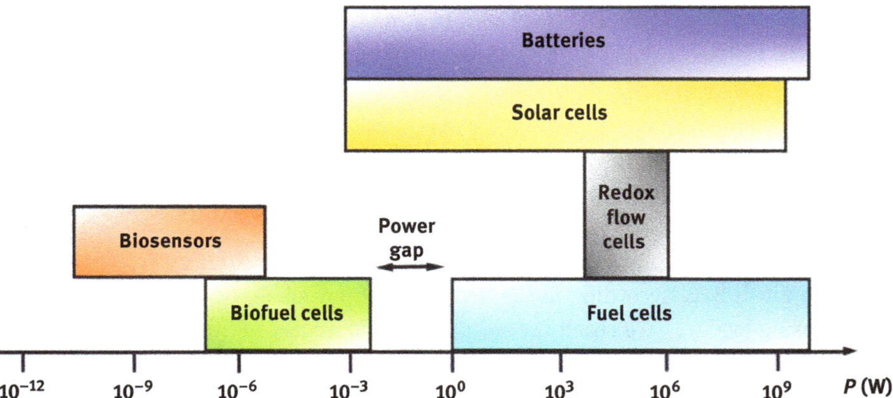

Fig. 9.8: A sketch of power output ranges of biofuel cells and other cells (including solar cells), illustrating the power gap between the fuel and biofuel cell range, challenging further biofuel cell research for practical applications (adapted after [46]).

However, one should take into account a tremendous development in this area, and therefore, even at the stage of writing this book, the above diagram seems to be obsolete (c.f. [27, 30, 33]). Due to the fact that both oxygen and glucose as biofuels are quite abundant in the environment, the development of BFCs is still progressing and the introduction of new electrode materials, especially conducting nanomaterials, constantly improves the performance of such power sources. Also, modern miniaturization strategies can provide charge pump systems boosting the final output voltage of BFCs to the levels compatible with current electronics [24, 30, 39, 40, 47].

Bibliography

[1] Suzuki H. Active Site Structure. In: Suzuki H., ed. How Enzymes Work: From Structure to Function. CRC Press, Taylor & Francis, 2015, pp. 117–140.

[2] Stryer L, Berg JM, Tymoczko JL. Biochemistry (5th ed.). San Francisco: W.H. Freeman. 2002.

[3] Petsko GA, Ringe D. Chapter 1: From sequence to structure. In: Lawrence E, Robertson M, eds., Protein structure and function. London: New Science Press Ltd. 2003, pp. 2–27.

[4] Boyer R. Enzymes I: Reactions, Kinetics, and Inhibition. In: Molloy K, ed., Concepts in Biochemistry. 3rd ed. John Wiley & Sons, Inc. 2006, pp. 131–162.

[5] Rusling JF, Wang B, Yun S. Electrochemistry of Redox enzymes. In: Bartlett PN. ed., Bioelectrochemistry. Fundamentals, Experimantal Techniques and Applications, John Wiley & Sons Ltd. 2008, pp. 39–76.

[6] Masa J, Schuhmann W. Electrocatalysis and bioelectrocatalysis – Distinction without a difference. Nano Energy. 2016, 29, 466–475.

[7] Tarasevich MR. Bioelectrocatalysis. In: Srinivasan S, et al., eds. Comprehensive Treatise of Electrochemistry. Springer, 1985, vol. 10, pp. 231–295.

[8] Willner B, Willner I. Reconstituted redox enzymes on electrodes: From fundamental understanding of electron transfer at functionalized electrode interfaces to biosensor and biofuel cell applications. In: Willner I, Katz E, eds. Bioelectronics. From Theory to Applications, Wiley VCH, 2005, pp. 35–123.

[9] Leech D, Kavanagh P, Schuhmann W. Enzymatic fuel cells: Recent progress Electrochimica Acta. 2012, 84, 223–234.

[10] Putzbach W, Ronkainen NJ. Immobilization Techniques in the Fabrication of Nanomaterial-Based Electrochemical Biosensors: A Review. Sensors. 2013, 13, 4811–4840.

[11] Tsujimura S. From fundamentals to applications of bioelectrocatalysis: bioelectrocatalytic reactions of FAD-dependent glucose dehydrogenase and bilirubin oxidase. Biosci. Biotechnol. and Biochem. 2019, 83, 39–48.

[12] Jackowska K, Krysinski P. New trends in the electrochemical sensing of dopamine. Anal. Bioanal. Chem. 2013, 405, 3753–3771.

[13] Florescu M, David M. Tyrosinase-Based Biosensors for Selective Dopamine Detection. Sensors. 2017, 17, 1314–1330.

[14] Kizling M, Rekorajska A, Krysinski P, Bilewicz R. Magnetic-field-induced orientation of fructose dehydrogenase on iron oxide nanoparticles for enhanced direct electron transfer. Electrochem. Comm. 2018, 93, 66–70.

[15] Majdecka D, Draminska S, Janusek D, Krysinski P, Bilewicz R. A self-powered biosensing device with an integrated hybrid biofuel cell for intermittent monitoring of analytes. Biosens. Bioelectron. 2018, 102, 383–388.

[16] c.f., Bollella P, Ludwig R, Gorton L. Cellobiose dehydrogenase: Insights on the nanostructuration of electrodes for improved development of biosensors and biofuel cells. Appl. Mat. Today. 2017, 9, 319–332.

[17] Mahshid S, Li C, Mahshid SS, Askari M, Dolati A, Yang L, Luo S, Cai Q. Sensitive determination of dopamine in the presence of uric acid and ascorbic acid using TiO_2 nanotubes modified with Pd, Pt and Au nanoparticles analyst. 2011, 136, 2322–2329.

[18] Gregg BB, Heller A. Cross-linked redox gels containing glucose-oxidase for amperometric biosensor applications. Anal. Chem. 1990, 62 258–263.

[19] Njagi J, Ispas C, Andreescu S. Mixed ceria-based metal oxides biosensor for operation in oxygen restrictive environments. Anal. Chem. 2008, 80, 7266–7244.

[20] Zhu Z, Garcia-Gancedo L, Flewitt AJ, Xie H, Moussy, F, Milne WI. A critical review of glucose biosensors based on carbon nanomaterials: Carbon nanotubes and graphene. Sensors. 2012, 12, 5996–6022.

[21] Karyakin AA, Karyakina EE, Schmidt HL. Electropolymerized azines: A new group of electroactive polymers. Electroanalysis. 1999, 11, 149–155.

[22] Karyakin AA, Ivanova YuN, Karyakina EE. Equilibrium (NAD(+)/NADH) potential on poly(Neutral Red) modified electrode. Electrochem. Commun. 2003, 5, 677–680.

[23] Wang J. Electrochemical glucose biosensors. Chem. Rev. 2008, 108, 814–825.

[24] Harper A, Anderson MR. Electrochemical glucose sensors – developments using electrostatic assembly and carbon nanotubes for biosensor construction. Sensors. 2010, 10, 8248–8274.

[25] Heller A. Miniature biofuel cells. Phys. Chem. Chem. Phys. 2004, 6, 209–216.

[26] Barton SC, Pickard M, Vasquez-Duhalt R, Heller A. Electroreduction of O-2 to water at 0.6 V (SHE) at pH 7 on the 'wired' Pleurotus ostreatus laccase cathode. Biosens. Bioelectron. 17, 2002, 1071–1074.

[27] Santoro C, Arbizzani C, Erable B, Ieropoulos I. Microbial fuel cells: From fundamentals to applications. A review. J. Power Sources. 2017, 356, 225–244.

[28] Rinaldi A, Mecheri B, Garavaglia V, Licoccia S, Di Nardo P, Traversa E. Engineering materials and biology to boost performance of microbial fuel cells: a critical review. Energy Environ. Sci. 2008, 1, 417–429.

[29] Katz E, Willner I, Kotlyar AB. A non-compartmentalized glucose vertical bar O-2 biofuel cell by bioengineered electrode surfaces. J. Electroanal Chem. 1999, 479, 64–68.

[30] Rasmussen M, Abdellaoui S, Minteer SD. Enzymatic biofuel cells: 30 years of critical advancements. Biosens. Bioelectron. 2016, 76, 91–102.

[31] Ohara TJ, Rajagopalan R, Heller A. Wired enzyme electrodes for amperometric determination of glucose or lactate in the presence of interfering substances. Anal. Chem. 1994, 66, 2415–2457.

[32] Rodriguez-Delgado MM, Aleman-Nava GS, Rodriguez-Delgado JM, Dieck-Assad G, Martinez-Chapa SO, Barcelo D, Parra R. Laccase-based biosensors for detection of phenolic compounds. Trends Anal. Chem. 2015, 74, 21–45.

[33] Xiao XX, Xia HQ, Wu RR, Bai L, Yan L, Magner E, Cosnier S, Lojou E, Zhu ZG, Liu AH. Tackling the challenges of enzymatic (bio)fuel cells. Chem. Rev. 2019, 119, 9509–9558.

[34] Suzuki A, Ishida K, Muguruma H, Iwasa H, Tanaka T, Hiratsuka A, Tsuji K, Kishimoto T. Diameter dependence of single-walled carbon nanotubes with flavin adenine dinucleotide glucose dehydrogenase for direct electron transfer bioanodes. Jap. J. Appl. Phys. 2019, 58, art. no. 051015.

[35] Sakthivel M, Ramarai S, Chen SM, Chen TW, Ho KC. Transition-metal-doped molybdenum diselenides with defects and abundant active sites for efficient performances of enzymatic biofuel cell and supercapacitor applications. ACS Appl. Mat. Interf. 2019, 11, 18483–18493.

[36] Bollella P, Ludwig R, Gorton L. Cellobiose dehydrogenase: Insights on nanostructuration of electrodes for improved development of biosensors and biofuel cells. Appl. Mat. Today. 2017, 9, 319–332.

[37] Bollella P, Fusco G, Stevar D, Gorton L, Ludwig R, Ma S, Boer H, Koivula A, Tortolini C, Favero G, Antiochia R, Mazzei F. A glucose/oxygen enzymatic fuel cell based on gold nanoparticles modified graphene screen-printed electrode. Proof-of-concept in human saliva. Sens. Actuat. B: Chem. 2018, 256, 921–930.

[38] Leech D, Kavanagh P, Schuhmann W. Enzymatic fuel cells: Recent progress. Electrochim. Acta. 2012, 84, 223–234.

[39] Wey AT, Southcott M, Jemison WD, MacVittie K, Katz E. Electrical circuit model and dynamic analysis of implantable enzymatic biofuel cells operating *in vivo*. Proc. IEEE 2014, 102, 1795–1810.

[40] Majdecka D, Draminska S, Janusek D, Krysinski P, Bilewicz R. A self-powered biosensing device with an integrated hybrid biofuel cell for intermittent monitoring of analytes. Biosens. Bioelectron. 2018, 102, 383–388.

[41] Montes-Cebrián Y, Álvarez-Carulla A, Colomer-Farrarons J, Puig-Vidal M, Miribel-Català PL. Self-powered portable electronic reader for Point-of-Care amperometric measurements. Sensors. 2019, 19, 3715.

[42] Shitanda I, Momiyama M, Watanabe N. Toward wearable Energy storage devices: Paper-based biofuel cells based on a screen-printing array structure. ChemElectroChem. 1017, 4, 2460–2463.

[43] Hou C, Liu A. An integrated enzymatic biofuel cells and supercapacitor for both efficient electric energy conversion and storage. Electrochim. Acta. 2017, 245, 295–300.

[44] Kang ZP, Jiao KL, Cheng J, Peng RY, Jiao SQ, Hu ZQ. A novel three-dimensional carbonized PANI (1600)@CNTs network for enhanced biofuel cell. Biosens. Bioelectron. 2018, 101, 60–65.

[45] Reuillard B, Le Goff A, Agnes C, Holzinger M, Zebda A, Gondran C, Elourzaki K, Cosnier S. High power enzymatic biofuel cell based on naphthoquinone-mediated oxidation of glucose oxidase in a carbon nanotube 3D matrix. Phys. Chem. Chem. Phys. 2013, 15, 4892–4896.i PCCP 15 2013 4892

[46] Bullen RA, Arnot TC, Lakeman JB, Walsh FC. Biofuel cells and their development. Biosens. Bioelectron. 2006, 21, 2015–2045.

[47] Cosnier S, Gross AJ, Giroud F, Holzinger M. Beyond the hype surrounding biofuel cells: What's the future of enzymatic fuel cells? Curr. Opin. Electrochem. 2018, 12, 148–155.

Part III: **Photoelectrochemistry in materials science – selected topics**

Generally, the photoelectrochemistry deals with an influence of electromagnetic irradiation (light illumination) on the electrode properties and on the electrochemical reactions taking place at the electrode. We can distinguish mainly two types of photoelectrochemical behavior:

1. Illumination of the electrode and its photoexcitation results in the subsequent electrochemical reaction of neutral reactants at electrode. Such behavior is characteristic of the semiconductor (SC)–electrolyte interface.
2. Illumination of nonphotoactive electrode and its vicinity results in photoexcitation of reactants molecules and their further reactions. Such behavior is mostly characteristic for photocatalysis and photoelectrocatalysis, but also may be observed at SC–electrolyte interface with some dyes as reactants.

This part of the book we dedicate to the photoelectrochemistry of SCs and its applications in solar energy conversion. In the beginning, we will briefly describe the nature of SCs, unusual properties of SC–electrolyte interface, simple charge transfer reactions, and the behavior of SC under illumination (Chapter 10). Photoelectrochemical cells and their application in solar energy conversion will be discussed in Chapter 11. Going further, we will also characterize some photocatalytic processes at SC particles (Chapter 12).

https://doi.org/10.1515/9783111160986-010

10 Semiconductors electrochemistry and photoelectrochemistry: fundamentals

10.1 Basic characteristics of semiconductors

Semiconductors (SCs) are materials with intermediate electronic conductivity between the conductivity of metals and insulators. Typical SCs are germanium (Ge) and silicon (Si), elements of group IV. SC compounds consist of the elements of other groups, such as (i) $A^{III}B^{V}$ (e.g., GaAs, GaP, InSb) and (ii) $A^{II}B^{VI}$ (e.g., ZnTe, CdTe, CdSe, CdS, ZnO). The compounds containing rare-earth metals and elements of group VI (e.g., TiO_2, WS_2, $MoSe_2$) also have properties of SCs. The efforts are made to synthesize the more complicated compounds such as $ZnSeP_2$ or perovskites.

For the description of SC properties, the band theory of solid state is applied [1–3]. It is beyond the scope of this book to present in detail this theory; however, some information is necessary to understand this chapter. For the solid state, it is assumed that the energy of electron levels of corresponding atoms constitute the bands. Between the bands with energy of the so-called allowed electron state is a band with a forbidden energy region. The most important are the bands called the valence (VB) and the conduction band (CB). The VB is formed with occupied energy levels of valence electrons and energetically is characterized by the highest occupied level of energy E_V. The CB is unoccupied at T = 0 K and can be filled with electrons when additional energy is supplied to solids (heat, electric, or electromagnetic field). This band is characterized by the energy of the lowest unoccupied energy level E_C. In metals, the VB overlaps with the CB. In SCs and insulators (dielectrics), these bands are separated by the band of forbidden energy, characterized by band gap energy E_g

$$E_g = E_C - E_V$$

SCs can be divided into two categories: intrinsic and doped SC. In the intrinsic SCs at $T > 0$, the current carriers are generated as a result of the thermal excitation of electrons from the VB to the CBs. At the same time, the equal number of holes, positively charged carriers, is created in the VB. When another external stimulus (electric field, light) is absent, the equilibrium concentration of electrons n_0 and holes p_0 is established, so in the intrinsic SC: $n_0 = p_0 = n_i$, and $n_0 p_0 = n_i^2$.

Both electrons and holes participate in the current flow; their concentration in room temperature in intrinsic SC is low 10^{10}–10^{11} cm^{-3}. To increase the conductivity, the foreign atoms (impurities) are induced in the SC lattice. These foreign atoms can work as donors or acceptors of electrons. They have very low ionization energy (0.01 eV), and so they are completely ionized at room temperature and become charged. If ionized, the donors of concentration N_D supply electrons to the CB, creating electron conductivity and becoming positively charged N_D^+. In this type of SC (n-SC), the concentration of

https://doi.org/10.1515/9783111160986-011

electrons is much higher in the CB than holes in the VB: $n_0 \gg p_0$ The electric current in bulk of SC is then carried mostly by electrons, and because of that, the electrons are called the "majority carriers" and the holes "minority carriers." When acceptor atoms are induced in the lattice, they withdraw the valence electrons creating holes. In this type of SC, the concentration of positively charged holes in the VB is much higher than the concentration of electrons in CB: $p_0 \gg n_0$. It is the so-called p-type semiconductor (p-SC), where "majority carriers" are holes and "minority carriers" are electrons. The acceptors in the SC lattice are charged negatively, and their concentration $N_A = N_A^-$.

For the doping of Ge or Si, the elements of group V (P, As, Sb) are used as donors and the elements of group III (B, Al, Ga, In) as acceptors are utilized.

There are some parameters resulting from the band theory of solid state, which are important for the description of SC properties. We will briefly describe some of them.

The probability of the occupancy of the state of energy E by electron is given by Fermi–Dirac function:

$$f(E) = \frac{1}{\exp\left(\frac{E - E_F}{k_B T}\right) + 1} \tag{10.1}$$

where E_F is energy of Fermi level, k_B is Boltzman constant, and T is temperature (K).

The occupancy of electron levels and their distribution in an intrinsic SC is shown in Fig. 10.1.

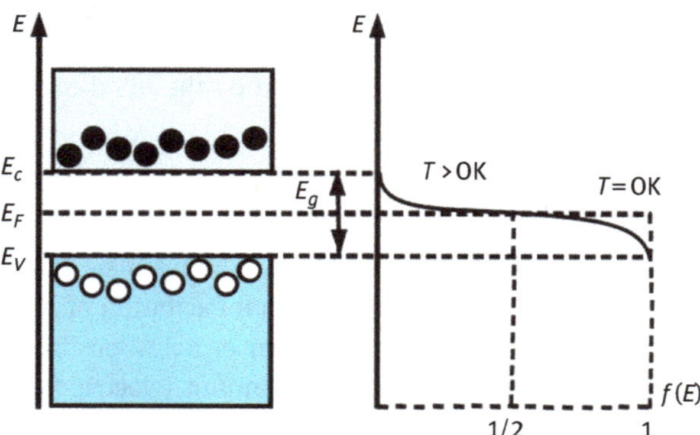

Fig. 10.1: Occupancy of electron levels and Fermi–Dirac function for intrinsic semiconductor in thermal equilibrium at $T > 0$ K. The Fermi energy remains approximately at the center of the bandgap.

As one can see that at T = 0 K, the $f(E) = 1$ if $E < E_F$ and $f(E) = 0$, if $E > E_F$. It means that at T = 0, all electrons are located in the VB. At the higher temperatures, there are as many electrons excited into the CB as there are empty states in the VB, and so E_F remains at $\frac{1}{2}E_g$. When the energy of electron level E is much higher than E_F and

$(E - E_F) \gg k_B T$, the Fermi–Dirac distribution function became Boltzman distribution function:

$$f(E) = \exp\frac{E_F - E}{k_B T} \tag{10.2}$$

Another important term is the function of density electron states in the conduction and VB:

$$D_C(E) = 4\pi\left(\frac{m_e^{ef}}{h^2}\right)^{\frac{3}{2}} \cdot (E - E_c)^{\frac{1}{2}}, \qquad \text{if } E > E_F \tag{10.3a}$$

$$D_V(E) = 4\pi\left(\frac{m_p^{ef}}{h^2}\right)^{\frac{3}{2}} \cdot (E_V - E)^{\frac{1}{2}}, \qquad \text{if } E < E_F \tag{10.3b}$$

m_e^{ef} is the effective mass of an electron at the bottom of the CB, m_p^{ef} is the effective mass of a hole at the top of the VB, and h is Planck's constant.

Based on both functions, Boltzman function and the density state function, the concentration of the charge carriers can be described in the thermal equilibrium by the following equations:

$$n_0 = N_C \exp\left(\frac{E_F - E_c}{k_B T}\right) \tag{10.4a}$$

$$p_0 = N_V \exp\left(\frac{E_V - E_F}{k_B T}\right) \tag{10.4b}$$

where N_C, N_V are the density of electron state in the CB and the VB, respectively.

The term *Fermi level* is very important and is defined as the energy of level for which the probability of occupation by electron is ½. In electrochemistry, the Fermi level is identical with an electrochemical potential of electron (Chapter 1):

$$E_F = \tilde{\mu}_e = \mu_e - F\varphi \tag{10.5}$$

μ_e is a chemical potential of electron, and φ is inner potential in particular phase.

The E_F value depends on the choice of the reference state. In physics, the energy of electron in vacuum is taken as reference and is accepted to be zero.

For intrinsic SC, the Fermi level is located near the center of the forbidden band:

$$E_F = \frac{E_V + E_C}{2} + \frac{k_B T}{2}\ln\frac{N_V}{N_C} \tag{10.6a}$$

For doped SCs, the Fermi level is described by the following expressions:

$$E_F = E_C + k_B T \ln\frac{N_D}{N_c} = E_C + k_B T \ln\frac{n_0}{N_c}, \qquad n_0 \approx N_D, \qquad \text{n–SC} \tag{10.6b}$$

$$E_F = E_V + k_B T \ln\frac{N_V}{N_A} = E_V + k_B T \ln\frac{N_V}{p_0}, \qquad p_0 \approx N_A, \qquad \text{p–SC} \tag{10.6c}$$

where N_D, N_A are the concentrations of donors or acceptors in the bulk of SC.

The scheme of the band energy and the location of energy levels E_F, E_D, E_A for intrinsic, n-SCs, and p-SCs is presented in Fig. 10.2(a–c).

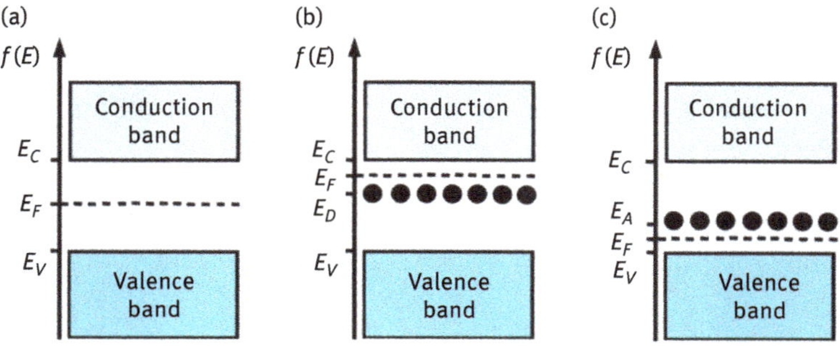

Fig. 10.2: Scheme of the energy bands for the intrinsic (a), n-type (b), and p-type (c) semiconductors. The Fermi energy level E_F and energy levels of donors E_D and acceptors E_A are marked.

10.1.1 Semiconductor under illumination

Two types of bandgap in SCs may be considered: direct and indirect band. According to this, we distinguished SCs with direct and indirect bandgap energy. In the direct band SC, the top of the VB occurs at the same value of the crystal electron momentum k, as the bottom of the CB. In indirect band SC, the top and the bottom of SC bands are shifted and electron momentum is not for them the same. Figure 10.3(a, b) illustrates both types of SC and optical transitions from the valence to CB during illumination.

Let us briefly consider phenomena during the illumination of SC. When the energy of photons $E_{ph} = h\nu < E_g$, the photons may be absorbed only by lattice or free carriers, then the absorption coefficient α is low. Absorption coefficient α is the rate of decay of light incident intensity I_0 and is a characteristic parameter, appearing in Lambert–Beer's law:

$$I(x) = I_0 \exp(-\alpha x) \tag{10.7a}$$

Here, I_0 is the intensity of incident illumination, and $I(x)$ is the illumination intensity at the distance x, at which the absorption is considered.

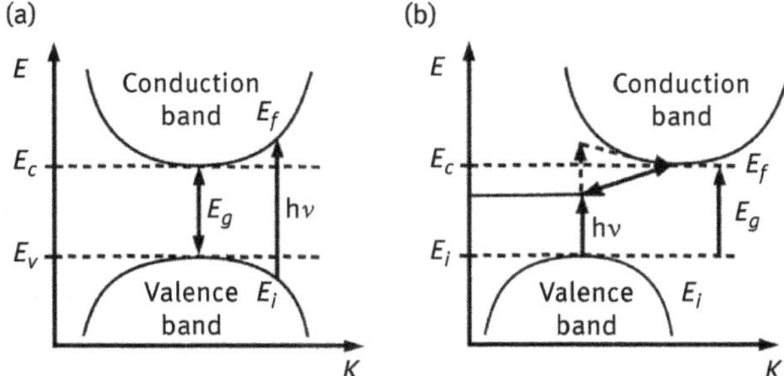

Fig. 10.3: Excitation of electrons across the bandgap by photon absorption. (a) Direct process, no phonon is required. (b) The indirect process necessitates a phonon interaction.

If the photon of energy $(h\nu \approx E_g)$ is absorbed in the "direct" SC, electron–holes pairs are generated: electrons at the bottom of the CB and holes at the top of the VB. The excited electrons and holes are not stable and their annihilation results in photon emission (the so-called radiative recombination). The increase of photon energy $(h\nu > E_g)$ results in an increase in kinetic energy of generated electron and holes. This excess energy is lost to the lattice, such process is called *thermalization*. In direct transition from band to band (interband transition), the absorption coefficient shows a step rise with the photon energy increase, up to 10^4 cm^{-1}. The penetration depth $x = \frac{1}{a}$ is then about 1 µm. Examples of "direct" SCs are GaAs, CdTe, CdSe, InP, ZnO. When the "indirect" SC is irradiated with photon energy close to E_g, the electron transition between the maximum of the VB and the minimum of the CB is not possible. The transitions occur only with the absorption of a photon and simultaneous absorption or emission of a phonon (interaction with the lattice vibrations). The band-to-band transition, in the "indirect" SC, occurs when $(h\nu > E_g)$. Absorption coefficient for "indirect" SC is much smaller than that for "direct" SC; it means that the penetration depth of photon is higher in "indirect" SC. Examples of such SCs are Si, Ge, GaP, and a-SiC.

The absorption coefficient a for direct and indirect transitions is given by

$$a = A\left(h\nu - E_g\right)^n \tag{10.7b}$$

where A is a proportionality coefficient and n = ½ or n = 2 for direct and indirect transition, respectively.

Before going further, we introduce the other term applied in the SC physics: *exciton*. In the interband transitions when electrons and holes are generated they formed the electrically neutral, two particles state electron–hole referred to *exciton*. The excitons result from electron–hole Coulomb attraction. They move through a crystal as a single entity. The lifetime of excitons is limited. They may annihilate in the recombination process or dissociate in the electric field to the "free" electron and holes, moving in the opposite direction.

When the SC is no longer in thermal equilibrium being perturbed by an external stimulus, for example, illumination, the concentration of electrons n and holes p are not equal to the equilibrium value n_0, p_0. In this case, the concentrations of carriers in the bands are as follows:

$$n(x) = n_0 + \Delta n \tag{10.8a}$$

$$p(x) = p_0 + \Delta p \tag{10.8b}$$

Under illumination, the relative changes in the concentration of minority carriers are much higher than the changes in the concentration of majority carriers. These expressions illustrate the relative changes in the concentration:

$$\text{for n-SC:} \quad \frac{\Delta p}{p_0} \gg \frac{\Delta n}{n_0} \qquad \text{and} \qquad \text{for p-SC:} \quad \frac{\Delta n}{n_0} \gg \frac{\Delta p}{p_0}$$

The changes in the concentration of carriers under illumination disturb the equilibrium. The terms named *quasi-Fermi levels* are introduced for the characterization of the energy of charge carriers in non-equilibrium conditions. The quasi-Fermi levels for electrons ${}_{\mathrm{n}}E_F^*$ and holes ${}_{\mathrm{p}}E_F^*$ are not equal as can be seen from the following equations:

$${}_{n}E_F^* = E_c + k_B T \ln \frac{n}{N_C} \tag{10.9a}$$

$${}_{p}E_F^* = E_V + k_B T \ln \frac{N_V}{p} \tag{10.9b}$$

The changes in concentration and energy of electrons cause the split of Fermi level as is shown schematically in Fig. 10.4(a, b).

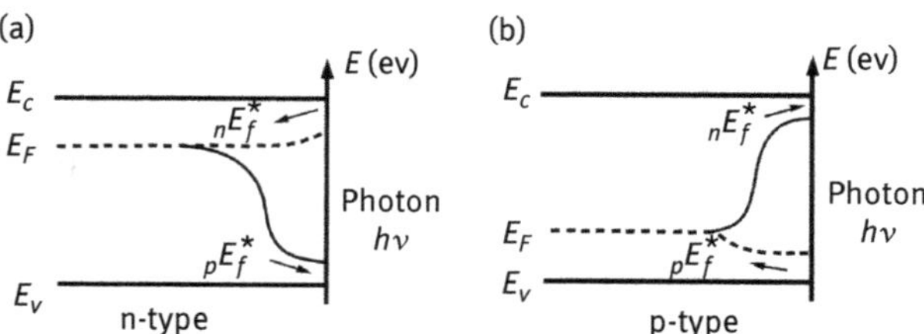

Fig. 10.4: Quasi-Fermi levels for semiconductors under illumination showing significant changes in Fermi energy levels of minority carriers (a) holes (n-SC) and (b) electrons (p-SC).

As you can see, the changes in Fermi energy levels are much higher for minority carriers because of the significant increase in their concentration under illumination. The bands in both Figs. 10.2 and 10.4 are flat. This is a strong simplification and means that the surface of the SC is not charged. In practice, we described, investigated, and used junctions (interfaces) such as Me|SC, n-SC|p-SC, SC|electrolyte. They are charged and vary in their properties from the bulk properties. The behavior and proper energy scheme of the SC–electrolyte interface will be described later.

10.1.2 Recombination

As was mentioned above, when the SC is illuminated with the energy $E \geq E_g$, a certain amount of the electron–hole (e–h$^+$) pairs is generated: electrons in the conduction and holes in the VB, resulting in the perturbation of equilibrium. The recombination is the process of the equilibrium restoration as a result of the annihilation of e–h$^+$ pairs.

The nonequilibrium concentration of electrons n and holes p is determined in the bands by equations:

$$\frac{d(\Delta n)}{dt} = G_n - R_n \tag{10.10a}$$

$$\frac{d(\Delta p)}{dt} = G_p - R_p \tag{10.10b}$$

G_n, G_p are generation rates, and R_n, R_p are recombination rates of electrons and holes, respectively.

The recombination rates are described by equations:

$$R_n = \frac{n - n_0}{\tau_n} \tag{10.11a}$$

$$R_p = \frac{p - p_0}{\tau_p} \tag{10.11b}$$

Equations (10.11a, b) describe the mean lifetime of electrons τ_n generated in the CB and of holes τ_p generated in the VB. If the deviations from equilibrium are small, we can assume that τ_n, τ_p are constant. Under this condition when the external stimulus is switched off, the lifetime of carriers is defined as the time at which the excess concentration of electrons or holes decreases to $\frac{1}{\exp}$ its initial value.

We can distinguish several types of recombination:

(i) the direct band-to-band recombination,
(ii) the recombination with participation of impurities, defects that work as the traps, often called *trap-assisted recombination* or *Shockley–Read–Hall* (SRH) *recombination*,
(iii) the surface recombination.

In direct band-to-band recombination, the generated free electrons go back from CB to VB. It may occur with photon emission (radiative recombination), or the energy excess may be transferred to SC lattice as phonons. The recombination rate for band-to-band radiative recombination is described by equation $R_r = B_r(np - n_0p_0)$. Since during illumination, the concentration of majority carriers is slightly changed, the radiative recombination depends only on the concentration of minority charge carriers. For n-SC, the recombination rate is described by R_p; for p-SC by R_n, eqs. (10.11b) and (10.11a), respectively. Such type of recombination is dominant in narrow bandgap SCs and in wide bandgap SCs with direct transition (CdS, GaAs).

In the second case, SRH recombination occurs via the intermediate bulk local energy level created within the bandgap by dopants and defects in the crystal lattice. Such energy levels are named *traps*. If their energy lays close to the CB they are electron traps; if it lays close to the VB they become the hole traps. This type of recombination is non-radiative; the excess of energy is lost in the form of thermal energy in the lattice vibrations. SRH recombination is dominant in indirect bandgap SCs. Much more about the trap-assisted recombination and the equations describing the recombination rate and lifetime of charge carriers can be found in [3]. The bulk recombination decreases the efficiency of solar energy conversion and should be limited. Among the methods that are applied is the reduction of the SC bulk size by nanostructuring. The surface recombination will be described later in Section 10.5.

10.1.3 Carriers and electric field

When the electric field is not applied to SCs, the concentration of charge carriers is determined by equilibrium values n_0 and p_0 eq. (10.4); the average velocity of the movement of the carriers is zero. In the electric field, the velocity (the so-called drift velocity) of the electrons and holes differs resulting from differences in mobility of electrons μ_n and holes μ_p. The external electric field causes the changes in the energy of carriers and in consequence influences their concentration:

$$n = n_0 \exp \frac{e(\varphi - \varphi_B)}{k_B T} \tag{10.12a}$$

$$p = p_0 \exp \frac{-e(\varphi - \varphi_B)}{k_B T} \tag{10.12b}$$

Here, $\varphi - \varphi_B$ is the potential drop in SC; further in the text it will have the meaning of potential drop in the space charge layer of SC.

The ordered movement of electrons and holes in the electric field is called *drift current* i_{drift} or in electrochemistry – migration current. The other kind of current that flows results from variation in concentration of carriers. It is diffusion current i_{diff} similar to that described for electrochemical kinetics (Chapter 1). Generally, the density of total current j is the sum of drift current density and diffusion current density of both carriers:

$$j = j_n + j_p \tag{10.13a}$$

$$j_n = j_{n,\,\text{drift}} + j_{n,\,\text{diff}} = en\mu_n \in + eD_n \frac{dn}{dx} \tag{10.13b}$$

$$j_p = j_{p,\,\text{drift}} + j_{p,\,\text{diff}} = ep\mu_p \in - eD_p \frac{dp}{dx} \tag{10.13c}$$

where e is elementary charge (e = 1.6 * 10^{-19} C), D_n and D_p are diffusion coefficients of electrons and holes, μ_n, μ_p are mobility of electrons and holes, and $\in$ is the intensity of the electric field. As there is a large difference between the concentration of majority and minority carriers, only one part of eq. (10.13a) is important.

10.2 Semiconductor–electrolyte interface

Before going further, we should correlate the energy scale used by physicists and electrochemists. In thermodynamic electrochemical scale, all oxidation–reduction potentials or equilibrium potentials are referred to the *standard hydrogen electrode* (SHE); its standard electrode potential is accepted to be 0 V ($E^0 = 0$) in an ideal 1 M solution of acid, at any temperature. In practice, the values of potentials are referred to different reference electrodes: normal hydrogen electrode (NHE), standard potential of which is accepted to be 0 V ($E^0 = 0$) in 1 N acid solution, SCE – saturated calomel electrode, silver chloride electrode $Ag|AgCl|Cl^-$, and others.

It is clearly seen that for the description of the SC properties more convenient is to use physical scale, where energy of electron level $E_F = \tilde{\mu}_e$, E_c, E_v and other energy levels are referred to as the value 0 eV in vacuum. If we want to consider the properties of a SC electrode from an electrochemical point of view, we should correlate both scales. Such a correlation is shown in Fig. 10.5. The correlation is based on some experimental and theoretical results pointing out that the value of the Fermi energy level (electrochemical potential of electrons) for the SHE electrode is -4.5 ± 0.1 eV (more exactly -4.44 ± 0.2 eV) [4].

10.2.1 Model of semiconductor–electrolyte interface

When the metal is immersed in the electrolyte solution, the double layer is created spontaneously at the interface, which consists of the Helmholtz layer and the Gouy–Chapman layer (Chapter 1). The concentration of electrons in a metal is very high, ca. 10^{22} cm^{-3}; therefore the charge of accumulated ions at the metal surface is compensated within a few Å in the metal vicinity. In the SC, the electron concentration is much lower, in intrinsic SC is not higher than 10^{11} cm^{-3}, and in doped SC it is not higher than 10^{15}–10^{16} cm^{-3}. Because of that, the charge of electrolyte ions or other charged species existing on/at the

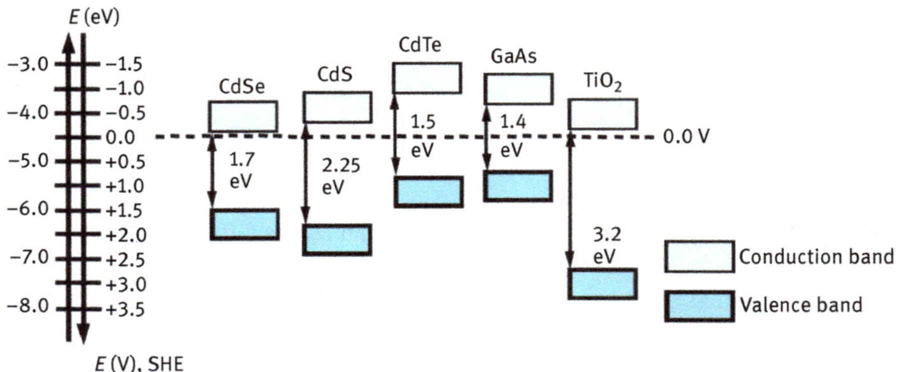

Fig. 10.5: The position of conduction and valence band edges for some semiconductors is shown on the vacuum scale and with respect to the SHE reference.

SC surface is compensated in the vicinity of SC. This charged region in SC is called *space charge layer* and is characterized by the potential drop φ_{SC}, charge Q_{SC}, capacity C_{SC}, and thickness L_{SC}. The existence of the space charge layer in SC results in the differences between the behavior of SC electrodes and metal electrodes. The electric field of the space charge layer leads to the changes in the concentration of electrons, to the changes in their potential energy resulting in the band bending. Since the electron is negatively charged, the bands are bent downward if $\varphi_{SC} > 0$ and upward when $\varphi_{SC} < 0$. The schematic model of SC–electrolyte, together with potential changes, is shown in Fig. 10.6. Note that the x-axis is directed toward the SC bulk and that at the interface SC–electrolyte $x = 0$.

The Galvani potential φ_G of SC–electrolyte interface is described as follows:

$$\begin{aligned}\varphi_G &= \varphi(SC) - \varphi(el) = [\varphi(\infty) - \varphi(0)] + [\varphi(0) - \varphi(-L_H) + \varphi(-L_H) - \varphi(-\infty)] \\ &= -\varphi_{SC} + \varphi_H + \varphi_{G-Ch}\end{aligned} \tag{10.14}$$

φ_{SC}, φ_H, φ_{GCh} are potential drops in the space charge layer of SC, in Helmholtz, and in Gouy–Chapman layers (ionic part of double layer), respectively.

Note that the potential of the ionic part of the double layer and space charge layer have different signs. It results from the assumptions that potential (electric field) decreases to 0 in the bulk of SC and for ionic part of the layer it drops to 0 in the bulk of electrolyte.

The Galvani potential between two conductors cannot be measured, as discussed in Chapter 1. What we can measure is the electrode potential E. In practice, the term "electrode potential" is used to denote the difference between two electrodes, the studied electrode, and the reference electrode. In reality, the electrode potential is the algebraic sum of all individual Galvani potentials of interfaces. If the electrode is in

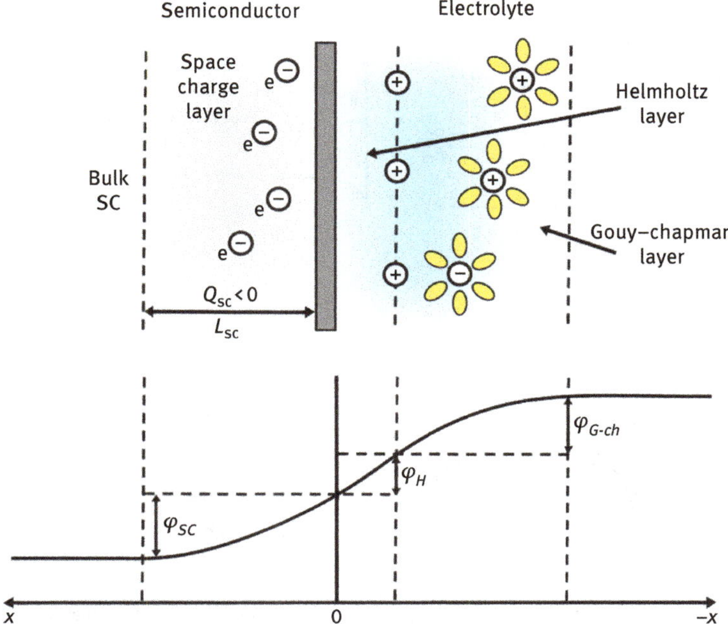

Fig. 10.6: Scheme illustrating the electric double layer at the semiconductor–electrolyte interface. The distribution of potentials within the layer is shown.

equilibrium with the redox system, the electrode potential can be calculated by the Nernst equation with respect to the NHE electrode.

In the part of the text that follows, we will denote the electrode potential as E, and potential applied from an external source as U.

10.2.2 Space charge layer

Three states of the space charge layer can be distinguished: depletion, inversion, and accumulation. They are presented in Fig. 10.7.

Depletion occurs when the charge in the space charge layer derives from the ionized atoms of donors or acceptors. In the case of n-SC, as the donors are positively charged, $Q_{SC} > 0$, the potential drop $\varphi_{SC} < 0$. In p-SC, acceptors are negatively charged then $Q_{SC} < 0$ and the potential drop $\varphi_{SC} > 0$ When in the n-SC, the positive charge Q_{SC} increases and the potential drop φ_{SC} became more negative; the space charge layer is in the inversion state. In this case, the charge is formed by mobile minority carriers – holes. In the p-SC, if the space charge layer is in the inversion state, the charge is also formed by mobile minority carriers, but in this case, electrons. Therefore, Q_{SC} became more negative and φ_{SC} more positive in comparison with the depletion state. The ac-

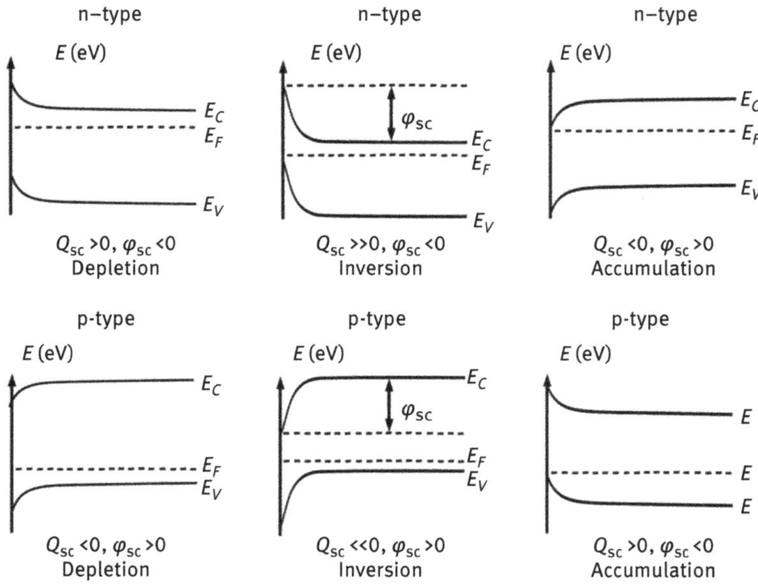

Fig. 10.7: Diagrams for different states of space charge region: depletion, inversion, and accumulation.

cumulation state in the space charge layer exists, when the charge is created by majority carriers. Then for n-SC Q_{SC} is negative, while for p-SC Q_{SC} is positive and so $\varphi_{SC} > 0$, $\varphi_{SC} < 0$ for n-SC and p-SC, respectively.

If $\varphi_{SC} = 0$, the bands are not bent; they are flat to the surface. Such a state can be forced by the external potential. Potential of SC electrode (measured against reference electrode) at which such state is reached is called the *flat band potential* E_{fb}. The positions of E_{fb} for some SCs are shown in Fig. 10.5.

The expressions describing Q_{SC}, φ_{SC}, L_{SC}, C_{SC} are complicated and their forms depend on the type of SC and level of doping. Their derivations are beyond the scope of this book and can be found in [3]. Here we describe the simple case for n-SC when the depletion state exists in the space charge layer. As you will find later, this depletion state is very useful when the conversion of light energy is considered.

In the depletion state of the space charge layer of n-SC, $Q_{SC} > 0$; the charge is determined by the concentration of ionized donors $N_D = N_D^+$. The potential distribution can be obtained in this case by solving eq. (10.15), setting the following boundaries for integration: $\varphi(L_{SC}) = 0$, $\varphi(0) = \varphi_{SC}$, $x = L_{SC}$

$$\frac{d^2\varphi}{dx^2} = -\frac{eN_D}{\epsilon_0 \epsilon_{SC}} \tag{10.15}$$

where ϵ_{SC} is relative dielectric permittivity of SC, and ϵ_0 is dielectric permittivity of vacuum.

The solution of this equation is as follows:

$$\varphi(x) = -\frac{eN_D}{2\epsilon_0\epsilon_{SC}}(x - L_{SC})^2 \tag{10.16}$$

at the interface where $x = 0$,

$$\varphi(x) = \varphi_{SC} = -\left(\frac{eN_D}{2\epsilon_0\epsilon_{SC}}\right)L_{SC}{}^2 \tag{10.16a}$$

$$L_{SC} = \left(\frac{2\epsilon_0\epsilon_{SC}|\varphi_{SC}|}{eN_D}\right)^{1/2} \tag{10.17}$$

It is clearly seen that the thickness of the space charge layer L_{SC} depends on the doping. In the lightly or moderately doped SC, L_{SC} is in the range of 10–1,000 nm.

The other term that is important is the so-called Debye length (screening length):

$$L_D = \left(\frac{\epsilon_0\epsilon_{SC}k_BT}{2e^2N_D}\right)^{1/2}, \qquad (\text{n} - \text{SC, depletion}) \tag{10.18}$$

This term is defined for all SC states and types, by equation:

$$L_D = \left(\frac{\epsilon_0\epsilon_{SC}k_BT}{e^2(n_0 + p_0)}\right)^{1/2} \tag{10.19}$$

and is also applied in the description of Gouy–Chapman layers in electrolyte (Chapter 1).

The differential capacity of the space charge layer can be delivered from the equation $C_{SC} = \frac{dQ_{SC}}{d\varphi_{SC}}$, and for the depletion state of n-SC is given by

$$C_{SC} = \frac{\epsilon_0\epsilon_{SC}}{L_D\sqrt{2}}\left(\left|\frac{\varphi_{SC}}{k_BT}\right| - 1\right)^{1/2} \tag{10.20}$$

Using eq. (10.18), we can write (10.20) in the form:

$$C_{SC}^{-2} = \frac{2}{\epsilon_0^2\epsilon_{SC}^2N_D}\left(|\varphi_{SC}| - \frac{k_BT}{e}\right) \tag{10.21a}$$

or

$$C_{SC}^{-2} = \frac{2}{\epsilon_0^2\epsilon_{SC}^2N_D}\left(E - E_{fb} - \frac{k_BT}{e}\right) \tag{10.21b}$$

where E_{fb}, E are flat band potential and electrode potential, measured against the reference electrode, respectively.

The dependence of C_{SC}^{-2} versus E is linear, and such a plot is called a *Mott Schottky plot* and measurements of C_{SC} as a function of SC electrode potential give us the possibility to determine E_{fb} and N_D. Mostly the impedance spectroscopy is applied for C_{SC} determination (Chapter 2).

The other important parameters are the concentration of electrons and holes in the space charge layer. They vary from equilibrium values in the bulk of SC and depend on the potential drop in the space charge layer:

$$\varphi_{\text{SC}} = \varphi_s - \varphi_b \tag{10.22}$$

Here φ_s is potential on the SC surface and φ_b is potential in the bulk of SC referred to as 0:

$$n = n_0 \exp\left(\frac{e\varphi_{\text{SC}}}{k_B T}\right) \tag{10.23a}$$

$$p = p_0 \exp\left(\frac{-e\varphi_{\text{SC}}}{k_B T}\right) \tag{10.23b}$$

10.2.3 Polarization of semiconductor–electrolyte interface

As was mentioned above, the Galvani potential of SC–electrolyte interface is the sum of three potential drops:

$$\varphi_G = \varphi(\text{SC}) - \varphi(\text{el}) = -\varphi_{\text{SC}} + \varphi_H + \varphi_{G-Ch}$$

In consequence, the differential capacitance of SC–electrolyte interface C also consists of three capacities $C_{\text{SC}}, C_H, C_{G-Ch}$ connected in series:

$$\frac{1}{C} = \frac{1}{C_{\text{SC}}} + \frac{1}{C_H} + \frac{1}{C_{G-Ch}} \tag{10.24}$$

If we compare the thickness of these three regions in the middle concentrated electrolyte, the inequalities $L_{\text{SC}} > L_H$ and $L_{\text{SC}} > L_{G-Ch}$ are fulfilled in the majority of cases. Note that $L_{G-Ch} = \lambda_D$ (Chapter 1). Therefore, the main interfacial potential drop φ_G is realized in the space charge layer of SC, $|\varphi_{\text{SC}}| \gg |\varphi_H|, |\varphi_{G-Ch}|$, and also C_{SC} has the main contribution to capacitance C of the electrical double layer.

When the potential of SC electrode E is changed by externally applied voltage, U, the potential profiles should vary in every part of double layer:

$$U - E = \Delta E = -\Delta\varphi_{\text{SC}} + \Delta\varphi_H + \Delta\varphi_{G-Ch} \tag{10.25}$$

Let us consider the case when $|\Delta\varphi_{\text{SC}}| \gg \Delta\varphi_H, \Delta\ \varphi_{G-Ch}$. It means that the change of SC electrode potential takes place in the vicinity of the space charge layer, so $\Delta E = |\Delta\varphi_{\text{SC}}|$. In this situation, the positions of the band edges at SC surface $E_{C,S}$, $E_{V,S}$ are constant with respect to the reference electrode and electrolyte. The band edges are "pinned" to the SC surface (Fig. 10.8(a)). If the SC is very strongly doped, the Fermi energy levels are close to the conduction or VB. The SC begins to behave as the metal. In this case, we referred to the term “metalized” SC. It results in a situation, $|\Delta\varphi_H| \gg |\Delta\varphi_{SC}$, and means that the changes in electrode potential take place in the Helmholtz layer. In the energy diagram, the energy levels of the surface are shifted with respect to the electrolyte and the bands are “unpinning.” However, with respect to the Fermi level of SC, the band edges $E_{C,S}$, $E_{V,S}$ maintain the same relative position (Fig. 10.8(b)).

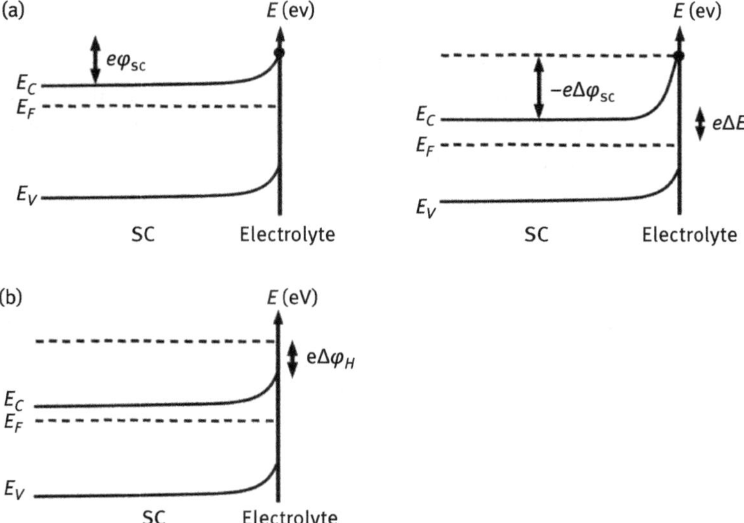

Fig. 10.8: (a) Energy schemes for semiconductor–electrolyte interface showing influence of applied external potential. Scheme illustrates the pining of band edges E_C, E_V at surface. The application of external potential change Fermi energy E_F and φ_{SC}. $\Delta E = \Delta\varphi_{SC}$. (b) Energy scheme for semiconductor–electrolyte interface to which the external potential is applied illustrates the pining of Fermi level and unpinning of E_C, E_V. The polarization changes the position of E_C, E_V and φ_H. $\Delta E = \Delta\varphi_H$.

The second case, in which the changes in electrode potential take place mostly in Helmholtz layer $|\Delta\varphi_H| > |\Delta\varphi_{SC}|$, is when the concentration of the surface state is large. The influence of the surface state on the behavior of the SC–electrolyte interface will be described in Section 10.5.

Much more about features of SC–electrolyte interface can be found in [3, 5, 6].

10.3 Fundamentals of electrochemical reactions on the SC electrode

In the previous sections, the specific features of SC and their influence on SC–electrolyte interface behavior were described. These features such as the bandgap energy, two types of charge carriers, the existence of space charge layer in SC, and the surface states influence the electrode kinetics as well and differ it from electrochemical reaction at the metal electrodes. More about the kinetics of the reactions at SC electrodes can be found in the books [3, 5, 7–9]. Here we will concentrate on the reactions occurring at SC electrodes in the solution containing the redox couple. The properly selected redox systems are used in electrochemical solar cells to create the depletion state in the space charge layer of SC and also to protect the SC electrodes against corrosion and photocorrosion.

Semiconductors in the solution with the redox system

If we immerse SC in the electrolyte containing redox couple, the equilibrium is established. This equilibrium is described by equality of electrochemical potentials of electrons in SC and in redox couple in electrolyte:

$$\tilde{\mu}_e(SC) = \tilde{\mu}_e \ \ (\text{Redox}) \tag{10.26a}$$

Since the electrochemical potential of electrons in solid state is equal to the Fermi energy level $E_F = \tilde{\mu}_e$, the concept of the introduction of the Fermi energy level of redox couple and description of the redox system in terms of occupied and unoccupied energy levels was developed by Gerisher [7]. This model will be described very briefly later. The equilibrium between SC and electrolyte solution with redox couple can be described then by relation:

$$E_F = E_{F,\text{Redox}}, \ \ E_{F,\text{Redox}} = -4.5eV - eE_{\text{Redox}} \tag{10.26b}$$

$$E_{F,\text{Redox}} = E^0_{F,\text{Redox}} + RT\ln\frac{c_{\text{Red}}}{c_{\text{Ox}}} \tag{10.26c}$$

$E_{F,\text{Redox}}$ is the Fermi energy level of the redox couple, and E_{Redox} is the equilibrium potential of the redox couple.

If $E_F > E_{F,\text{Redox}}$, the equilibrium is established by the movement of electrons from n-SC to Ox form of the redox system. As a result, in the space charge layer of SC, the depletion state is formed ($Q_{SC} > 0$). It is also true, when $E_{fb} < E_{\text{Redox}}$.

In p-SC the depletion state is formed, when $E_F < E_{F,\text{Redox}}$ or $E_{fb} > E_{\text{Redox}}$ In this case, $Q_{SC} < 0$. Described situations are illustrated in Fig. 10.9(a, b). To obtain in the space charge layer of SC, the inversion state, or accumulation state, one has to choose redox couple with proper equilibrium potential E_{Redox}.

To create the inversion state in the space charge layer, we have to use the redox couples which fulfill the relations $E_F \gg E_{F,\text{Redox}}$ for n-SC and $E_F \ll E_{F,\text{Redox}}$ for p-Sc, respectively. The accumulation state is obtained when $E_F < E_{F,\text{Redox}}$ for n-SC and $E_F > E_{F,\text{Redox}}$ for p-SC.

Note that the units of $E_F, E_{F,\text{Redox}}$ are eV but the units of E_{fb}, E_{Redox} are V.

Let us now consider the charge transfer processes in the dark taking place at n-SC electrode immersed in the electrolyte with a redox system [8]. The established equilibrium is dynamic. It means that the electrons flow from the CB of n-SC to the oxidized form of redox couple in the electrolyte solution and electrons flow from the reduced

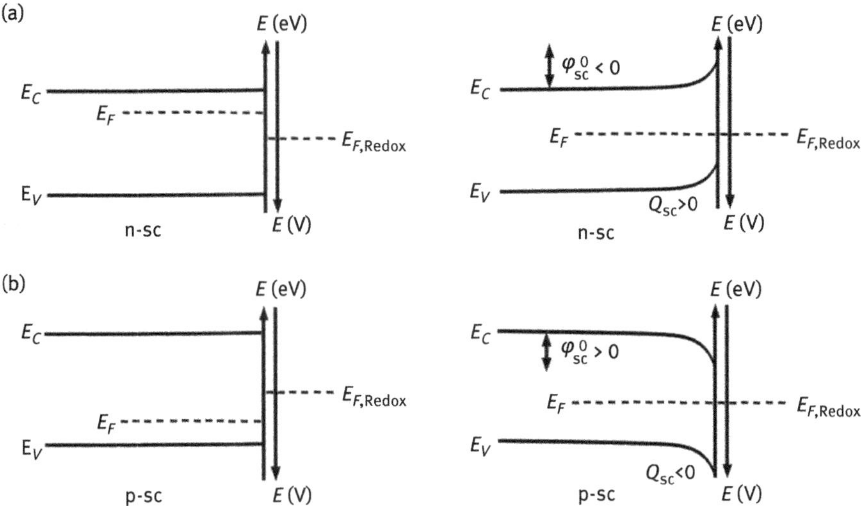

Fig. 10.9: Influence of redox system in electrolyte on the state of space charge layer in semiconductor: (a) n-SC and (b) p-SC. The redox system generates the depletion state in the space charge layer of semiconductors.

form of redox system in solution to the CB of n-SC. These are described by the reactions (i) and (ii):

i) $e(SC) + Ox(el) \xrightarrow{k_f} Red(el)$, forward reaction

ii) $e(Red, el) + v(SC) \xrightarrow{k_f} Ox(el)$, reverse (backward) reaction

Here v represents the empty electron state in the CB of n-SC and is assumed to be constant, independent of potential, and e(SC) and e(Red, el) are electrons from SC and from Red form in the electrolyte solution, respectively.

The rate of both reactions can be expressed by simple equations:

$$\text{rate(i)} = k_f n_{S0} c_{Ox}, \qquad \text{rate(ii)} = k'_r c_{Red}, \qquad k'_r = k_r v$$

n_{S0} is the surface (space charge layer) concentration of electron in the CB of n-SC in equilibrium, $n_{S0} = n\exp(-\frac{e\varphi^0_{SC}}{k_B T})$, and φ^0_{SC} is a potential drop in the space charge layer of SC when equilibrium is established between SC and the redox system in the electrolyte.

The flux of electron in forward and reverse reaction creates the cathodic and anodic current, in CB, respectively. The currents are very low and the net current in equilibrium is zero like for the case of a metal electrode in the electrolyte with a redox system. Considering the above, it is clearly seen that for n-SC only forward reaction (cathodic current) will be a potential-dependent one since the electron concentration in the space charge layer is also potential dependent. Applying the external potential (bias), U, we can write the equation describing the cathodic current density:

$$j_{C,\mathrm{CB}} = -eFk_f c_{\mathrm{Ox}} n_s, \qquad \text{where} \qquad n_s = n\exp\left(-\frac{e(\varphi_{\mathrm{SC}}^0 + U)}{k_B T}\right) \tag{10.27a}$$

$$j_{C,\mathrm{CB}} = -eFk_f c_{\mathrm{Ox}}\ n\exp\left(-\frac{e(\varphi_{SC}^0 + U)}{k_B T}\right) = -eFk_f c_{\mathrm{Ox}} n_{S0}\exp\left(-\frac{eU}{k_B T}\right) \tag{10.27b}$$

$$j_{C,\mathrm{CB}} = j_{0,\mathrm{CB}}\ \exp\left(-\frac{eU}{k_B T}\right) \tag{10.27c}$$

The anodic current density is constant, not U-dependent, and in equilibrium it is

$$j_{A,\mathrm{CB}}^{\mathrm{eq}} = -eFk'_r c_{\mathrm{Red}} \cong j_{C,\mathrm{CB}}^{\mathrm{eq}} = -eFk_f c_{\mathrm{Ox}} n_{S0} = j_{0,\mathrm{CB}} \tag{10.28}$$

Therefore, the current that flows in the dark across the CB of n-SC is described by the equation:

$$J_{\mathrm{CB}} = j_{A,\mathrm{CB}} - J_{C,\mathrm{CB}} = j_{0,\mathrm{CB}}\left(1 - \exp\frac{-eU}{k_B T}\right) \tag{10.29a}$$

$$J_{\mathrm{CB}} = j_{0,\mathrm{CB}}\left(1 - \exp\frac{-FU}{RT}\right) \tag{10.29b}$$

Please note that $j_{A,\mathrm{CB}} \cong j_{A,\mathrm{CB}}^{\mathrm{eq}}$.

Now we consider the p-SC. Here, the VB is practically full, and so the electrons transfer to the solution is constant and independent of potential. However, the electron transfer from the reduced form of redox couple in the solution to the VB depends on the density of holes, empty space at the surface (space charge layer). This density is the potential dependent $p_{S0} = p\exp(-\frac{\varphi_{\mathrm{SC}}^0}{k_B T})$, and so using a similar procedure as above we can obtain the density of current across the VB:

$$J_{\mathrm{VB}} = j_{0,\mathrm{VB}}\left(\exp\frac{eU}{k_B T} - 1\right) \tag{10.30a}$$

$$J_{\mathrm{VB}} = j_{0,\mathrm{VB}}\left(\exp\frac{FU}{RT} - 1\right) \tag{10.30b}$$

The $j_{0,\mathrm{CB}}$ and $j_{0,\mathrm{VB}}$ have a meaning of exchange current density (Chapter 1); however, they do not depend only on rate constant k and c_{Ox}, c_{Red} but also on n_{S0} or p_{S0}, respectively. One can change these concentrations by changing the equilibrium condition. For instance in the case of n-SC, the application of a redox system with $E_{F,\mathrm{Redox}}$ much more negative than E_F of n-SC (E_{Redox} is more positive) causes the φ_{SC}^0 increase and thus a decrease of n_{S0}.

From eqs. (10.29a) and (10.30a), or (10.30b) and (10.30b), it is clearly seen that significant currents can flow only in one direction. This behavior is shown in Fig. 10.10 for both types of SC.

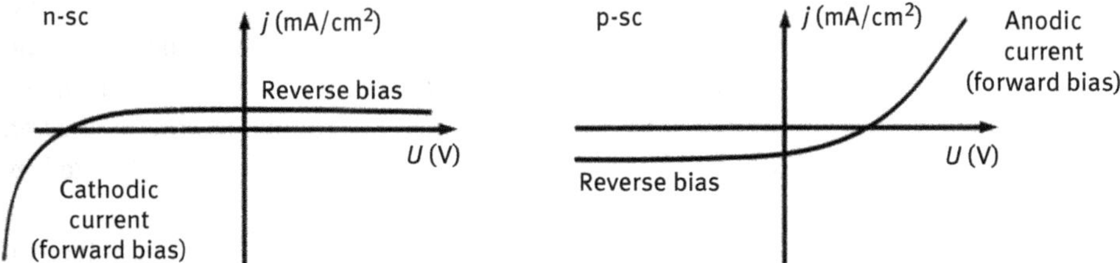

Fig. 10.10: Current density–potential characteristics of semiconductor electrodes in an electrolyte solution, in dark.

It is the current of majority carriers, electrons in n-SC and holes in p-SC. For n-SC, the cathodic current increases exponentially for negative U. For positive U the current is much lower in magnitude and almost independent of potential. For p-SC, the anodic current increases exponentially while U is positive and at negative U current is saturated and potential independent. From above it can be clearly seen that the SC–electrolyte interface has rectifying properties similar to p–n junctions. In SC electronics, when the current magnitude is large, the applied potential is called a *forward bias*, and when the current is small the notion reverse bias is used. The model discussed above is similar to that applied for the description of current–voltage behavior of n–p junction [8].

Assuming that the potential of SC electrode in the equilibrium with electrolyte solution containing the redox couple is determined by E_{Redox}, we can exchange the U in the above equations by the overpotential η. In Chapter 1, η was defined as $\eta = E - E_{\text{eq}}$. In this case, the equivalent description of overvoltage (overpotential) is $\eta = U - E_{\text{Redox}}$ and the current densities can be written:

$$J_{\text{CB}} = j_{0,\text{CB}}\left(1 - \exp\frac{-F|\eta|}{RT}\right) \tag{10.31a}$$

$$J_{\text{VB}} = j_{0,\text{VB}}\left(\exp\frac{F|\eta|}{RT} - 1\right) \tag{10.31b}$$

then Tafel equation and plot (Chapter 1) can be used to determine the $j_{0,\text{CB}}$ and $j_{0,\text{VB}}$.

In the electrochemistry of SCs, the Gerisher model is often used for the description of electrode kinetics with charge transfer mechanism [7]. The model can be called "macroscopic" since the charge transfer is considered between the energy state of electrons of the SC electrode and redox couple in the solution.

The basic concepts of this model are as follows: (i) the electrons can be exchanged only between the same energy levels of SC and redox couple, without any energy losses. (ii) The electron energy levels of redox couple are characterized by empty and occupied states; the empty, unoccupied states come from the oxidized form of redox couple, while occupied states come from reduced form. (iii) For the energy description of the redox couple and its electron states, the terms applied in solid state are used, such as energy of Fermi level $E_{F,\text{Redox}}$ (eq. 10.26b) and function of the density of electron states in the redox system $D_{\text{Redox}}(E)$ and $D_{\text{Ox}}(E)$, $D_{\text{Red}}(E)$ for the unoccupied and occupied energy electron levels, respectively. Due to the thermal fluctuation of the solvation shell, the occupation of these energy states in the redox electrolyte is given by Boltzman distribution function $f(E)$ (eq. 10.2). In this equation E_F is exchanged by $E_{F,\text{Redox}}$. Note that E means here the energy.

Density of states D_{Redox} is the sum of unoccupied and occupied states (electron energy levels):

$$D_{\text{Redox}}(E) = D_{\text{Ox}}(E) + D_{\text{Red}}(E)$$

they are equivalent to the density of state function in SC in CB and VB (eqs. 10.3a, b). For the redox system,

$$D_{\text{Ox}}(E) = c_{\text{Ox}} W_{\text{Ox}}(E) \tag{10.32a}$$

$$D_{\text{Red}}(E) = c_{\text{Red}} W_{\text{Red}}(E) \tag{10.32b}$$

$$W_{\text{Ox}}(E) = W^0 \exp\left[\frac{\left(E - E^0_{F,\text{Redox}} + \lambda\right)^2}{4k_B T\lambda}\right] \tag{10.32c}$$

$$W_{\text{Red}}(E) = W^0 \exp\left[\frac{\left(E - E^0_{F,\text{Redox}} - \lambda\right)^2}{4k_B T\lambda}\right] \tag{10.32d}$$

c_{Ox}, c_{Red} are concentration of oxidized and reduced forms of redox system, $E^0_{F,\text{Redox}}$ is the energy of standard Fermi level of redox system (eV), corresponding to standard potential E^0_{Redox} (V), and λ is the reorientation energy (Chapter 1, Marcus theory).

Above equations can be illustrated by Fig. 10.11(a, b), where the dependence of electron energies of a redox system versus density of state is shown for different concentration of Ox and Red forms.

Applying these equations and remembering that the flux of charge carriers electrons and holes depend not only on the redox system but also on the concentration of carriers and their distribution in SC, the equations for currents were delivered for SC electrode with redox system and under polarization. It is beyond this book; the interested reader will find much more about Gerischer's model and its application in [5, 7]. Here, using this model, we will only illustrate the correlation between the energy state of the redox system and SC when the electrode is in the equilibrium and polarized by an external potential source. Figure 10.12 illustrates the equilibrium established between n-SC electrode and the redox system in the electrolyte. The Fermi energy E_F and $E_{F,\text{Redox}}$ are equal and lay near the CB. The electrons transfer will occur

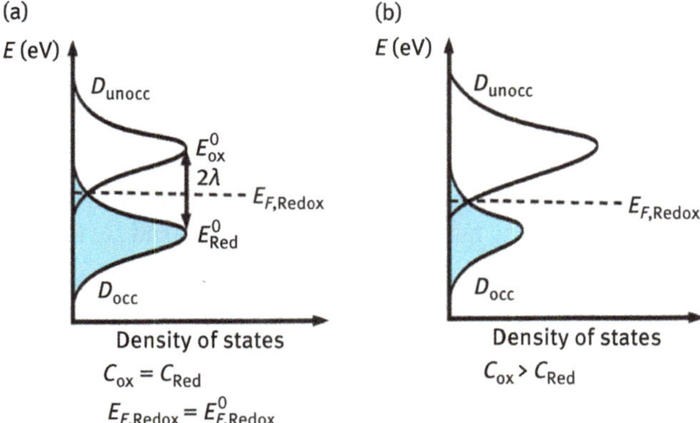

Fig. 10.11: Diagram showing the distribution of the electron energy levels of the oxidized and reduced forms of a redox couple: $c_{Ox} = c_{Red}$ (a) and $c_{Ox} > c_{Red}$ (b).

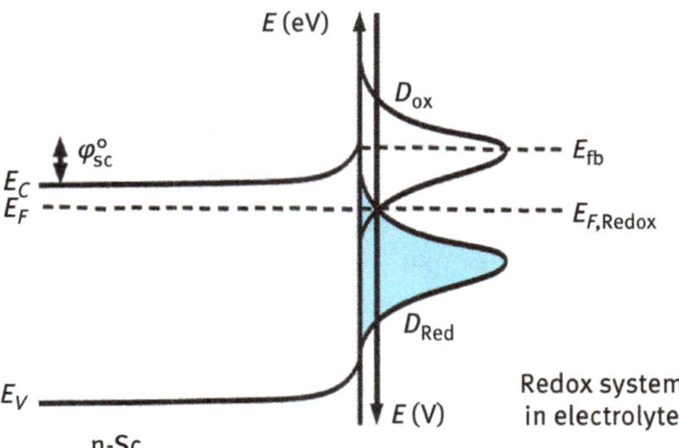

Fig. 10.12: Semiconductor–electrolyte interface at equilibrium showing overlap of the energy levels of Ox form of redox couple with conduction band. The space-charge layer is in a depletion state. The exchange current that flows between the conduction band and Ox form in the electrolyte is very small.

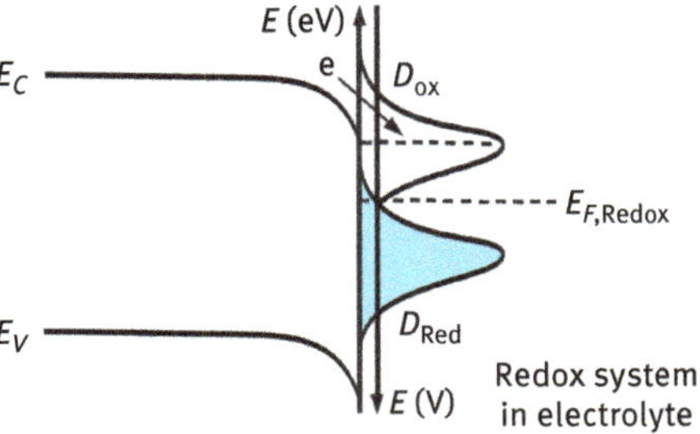

Fig. 10.13: Semiconductor electrode under cathodic polarization. Cathodic polarization generated the accumulation state in the space-charge layer rich in electrons. In this situation current that flows between the conduction band and Ox form in the electrolyte is large.

between the CB and oxidize form of redox couple. Additionally the position of flat band potential and φ^0_{SC} are marked. The exchange current will be very low since the concentration of electrons in the space charge layer of SC (depletion state) is low and energy levels of the oxidize states are empty. Figure 10.13 presents the changes in this scheme caused by the cathodic polarization. The cathodic polarization was used to change the state of the space charge layer from depletion to accumulation. In this case, the cathodic current (flow of electrons between the CB and oxidized form) will be large since the concentration of electrons is potential dependent.

10.4 Photoeffects at/in semiconductor electrode in electrolyte

It was mentioned above that when n-SC is immersed in electrolyte containing redox couple and Fermi energy E_F of SC is larger than Fermi energy of the redox system in electrolyte solution $E_{F,\text{Redox}}$ $(E_F > E_{F,\text{Redox}})$, the equilibrium is established and the depletion state is formed in the space charge layer of SC. Now we will illuminate n-SC, characterized by absorption light coefficient α, with the light energy $h\nu \geq E_g$. During the illumination, the pairs e–h$^+$ are generated and may be separated by the electric field of the space charge layer in the depletion state. The majority carriers (electrons) are directed to the bulk of SC and when the electrical circuit is closed to the opposite electrode. Minority carriers (holes) move to the interface and take place in the oxidation reaction of reduced species of redox couple: $\text{Red} + \text{h}^+ \rightarrow \text{Ox}$. This reaction is equivalent to the reaction of electrons of reduced forms with holes in SC. The changes in the energy of electron levels and in the potential of the space charge layer under illumination of n-SC surface are shown in Fig. 10.14.

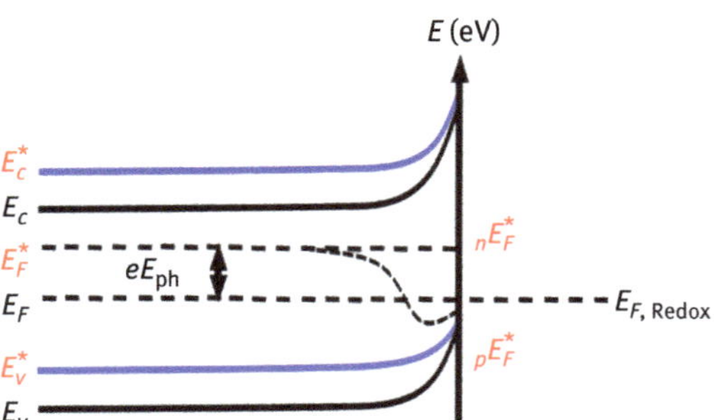

Fig. 10.14: Energy scheme of semiconductor–electrolyte interface, in dark and under illumination (red symbols). The conduction and valence band edges are pinned. During illumination the energy of Fermi level, E_F, is changed together with E_C, E_V in bulk of SC.

Since during illumination concentration of charge carriers increases, E_F, φ_{SC} are changed (Sections 10.1.1 and 2). The E_F or φ_{SC} changes are measured to the reference electrode as the open circuit potential under illumination:

$$E_{ph}^{OC} = E_{d}^{OC} + E_{ph} \tag{10.33}$$

E_d^{OC} is the OCP (open circuit potential) of electrode measured in the dark, and E_{ph} is the photopotential. If in the dark, the equilibrium is established between SC and redox couple present in electrolyte, the E_d^{OC} is equal to the equilibrium potential E_{Redox}. The photopotential is described by equation:

$$E_{ph} = -\frac{k_B T}{e}\ln(1 + bI_0) \tag{10.34}$$

b is parameter dependent on diffusion coefficient, diffusion length, the concentration of carriers, and absorption coefficient. The delivery of (10.34) equation and the equation determining b parameter can be found in the book [3]. Equation (10.34) points out that $E_{ph} < 0$ since b and $I_0 > 0$. If the depletion layer exists in SC and there is no surface recombination, the measured E_{ph}^{OC} is always more negative than E_d^{OC}. Equation (10.34) also shows the influence of the intensity illumination I_0 on E_{ph}^{OC}; at low intensities, this dependence should be linear, and at higher intensities it should be logarithmic.

During illumination, when the SC electrode is connected through an external circuit with another electrode in electrolyte solution, we can measure also changes in the current:

$$j_{light} = j_d + j_{ph} \tag{10.35}$$

j_{light}, j_d, j_{ph} are the current densities: under illumination j_{light}, in the dark j_d, and photocurrent j_{ph}.

To derive the expression for photocurrent, we should consider the depth of light penetration $x = \alpha^{-1}$. Two cases are shown in Fig. 10.15(a, b) for n-SC.

In the case, (i) $\alpha^{-1} < L_{SC}$, the light is absorbed in the narrow vicinity of the space charge layer. Photogenerated pair e–h$^+$ is separated and carriers moved in the electric field. The flux of holes to the interface SC–electrolyte with the redox system is constant and maximal. It means that the changes in φ_{SC} do not influence the photocurrent and the photocurrent should be independent of the potential. In the other case, (ii) $\alpha^{-1} > L_D$, where L_D is Debye length;

$$L_D = L_{SC} + L_{diff}, \quad L_{diff} \text{ is the diffusion length,}$$

the carriers are generated in all vicinity of deep light absorption. The direction of the electric field at the interface of SC–electrolyte is such that the minority carriers generated in the space charge layer move to the interface, participating in the reaction with

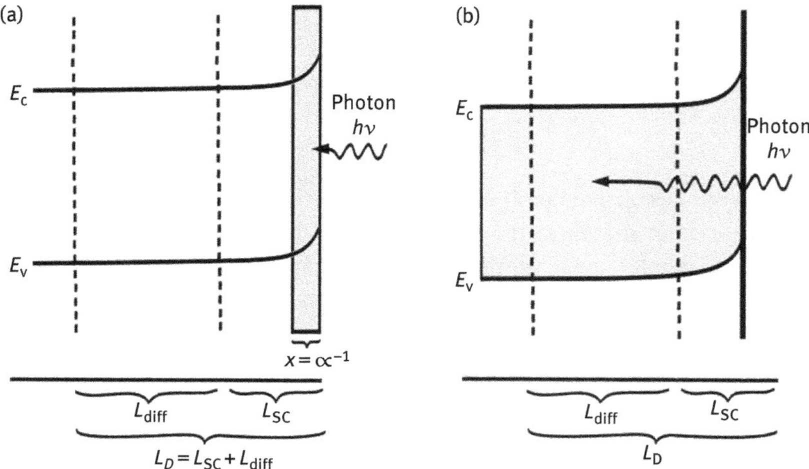

Fig. 10.15: Correlation between the depth of light absorption x, the thickness of the space-charge layer, L_{SC}, and diffusion length L_{diff}. In case (a) $x < L_{SC}$; in case (b) $x > L_D$. Adapted from [3].

a redox couple. The minority carriers generated beyond the depletion layer L_{SC}, but in the vicinity of L_D, move in the direction of interface by diffusion, and those that were generated even deeper than L_D recombine. The characteristic length L_{diff} (in short cut L_d) is equal to L_p for n-SC, or L_n for p-SC and is defined by the equations:

$$L_p = \sqrt{D_p \tau_p} = \sqrt{k_B T \mu_p \tau_p} \tag{10.36}$$

$$L_n = \sqrt{D_n \tau_n} = \sqrt{k_B T \mu_n \tau_n} \tag{10.37}$$

where D_p, D_n are diffusion coefficient of holes and electrons; τ_p, τ_n are lifetime of holes and electrons; and μ_p, μ_n are mobility of holes and electrons, respectively.

There are few models applied for the description of the photocurrent generated at the SC–electrolyte interface. We will apply the Gärtner model introduced first for the description of photocurrents generated under the illumination of SC–metal junction [9, 10]. In this model, the flux of minority carriers, generated under illumination, to the interface (junction) is the sum of two fluxes, one in the depletion region of SC, j_{SC}, and the other in the diffusion zone, j_{diff}, where the minority carriers near the depletion layer diffuse to this zone before the recombination. Then we can write an equation for the density of photocurrent:

$$j_{ph} = j_{SC} + j_{diff} \tag{10.38}$$

The photocurrent generated in the depletion state of the space charge layer is described by the equation:

$$j_{SC} = eI_0(1 - \exp^{-\alpha L_{SC}}) \tag{10.39}$$

The photocurrent originated from the diffusion is given for n-SC (minority carriers – holes) by the expression:

$$j_{\text{diff}} = -eI_0\left(\frac{\alpha L_p}{1+\alpha L_p}\exp^{-\alpha L_{SC}}\right) \tag{10.40}$$

The photocurrent (eq. 10.38) generated at n-SC–electrolyte interface:

$$j_{\text{ph}} = -eI_0\left(1 - \frac{\exp^{-\alpha L_{SC}}}{1+\alpha L_p}\right) \tag{10.41}$$

for p-SC, L_n should be used in eqs. (10.40 and 10.41).

The photocurrent depends upon L_{SC}. Taking into account the dependence of L_{SC} on φ_{SC} (eq. 10.17) and the relation between the potential drop in the space charge layer of SC and the electrode potentials $\varphi_{SC} = E - E_{\text{fb}}$, we obtain the relation between the photocurrent and n-SC electrode potential:

$$j_{\text{ph}} = -eI_0\left[1 - \frac{\exp^{-\alpha\left[C(E-E_{\text{fb}})^{1/2}\right]}}{1+\alpha L_p}\right] \tag{10.41}$$

where $$C = \left(\frac{2\varepsilon_0\varepsilon_{SC}}{eN_D}\right)^{1/2}$$

The total current that flows across SC electrode under illumination is the sum of photocurrent and the current of majority carriers. The latter is described by eq. (10.29b, 10.30b) since the changes of minority carriers concentration are very small.

The expressions are as follows:

$$j = j_{\text{ph}} + j_{0,\text{CB}}\left(1 - \exp\frac{-FU}{RT}\right), \qquad \text{n–SC} \tag{10.42a}$$

$$j = j_{\text{ph}} + j_{0,\text{VB}}\left(\exp\frac{FU}{RT} - 1\right), \qquad \text{p–SC} \tag{10.42b}$$

The photogenerated positive (anodic) current j_{ph} results from the flux of generated holes and dominates; when the n-SC electrode is reverse biased (polarized), at forward bias a small cathodic current is observed. For p-SC the photocurrent would be negative (cathodic), dominating at reverse bias and at the forward bias a small anodic current would be observed. The shape of the current–potential curve is shown in Fig. 10.16.

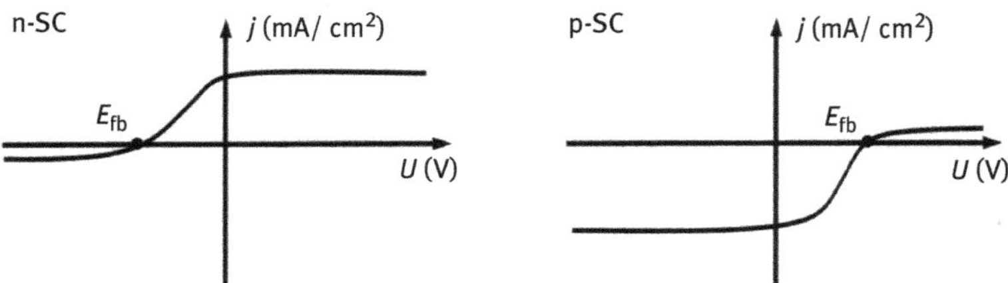

Fig. 10.16: Current density–potential characteristics for semiconductor electrode under illumination. Note that we used symbols E_{fb} and U_{fb} interchangeably; they have the same meaning and value.

In photoelectrochemistry the term "quantum yield" $\varnothing$ is applied. $\varnothing$ is defined as the ratio of photocurrent j_{ph} to the intensity of light I_0 illuminated the SC electrode:

$$\varnothing = \frac{j_{ph}}{I_0} \tag{10.43}$$

We can introduce this notation in eq. (10.41) and rewrite the expression in logarithmic form [10]:

$$-\ln(1-\varnothing) = \alpha C(E - E_{fb})^{1/2} + \ln(1 + \alpha L_d) \tag{10.44}$$

The equation pointed out that the graph $-\ln(1-\varnothing)$ versus $(E-E_{fb})^{1/2}$ should be linear and the intercept of the plot on the potential axis when $-\ln(1-\varnothing)=0$ gives the value of E_{fb}. The slope of the graph gives the value of αC, where $C=\left(\frac{2\varepsilon_0\varepsilon_{SC}}{eN_X}\right)^{1/2}$, N_X is N_D or N_A depending on the type of doping. If α is known from other measurements, the concentration of dopants can be determined. If the concentration of dopant is known from capacitance measurement (Section 10.2), we can determine the absorption coefficient. For such purpose, the SC–electrolyte interface should be illuminated with monochromatic light. Upon such measurements, using the light of different wavelengths λ, we can determine the dependence of α on λ. From the intercept on $-\ln(1-\varnothing)$ (y-axis), we can obtain a magnitude of $\ln(1+\alpha L_d)$ and calculate the diffusion length of minority carriers, L_d, and by using eqs. (10.36) and (10.37) the lifetime of carriers can be assessed.

The photocurrent measurement and application of the Gärtner model give us an opportunity to determine many important parameters. The main weakness of the Gärtner model is that it neglects the surface recombination.

10.5 Influence of surface states

So far, we have assumed that the free charge carriers are directly transferred from the conduction or VB to the redox couple in the electrolyte and that all the changes in the potential at the interface SC|electrolyte occur in the space charge layer of SC.

However, in these processes at the interface, the surface state can play an important role. Two types of surface states can be distinguished: intrinsic and extrinsic surface states. The intrinsic surface states are due to the SC solid-state nature and result from discontinuity and termination of crystal structure at the surface (Tamm state). The other intrinsic states (Shockley) arise as a result of the covalent bond breaking, the creation of dandling electron bonds and radical states. The extrinsic surface states issue from surface defects, adsorbates, oxide layer, and interface formation. In the case of the SC|electrolyte interface, the ions, reaction products, oxides, and other surface species may be responsible for such states. The distribution of surface states may be described usually by three models, two of them consider the fluctuation of distribution of energy (Gaussian model, exponential model). We will consider the simplest, monoenergetic model. In this model, the surface states are described by one localized energy level, E_{ss}, of electrons, which can lose or capture one electron. The energy level lay in the bandgap energy (Fig. 10.17) and can be filled with electron (donor level) or be empty (acceptor level).

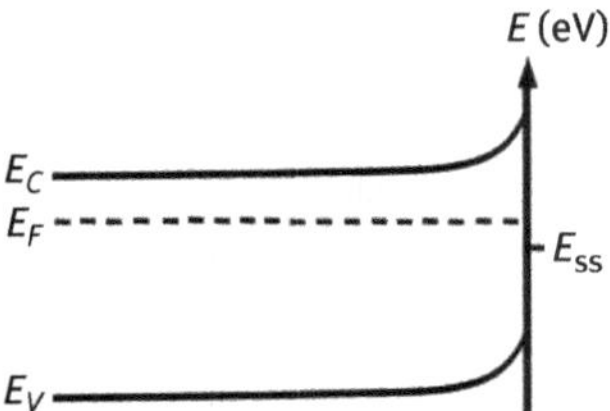

Fig. 10.17: Simplified energetic model of surface states characterized by one localized energy level E_{SS} of electrons.

The surface states are characterized by the density of surface states N_{SS}, the density of charge in the surface states Q_{ss}, and Fermi distribution function f_{ss}.The density charge is given by

$$Q_{ss} = eN_{ss}f_{ss} = \frac{N_{ss}}{1 + \exp\left(\frac{E_{SS} - E_F}{k_B T}\right)} \tag{10.45}$$

In the presence of surface states, the total charge of SC part of the double layer is the sum $= Q_{SC} + Q_{SS}$, so the resultant capacitance of SC part of the interface is equal to the two capacitors connected in parallel $C = C_{SC} + C_{SS}$. The scheme of the equivalent electrical circuit describing the SC|electrolyte capacitance is shown in Fig. 10.18.

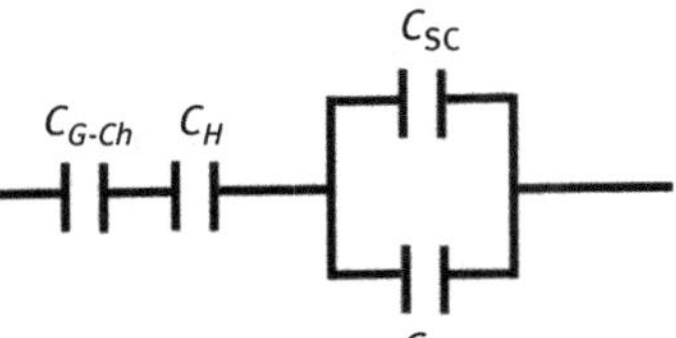

Fig. 10.18: Scheme of equivalent electrical circuit showing semiconductor–electrolyte capacitance.

It was mentioned earlier that a large amount of surface states change the potential distribution at SC–electrolyte interface and can cause the "metallization" of SC electrode ($|\Delta\varphi_H| \gg |\Delta\varphi_{SC}|$). The relation points out that any changes of the potential of SC electrode occur in the Helmholtz layer. To observe such changes in the potential distribution at SC | electrolyte interface, the density of surface states should be larger than 10^{13} cm^{-2} [3, 4]. Except for the changes in the potential distribution, the surface states can affect the kinetics of electrode reactions, changing the pathway of reaction. They can participate in the recombination and charge carriers trapping, decreasing the photocurrent and in consequence the efficiency of conversion of solar energy in photelectrochemical cells (PEC). In Section 10.4, the Gärtner model, which neglects the surface and space charge recombination, was applied for photocurrents determination. Many efforts have been made to improve the Gärtner model; here we will briefly describe Peter's model [11]. This model is presented in Fig. 10.19.

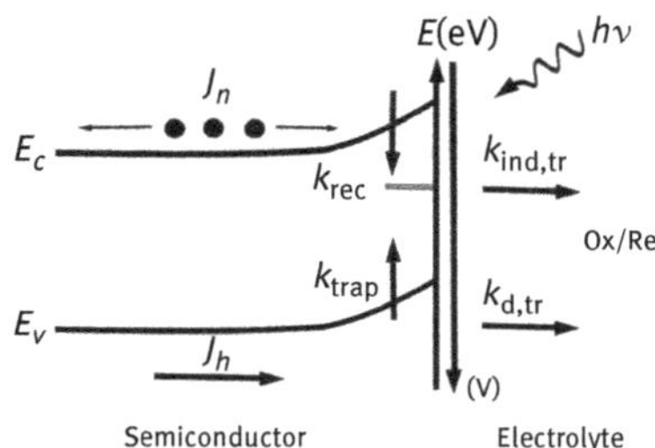

Fig. 10.19: Scheme illustrating the involvement of surface states in recombination, trapping and direct and indirect charge transfers. Semiconductor electrode under illumination. Adapted from [11].

Let us consider the n-SC electrode in an electrolyte solution with redox couple, which causes the formation of the depletion state in the space charge layer of SC. SC has the surface states with energy levels in the vicinity of bandgap energy. We illuminate the electrode (such electrode works as photoanode in the PEC) and generate charge carriers: electron and holes. The flux of minority carriers (holes) photogenerated within Debye's length, L_D, ($L_D = L_{SC} + L_{diff}$), arrives at the SC–electrolyte interface (more detailed description you can find in Section 10.4). Furthermore, there are few steps that should be considered: (1) direct transfer of photogenerated holes from VB to Red form of a redox couple, and (2) indirect transfer of photogenerated holes from the VB to Red form of a redox couple via the surface states, which in this case work as traps. The photogenerated holes can also recombine via the surface states with electrons from the CB. In this case, the surface states play the role of recombination centers. All these processes: charge transfer, trapping, and recombination occur with different rates, characterized by rate constants: $k_{d,\mathrm{tr}}$, $k_{\mathrm{ind,tr}}$, k_{trap}, k_{rec} and compete with each other. Let us consider two limiting cases. In the first case, all photogenerated holes that arrive at the surface participate in direct charge transfer between the VB and reduced species in the solution. The rate of direct charge transfer is characterized by the rate constant $k_{d,\mathrm{tr}}$ and the photocurrent is determined by flux of holes $\frac{dp}{dt} = G_p$ (generation rate of holes) and expressed by the Gärtner equation. In the second case, considered by Peter [11], all holes that arrive at the surface are rapidly trapped by surface states and further participate in the indirect charge transfer or in the recombination. The flux of holes across the interface is then given by

$$\frac{dp_s}{dt} = G_p - k_{\text{ind, tr}} p_s - k_{\text{rec}} p_s \tag{10.46a}$$

where p_s is the density of holes at surface equal to the density of trapped holes, $k_{\text{ind, tr}}$ is rate constant of indirect charge transfer from traps (surface states) to reduced form in solution, $k_{\text{ind, tr}} = k_{\text{trap}} c_{\text{Red}}$, k_{rec} is the recombination rate constant, and k_{trap} is rate constant of trapping.

Under the condition of steady-state illumination, $\frac{dp_s}{dt} = 0$, taking into account eq. (10.46a)

$$p_s = \frac{G_p}{k_{\text{ind, tr}} + k_{\text{rec}}} \tag{10.46b}$$

$$j_{\text{ph}} = e k_{\text{ind, tr}} p_s = e G_p \frac{k_{\text{ind, tr}}}{k_{\text{ind, tr}} + k_{\text{rec}}} \tag{10.46c}$$

This equation points out that trapping and recombination at the surface states decrease the photocurrent.

The above models do not consider the recombination, which can occur in the vicinity of the space charge layer of SC as a result of dopants and lattice defects.

Taking into account all recombination processes (bulk and surface) it is clearly seen how important is (i) the determination of parameters characterizing the minority charge carriers, such as the lifetime, mobility, and the diffusion length, and (ii) the determination of kinetic rate constants of recombination and charge transfer processes.

Some methods were lately developed: the intensity-modulated photocurrent spectroscopy and open-circuit voltage-decay, which provide some insights into the surface carrier dynamics. Interested readers will find more on this topic in references [12, 13]. In the last years, many efforts have been made to avoid recombination processes. Few types of approaches were developed: (i) nanostructuring of SC material on a scale comparable, or even smaller than, the width of space charge layer, (ii) application of cocatalyst to remove the kinetic limitations, and (iii) optimization of deposition procedure and surface treatment (etching, thermal treatment). However, the electrochemical treatment can, on the one hand, decrease the surface recombination but, on the other hand, can facilitate the photocorrosion. The photocorrosion is another process that should be eliminated in photoelectrochemical cells. The description of the photocorrosion processes and methods of protection against them will be addressed in the following section.

10.6 Corrosion, photocorrosion, and stability

Besides the trapping and recombination of minority carriers via surface state, the stability of the SC electrode is a crucial problem in photoelectrochemical cells. The SCs like the metals can go spontaneous decomposition (oxidation, dissolution) as a result of chemical, electrochemical reactions, that is, corrosion. The approach used for the description of metal corrosion can be also applied to the SC materials (Chapter 3). The main difference in corrosion of metals and SCs results from the possibility of participation of two charge carriers: electron and holes in corrosion of SC [3]. Let us consider the decomposition of MX SC. We can write the reactions of decomposition with participation of electrons from the CB (cathodic reaction) and holes from the VB (anodic reaction) as follows:

$$MX + ne + \text{solv} \rightarrow M + X^{n-}_{\text{solv}} \tag{10.47a}$$

$$MX + nh^+ + \text{solv} \rightarrow M^{n+}_{\text{solv}} + X \tag{10.47b}$$

In the corrosion, the conjugated reactions with those above are: $H_2 - 2\,e \rightarrow 2\ H^+$ (with 10.47a) and $2\ H^+ + 2\,e \rightarrow H_2$ (with 10.47b).

Based on the thermodynamic data and calculation of Gibbs free energy changes of reactions (10.47a, b), we can determine the standard equilibrium potential of decomposition of MX with participation of electrons $E^0_{\text{dec},n}$ and holes $E^0_{\text{dec},p}$. These reactions can be also described in energy scale by electrochemical potential (Fermi energy) $E_{F,\text{dec},n}$ and $E_{F,\text{dec},p}$, respectively.

The thermodynamic criteria point out that corrosion or the cathodic decomposition of the SC compound occurs, with the participation of electron from the CB, if the potential of the electrode will be lower than $E^0_{\text{dec},n}$. In the case of decomposition with holes from the VB, the corrosion or anodic decomposition is expected, if the potential of the electrode is higher than $E^0_{\text{dec},p}$:

$$E < E^0_{\text{dec},n}, \qquad E > E^0_{\text{dec},p}$$

Schematically, the thermodynamic stability related to the decomposition of SC, in the condition of flat band potential, is shown in Fig. 10.20.

In this figure, four cases are presented, the SC is (a) stable with respect to cathodic and anodic decomposition, (b) stable to cathodic decomposition, (c) stable to anodic decomposition, and (d) unstable and can undergo cathodic and anodic decomposition.

The decomposition of material significantly increases, if the SC electrode is illuminated. In Fig. 10.21, the influence of illumination on SC corrosion rate is presented.

It is clearly seen that illumination results in an increase in corrosion current density, $(\ln j_{\text{corr},d}, \ln j_{\text{corr},l})$ – means corrosion current density in dark and under illumination.

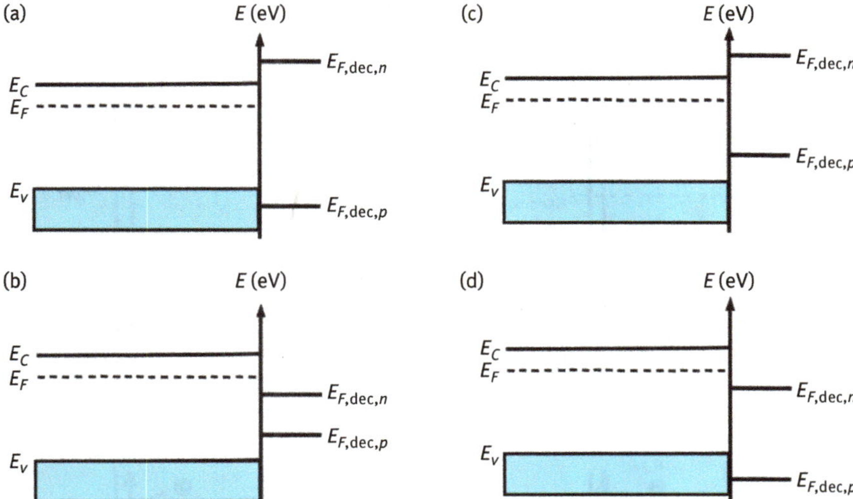

Fig. 10.20: Correlation between the position of band energies E_C, E_V and decomposition potentials in energetic scale $E_{F,\text{dec},n}$, $E_{F,\text{dec},p}$ for n-type semiconductor: (a) electrode is resistant to decomposition by electrons and holes, (b) electrode is not resistant to decomposition by electrons and holes, (c) electrode is not resistant to decomposition by holes, and (d) electrode is not resistant to decomposition by electrons.

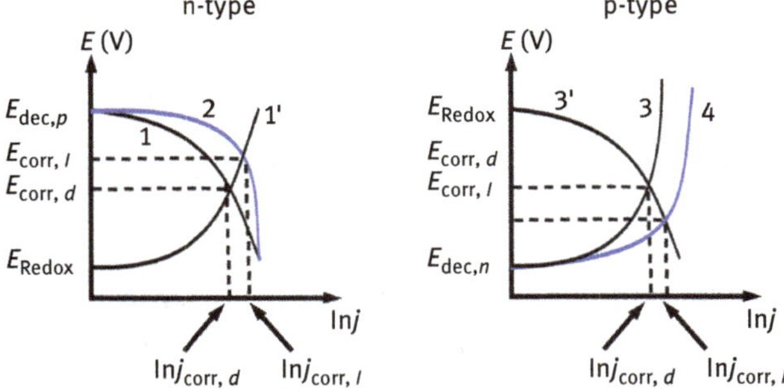

Fig. 10.21: Influence of illumination on the corrosion of n-type and p-type of SC. Curves 1 and 3 show corrosion of both semiconductors in dark. Curves 2 and 4 show corrosion of semiconductor under illumination. Note that corrosion current density under illumination is large than corrosion current density in dark. Adapted from [3].

Let us consider the few cases concerning the n- and p-SCs immersed in the electrolyte solution with the redox system. The redox system was chosen in such a way that the depletion layer was formed in the space charge layer of SC. Such a situation is specially created in SC electrodes working as photoanodes and photocathodes in the photoelectrochemical cells. Note that the behavior of SC electrodes in redox electrolyte and under illumination was described Sections 10.3 and 10.4.

In Fig. 10.22 four cases are presented for n- and p-SC.

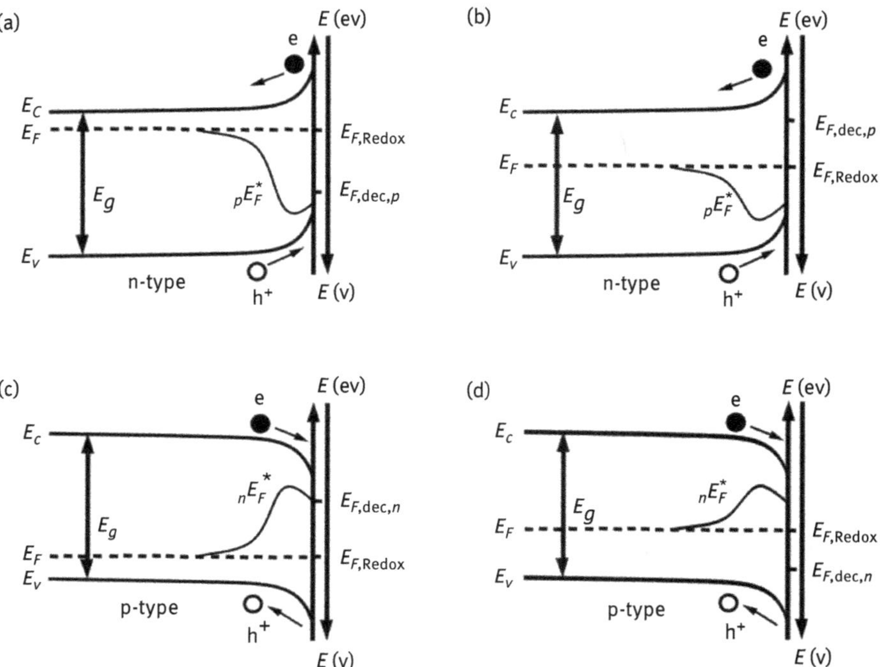

Fig. 10.22: Energetic diagrams presenting the thermodynamic stability of n- and p-SCs and their protection against photocorrosion by the redox system. The n-SC is protected against photocorrosion when $E_{F,\,\mathrm{Redox}} > E_{F,\,\mathrm{dec},p}$ (a); p-SC is protected when $E_{F,\,\mathrm{Redox}} < E_{F,\,\mathrm{dec},n}$ (b). Parts c, d point out that semiconductors undergo photocorrosion.

In the case of n-SC (a, b) only holes can participate in the reactions. They oxidize either the SC or the Red form of a redox couple in the solution. The reaction with the participation of electrons is not possible since their concentration in the CB, under such condition, is low. Which reaction dominates at SC–electrolyte interface depends on the relations between Fermi energies of redox systems and decomposition of SC, or on the relations between the corresponding potentials. In the case (a) and (b) the following reactions can occur:

$$MX + nh^{+} + \mathrm{solv} \longrightarrow M^{n+}_{\mathrm{solv}} + X$$
$$\mathrm{Red} + nh^{+} \longrightarrow \mathrm{Ox}$$

If $E_{F,\,\mathrm{Redox}} > E_{F,\,\mathrm{dec},\,p}$, or $E_{\mathrm{Redox}} < E^{0}_{\mathrm{dec},\,p}$, the oxidation of reduced specious is preferable (a), and if $E_{F,\,\mathrm{dec},\,p} > E_{F,\,\mathrm{Redox}}$, or $E^{0}_{\mathrm{dec},\,p} < E_{\mathrm{Redox}}$, a decomposition of n-SC takes place (b).

For p-SC, since the concentration of holes is very low in the depletion layer, the electrons participate in the reduction of the oxidized form of redox couple or in the reduction of SC materials:

$$MX + ne + \text{solv} \rightarrow M + X^{n-}_{\text{solv}}$$

$$\text{Ox} + ne \rightarrow \text{Red}$$

If $E_{F,\text{Redox}} < E_{F,\text{dec},n}$, or $E_{\text{Redox}} > E^0_{\text{dec},n}$ the reduction of Ox form takes place (c); if $E_{F,\text{dec},n} < E_{F,\text{Redox}}$, or $E^0_{\text{dec},n} > E_{\text{Redox}}$ the decomposition of p-SC occurs (d).

Thus, we can conclude that for the protection of SC against photocorrosion, we have to apply for photoanode the redox systems: $E_{\text{Redox}} < E^0_{\text{dec},p}$ and for photocathode the redox system: $E_{\text{redox}} > E^0_{\text{dec},n}$.

The diagram of electrochemical potentials for few SCs with marked decomposition potentials is shown in Fig. 10.23.

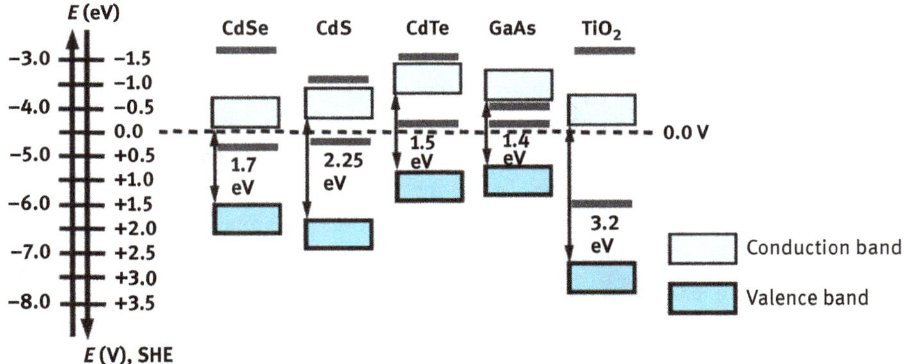

Fig. 10.23: Correlation between the position of conduction and valence band edges E_C,E_V and the decomposition potentials (black short line). All presented semiconductors are unstable and may be oxidized by holes during illumination.

As examples of the thermodynamic approach, we will consider the CdX (X = S, Se, Te) electrodes that are applied as photoanodes in photovoltaic liquid cells (Chapter 11). The data used here are taken from books [3, 14] and references therein. The photocorrosion reaction can be written:

$$\text{CdX} + 2\,\text{h}^+ \rightarrow \text{Cd}^{2+} + \text{X}, \qquad 2\,\text{H}^+ + 2\,\text{e} \rightarrow \text{H}_2$$

The calculated decomposition potentials $E^0_{\text{dec},p}$ are equal to 0.32 V (n-CdS), 0.12 V (n-CdSe), −0.08 V (n-CdTe) versus NHE. It was proved experimentally that such redox systems such as $S^{2-}|S^{2-}_n$, $Se^{2-}|Se^{2-}_n$,$Te^{2-}|Te^{2-}_n$ with standard redox potentials −0.48 V, −0.71 V, −0.81 V versus NHE almost fully protect the electrodes against photocorrosion. In these redox systems, the values of flat band potentials, E_{fb} are very negative and lay in the potential range from −1.3 V (n-CdS) to −1.7 V (n-CdTe) versus NHE. Since the difference between redox potentials of these couples and the flat potentials of CdX is large, the φ_{SC} and in the consequence E_{ph}^{oc} should be large, too. However, the solutions of these redox systems are colorful, and so a significant part of light energy is lost. Other solutions containing SO_3^{2-}, $S_2O_3^{2-}$, $Fe(CN)_6^{4-}$ ions were tested. It was found that their efficiency in the protection of the CdS photoanode is about 95–85%.

The quasi-thermodynamic model can be used for description of all types of the SC decompositions: corrosion in dark, dissolution under the external applied potential and photocorrosion. However, this approach does not provide full insight into the decomposition phenomena. We have to keep in mind the kinetics of these processes. It is difficult to make some generalization since the kinetics depends on the crystal structure, electronic structure, and surface states, their influence should be determined individually.

Like in the case of metals, some SCs can protect themselves against corrosion undergoing passivation. A good example is silicon, which in the majority of solution oxidizes and is covered with a protective oxide films. What is interesting is its ability to go from a passive state to an active state under illumination [3]. Apart from usage of redox systems and passivation, we can also protect the SC electrodes against decomposition by depositing the thin layers of conducting polymers on their surface. Polymers such as polypyrrole, polythiophene, polyaniline, and others were applied. The main problem is the deposition of CP on the SC surface. Polymers can be deposited chemically, electroless-deposited, or electrochemically. In the two first methods, the strong oxidants have to be used for monomer oxidation and further polymerization. In electrochemical deposition, the anodic potentials of monomer oxidation are large and can oxidize and dissolve the SC. For instance, the oxidation potential of pyrrole, thiophene, and 3-methylthiophene are 0.8 V, 1.6 V, 1.35 V versus SCE in acetonitrile solution, respectively. This problem was partially solved by the application of soluble polymers such as polyaliphatic-thiophenes and polyalkoxy-thiophenes. Interested readers can find more about conducting polymers, their synthesis, and applications in [15]. All this fundamental information introduced in this chapter will be applied in the description of the electrochemical photocells and their applications in the solar energy conversion.

Bibliography

[1] Kittel C. Introduction to solid state physic. 8 ed. Hoboken, USA, John Wiley & Sons Inc, 2005.
[2] Rosenberg HM. The solid state. 2 ed. Oxford, UK, Clarendon Press, 1978.
[3] Pleskov YV, Gurevich YY. Semiconductor photoelectrochemistry. NY,USA, Plenum Publishing Corporation, 1986.
[4] Bard AJ, Faulkner LR. Electrochemical methods. Fundamentals and applications. NY, USA, John Wiley &Sons Inc, 2001.
[5] Meming R. Semiconductor electrochemistry. 2 ed. Weinhaim, FRG, Wiley VCH Verlag GmbH, 2015.

[6] Rajeshwar K. Fundamentals of semiconductor electrochemistry and photoelectrochemistry. In ed., Licht S. Semiconductor electrodes and photoelectrochemistry. Vol 6. In Bard AJ, Stratman M., ed. Encyclopedia of electrochemistry. Weinheim, FRG, Wiley VH Verlag GmbH, 2002, 1–53.

[7] Gerischer H. Semiconductor electrode reactions. In: ed., Delahay P, Tobias CW. Advances in electrochemistry and electrochemical engineering, Vol 1. NY, USA. Interscience, 1961, 139–232.

[8] Fuller TF, Harb JN. Electrochemical Engineering. Hoboken, USA. John Willey& Sons Inc, 2018.

[9] Peter L.M. Semiconductor photoelectrochemistry. In: Pletcher D. Electrochemistry. Vol.9, London, United Kingdom, RSC, 1984, 66–100.

[10] Sharon M. The photoelectrochemistry of semiconductor/electrolyte solar cell. In: Licht S., ed Semiconductor electrodes and photoelectrochemistry. Wiley-VCh. Verlag Gmbh, Weinheim, FRG, 2002, 287–316.

[11] Peter LM. Photoelectrochemical kinetics at semiconductor electrodes. In ed., Compton RG, Hancoock G. Application of kinetic modeling. Vol.37 Comprehensive chemical kinetics. Amsterdam, The Netherlands, Elsevier, 1999, 223–280.

[12] Gurudayal, Peter LM, Wong LH, Abda FF. Revealing the influence of doping and surface treatment on the surface carrier dynamics in hematite nanorod photoanodes. ACs Appl. Mater. Interfaces, 2017, 9, 41265–41272.

[13] Zaban A, Greenshtein M, Bisquert J. Determination of the electron lifetime in nanocrystalline dye solar cells by open-circuit voltage decay measurements. Chem. Phys, Chem., 2003, 4, 859–864.

[14] Bouroushian M. Photoelectrochemistry and application. In: Electrochemistry of metal chalcogenides. Berlin, FRG, Springer Verlag, 2010, 207–308.

[15] Skotheim TA, Reynolds JR ed Conjugated polymers. Theory, properties and characterization CRC Press USA 2007.

11 Solar energy conversion in photoelectrochemical cells

11.1 General remarks

There are two main reasons for the rapidly increasing interest in the application of solar energy in our everyday life: first, the exhaustion of fossil fuel resources (e.g., oil, coal, and gas), and second, the problem with environmental pollution resulting mostly from the application of natural fuels for electric energy production. The Sun is the cleanest, most abundant, and practically inexhaustible energy source. The power that the Sun continuously delivers to the Earth is about 1.2×10^5 TW (terawatts), while our civilization produces and uses currently 13 TW, and about 80% of it comes from fossil fuels. Sunlight energy can produce heat or can be converted to electricity (solar to electricity) by applying photovoltaic (PV) cells. The best PV commercial solar cell based on crystal silicon has a conversion efficiency of about 18%, and for the laboratory silicon cell 25% efficiency was reached, being close to the theoretical limit of 31%. This cell has a single p–n junction, so it can capture only a small part of the solar spectrum. The effort is made to construct stacked cells with different bandgaps that enable to capture a greater fraction of solar energy. The calculated efficiency limit is then 43% for two p–n junctions, and increases reaching 66% for infinite numbers of junctions [1]. The cheaper solar cells can be made from other materials but their conversion efficiency is not so satisfied. It is worth to mention the DSSC (dye-sensitized solar cells). The laboratory solar cells based on dye sensitization of oxide semiconductors are typically 10–11% efficient. We will describe this type of cell in Section 11.3. In other solar cells, the junctions of semiconductor nanoparticle – conducting polymers – are applied. Up till now, their efficiency is not so large, about 3–4%, but they are flexible, can have large active surface, and their formation is not expensive [2, 3]. The surprising increase of efficiency comes from quantum-dot phenomena resulting in the multiplication of electron–hole pairs for a single incident photon [4].

The cells for solar energy storage are also intensively developed (STF – solar to fuel). In these cells, solar energy is converted into chemical energy and stored in the reaction products. The photocells where the splitting of water or the reduction of CO_2 occurs are such examples.

We can distinguish two types of cells: solid cells and liquid cells. The solid cells are PV cells containing n-SC/p-SC junctions or Me/SC junctions or multijunctions of different types. They convert solar energy into electrical energy. The liquid cells are referred to as photoelectrochemical cells (PEC) and will be described in detail later. They can convert solar energy into electricity (PV cells) or can store the energy in reaction products (photoelectrosynthetic cells).

https://doi.org/10.1515/9783111160986-012

11.2 Efficiency and key parameters

There are plenty of factors that influence the efficiency ε of solar energy conversion in the solar cells, such as thermodynamic efficiency, threshold efficiency, charge carriers separation efficiency, and storage efficiency. It is very difficult to determine these factors experimentally. Because of that, other quantities that are measured and determined are open-circuit photopotential $E_{\text{ph}}^{\text{oc}}$, short-circuit current density under illumination $j_{\text{sc, c}}^{\text{ph}}$, fill factor ff, and external quantum conversion efficiency EQE. The basic parameter determining the optional energy conversion is the thermodynamic efficiency, coming from the Carnot cycle. For solar energy conversion, $\varepsilon_{\text{therm}}$ is approximately 0.7. The conversion of solar energy is the threshold process, which means that only a part of quantum energy $h\nu = E_g$ is absorbed by the direct semiconductor, and for indirect SC $h\nu > E_g$. The term threshold efficiency, $\varepsilon_{\text{thresh}}$, means the ratio of the number of absorbed light quanta of a proper energy to the total number of light quanta striking the surface of SC. The important quantity is quantum yield $\varnothing$, which characterizes the conversion efficiency of the incoming photon flux of energy $h\nu \geq E_g$ into electrical current (Section 10.4). Quantum yield is connected with the efficiency of separation of charge carriers generated under illumination. This ability manifests itself in the value of short-circuit current density, $j_{\text{sc, c}}^{\text{ph}}$, under illumination. The efficiency of the carriers collection is determined by the changes of electrochemical potential of minority carriers and manifests itself as open-circuit photopotential, $E_{\text{ph}}^{\text{OC}}$. Another important parameter is the external quantum conversion efficiency EQE. This value indicates the photocurrent produced by the cell when illuminated with photons of a particular wavelength and does not take into account the losses associated with reflection of incident photon and recombination of charge carriers. The quality of the solar cell is characterized by a fill factor:

$$\text{ff} = \frac{\left(j_{\text{ph}}^{\max} E_{\text{ph}}^{\max}\right)}{j_{\text{sc, c}}^{\text{ph}} E_{\text{ph}}^{\text{OC}}} \tag{11.1}$$

All these parameters can be obtained from the working characteristics of a photocell (j vs. U) as shown in Fig. 11.1(a). The fill factor is near unity when the working characteristic is rectangular (case 1, Fig. 11.1(b)). The shape of the characteristics is changed when the internal resistance of the cell is large (case 2, Fig. 11.1(b)).

The power conversion efficiency can be determined by the expression

$$\varepsilon = \frac{P_{\max}}{P_{\text{in}}} = \frac{j_{\text{ph}}^{\max} E_{\text{ph}}^{\max}}{P_{\text{in}}} = \frac{j_{\text{sc, c}}^{\text{ph}} E_{\text{ph}}^{\text{OC}}}{P_{\text{in}}} \text{ff} \tag{11.2}$$

To compare different solar cells, these crucial parameters (eq. (11.2)) should be determined in the standard conditions (STC – standard test conditions).

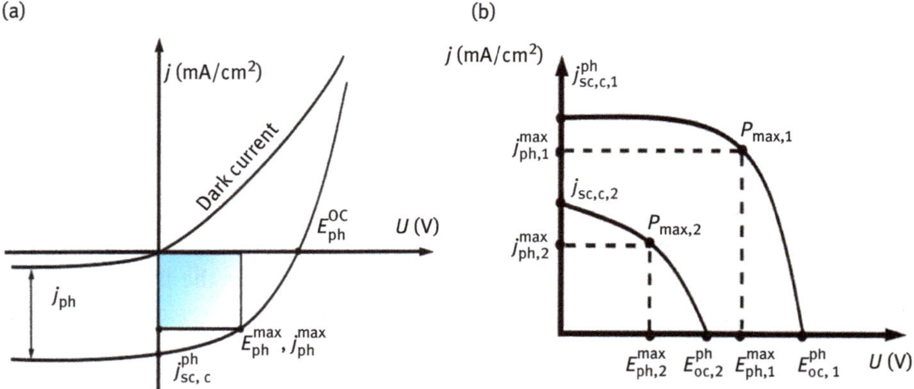

Fig. 11.1: (a) Current density–potential characteristics of a photocell in dark and under illumination. Parameters for fill factor determination are marked. (b) Scheme illustrates the influence of the internal resistance of photocell on the current–potential characteristics. If the internal resistance is low, the power $P_{max,1}$ is high. Increase of internal resistance of the photocell causes decrease of power $-P_{max,2}$.

The STC represents a temperature of 25 °C and an irradiance of 100 mW/cm^2 with an air mass 1.5 spectrum (AM 1.5). The standard reference solar spectrum AM 1.5 defined by the American Society for Testing and Materials characterizes the effect of the Earth's atmosphere on the solar radiation and corresponds to a solar zenith angle of 48.2°. Figure 11.2 shows the dependence of spectral irradiation on wavelength λ and energy.

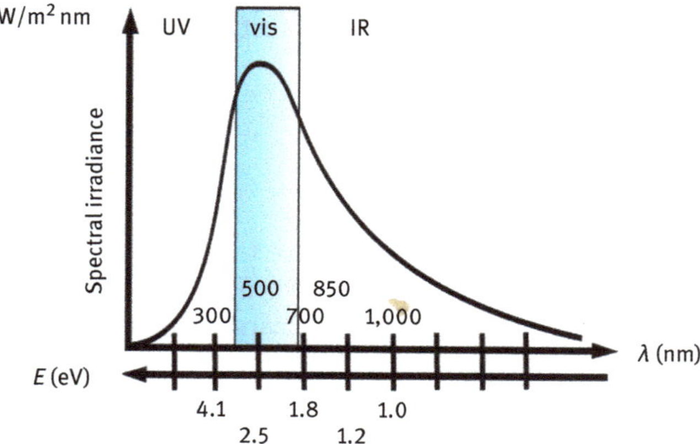

Fig. 11.2: Distribution of solar spectrum irradiance upon wavelength (λ, nm) and energy (eV).

It is clearly seen that semiconductors applied in a photocell should have the bandgap energies in the visible range of light. All these parameters described above characterize the PV cell, where solar energy is converted to pure electrical energy.

The solar energy can also be transformed into chemical energy and stored as fuel. If solar energy is used only for fuel formation (STF), the power efficiency can be determined by the equation:

$$\varepsilon_{STF} = \frac{P_{out}}{P_{in}} = \frac{j^{ph}_{sc,c}\Delta E^0 \eta_F}{P_{in}} \tag{11.3}$$

P_{out} is the output power of the photoelectrosynthetic cell and P_{in} the input solar power, $j^{ph}_{sc,c}$ the short-circuit photocurrent density, η_F Faradaic efficiency, and ΔE^0 the difference of the standard potentials of half-reactions occurring at the anode and cathode.

To determine the solar to hydrogen conversion efficiency ε_{STH} of photo-driven water splitting, we applied in eq. (11.3), the difference of standard potentials of the hydrogen and oxygen evolution ($\Delta E^0 = -1.23\,V$) or the Gibbs standard free energy of reaction: $H_2O \rightarrow H_2 + 1/2\ O_2$, $\Delta_r G^0 = 237$ kJ/mol [5].

11.3 Photoelectrochemical cells: classification – principle of operation

It was mentioned in Section 10.3 that when SC is immersed in an electrolyte solution containing a redox system, the space charge layer is created in SC at the interface. The state of this layer depended on the correlation between the Fermi level of SC and Fermi level of the redox system. Let us suppose that the depletion state is formed, which is the best state for energy conversion. Generated under illumination, the electron–hole pairs are separated in the electric field of space charge layer (n-SC), and electrons flow to the counter (auxiliary) metal electrode and holes to the interface SC/electrolyte, where electrons reduce the Ox form and holes oxidize the Red form of redox couple. Depending on the reactions that take place in the cells and their Gibbs free energy, $\Delta_r G$, we distinguish usually three types of photocells:
(i) Electrochemical PV cells, $\Delta_r G = 0$
(ii) Photoelectrosynthetic cells, $\Delta_r G > 0$
(iii) Photocatalytic cells, $\Delta_r G < 0$

11.3.1 Liquid photovoltaic cells

The electrochemical PV cell called sometimes “liquid junction photocell” or “regenerative photocell” may contain one or two SC electrodes. In the first case, when n-SC or p-SC is used under illumination, electrodes work as photoanode or photocathode. In the second case, two electrodes are photoactive, so the total open-circuit photopotential of the cell is the sum of two electrode photopotentials. The maximum value E^{OC}_{ph}

may be predicted from the differences of flat band potential of both SC electrodes if the band edges are pinned:

$$E_{\mathrm{ph}}^{\mathrm{OC}} = E_{\mathrm{fb}}(\text{n-SC}) - E_{\mathrm{fb}}(\text{p-SC}) \tag{11.4}$$

In Fig. 11.3(a) and (b), the energy diagram for PV cell in dark and under illumination is shown.

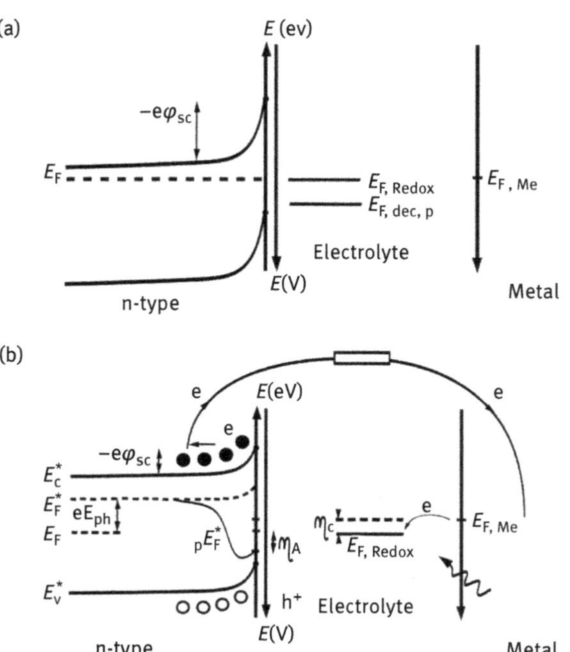

Fig. 11.3: Energy diagram for liquid photovoltaic cell (regenerative cell, $\Delta_r G = 0$) in the dark (a) and under illumination (b). The band of SC is "pinned." During illumination, the Fermi energy level (E_F) and potential drop in the space-charge layer of SC (φ_{SC}) are changed; η_A, η_C mean the overpotentials for anodic and cathodic reactions, respectively.

In dark, equilibrium is established when Fermi energy levels of semiconductor, metal electrodes, and redox system become equal: $E_{\mathrm{F,SC}} = E_{\mathrm{F,Me}} = E_{\mathrm{F,Redox}}$. As a result of equilibration of Fermi energy levels, the depletion state of the space charge layer is created, and the energy bands are bent (look in Sections 10.3 and 10.4). Under illumination, the bands unbend, the potential drop, φ_{SC}, decreases, and the Fermi level of SC is shifted, which leads to the change in electrode potential. This change is equal to photopotential, E_{ph} (marked in Fig. 11.3(b) as eE_{ph}, in energy scale). When the edges of the energy band are pinned and φ_H remains constant, the maximum $E_{\mathrm{ph}} = E_{\mathrm{ph}}^{\mathrm{OC}}$ and is determined by difference ($E_{\mathrm{Redox}} - E_{\mathrm{fb}}$) or φ_{SC} in dark. To obtain the large values of $E_{\mathrm{ph}}^{\mathrm{OC}}$ in photocell, we should choose the proper redox couple and semiconductor to make this difference as large as possible. The position of decomposition potential of SC ($E^0_{\text{dec,p}}$) or Fermi energy level of decomposition by minority carriers ($E_{\text{F,dec,p}}$) is shown in the diagram, η_A, η_C mean the overpotential for anodic and cathodic reactions. As mentioned in Section 10.6, the redox system can protect n-SC against the photocorrosion when:

$$E_{\text{F,Redox}} > E_{\text{F,dec,p}}\ (eV) \quad \text{or} \quad E_{\text{Redox}} < E^0_{\text{dec,p}} \quad (\text{V})$$

The general schemes of regenerative photocells can be written as

$$1)\ \text{n}-\text{SC}\left|\frac{\text{electrolyte}}{\text{Ox, Red}}\right|\text{Me(Pt)};\ 2)\ \text{p}-\text{SC}\left|\frac{\text{electrolyte}}{\text{Ox, Red}}\right|\text{Me(Pt)};\ 3)\ \text{n}-\text{SC}\left|\frac{\text{electrolyte}}{\text{Ox, Red}}\right|\text{p}-\text{SC}.$$

The reactions that are going under illumination are as follows:

$$\begin{array}{llll} 1) & \text{n}-\text{SC}^*; \ \text{Red}_1 + \text{h}^+(\text{SC}) \rightarrow \text{Ox}_1; & \text{Pt}; \ \text{Ox}_1 + \text{e} \rightarrow \ \text{Red}_1 \\ 2) & \text{p}-\text{SC}^*; \ \text{Ox}_1 + \text{e(SC)} \rightarrow \ \text{Red}_1; & \text{Pt}; \ \text{Red}_1 - \text{e} \rightarrow \ \text{Ox}_1 \\ 3) & \text{n}-\text{SC}^*; \ \text{Red}_1 + \text{h}^+(\text{SC}) \rightarrow \text{Ox}_1; & \text{p}-\text{SC}^*; \ \text{Ox}_1 + \text{e(SC)} \rightarrow \ \text{Red}_1 \end{array}$$

In all cases, the overall reaction is: $\text{Red}_1 + \text{Ox}_1 \leftrightarrow \text{Ox}_1 + \text{Red}_1, \Delta_r G = 0$.

The typical examples of such cells are chalcogenide electrodes such as n-CdTe, n-CdSe, and n-CdS immersed in the solution containing redox couples: Te^{2-}/Te_n^{2-}, Se^{2-}/Se_n^{2-}, or S^{2-}/S_n^{2-}. These redox couples not only form the depletion state in the space charge layer of SC, but also protect the semiconductor against photocorrosion (Section 10.6).

The highest reported conversion efficiency of about 16% was obtained for single crystals of n-CdSe in solution containing $Fe(CN)_6^{3-}$ ions. The efficiencies of 12% and 11% were determined for n-GaAs in Se^{2-}/Se_n^{2-} solution and n-CdTe in Te^{2-}/Te_n^{2-} solution, respectively. However, the application of single crystals is not economical; therefore, plenty of methods such as vapor deposition, chemical deposition, electrochemical, deposition, and sol–gel are applied for the formation of thin solid polycrystalline or amorphous semiconducting films. These methods have been used to prepare thin polycrystalline films of n-Cd(S, Te, Se), p-CdTe, p-InP, n-GaAs, and others. We may treat the liquid PV cell described earlier as the classical cell. More broad information about such cells can be found in [6–8] and references therein.

Let us now concentrate on the new generation of electrochemical PV cells:

(i) DSSC
(ii) Quantum dot-sensitized solar cells (QDSSC)

11.3.2 Dye-sensitized solar cells

These types of cells were first invented by Grätzel in the late 1990s of twentieth century [9]. The first laboratory cell contained a titanium sheet covered with the "fractal" TiO_2 film of a high surface area. The surface of the film was further covered with the ruthenium dye RuL_3 (L-2,2′-bipirydyl-4,4-dicarboxylate). All system $Ti/TiO_2/RuL_3$ worked as photoelectrode. As a counter electrode, Pt wire was used. Cell was filled with an aqueous, slightly acidic solution containing bromide and a small amount of bromine. The open-circuit photopotential reached 1 V, and the overall conversion efficiency in full sunlight was 1–2%. It was the beginning. At present, the efficiency of conversion in dye-sensitized PV cells is 10–12% [10, 11].

How this cell operates and what was changed to reach such efficiency? Figure 11.4 presents the principle of cell operation.

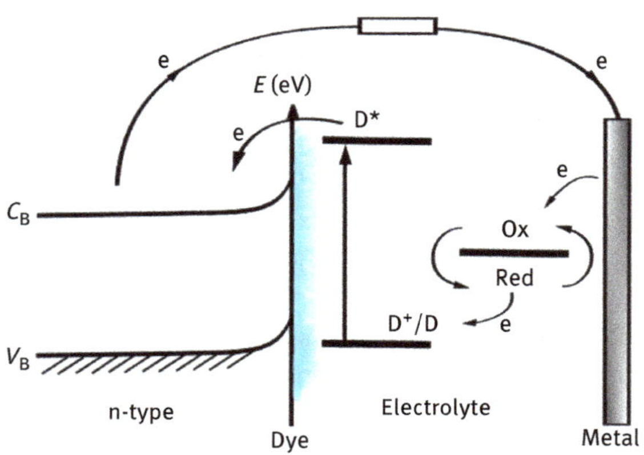

Fig. 11.4: Operating principle and energy diagram of a dye-sensitized solar cell. D^+/D, dye in a ground and oxidized state; D^* dye in an excitation state; Ox/Red, mediator in electrolyte.

When the SC with high value of E_g is illuminated with visible light, the dye (sensitizer) molecules absorbed the light and become excited.

If the energy of the excited state of the dye lies above the energy E_c of conduction band (CB) of SC, the photoexcitation of dye is followed by electron injection into CB. The oxidized dye molecules are regenerated by a reduced form of redox couple in the electrolyte solution. The redox system itself is regenerated at a counterelectrode by electrons flowing by the external circuit from the SC electrode. Let us consider the cell containing the I^-/I_3^- redox couple in the organic solvent such as acetonitrile or propylene carbonate with lithium salt as a supporting electrolyte.

The basic reactions that occur during the illumination are as follows:

$$D + h\nu \rightarrow D^*; \qquad D^* \rightarrow D^+ + e(\mathrm{SC})$$

$$\frac{3}{2}I^- + D^+ \rightarrow D + \frac{1}{2}I_3^-; \qquad I_3^- + 2e\ (\mathrm{Me}) \rightarrow 3I^-$$

The other reactions that take place are: (i) the radiative deactivation of excited dye, (ii) recapture of electrons from CB by oxidized dye molecules, and (iii) recapture of

CB electrons by oxidized species of the redox couple. All these additional reactions decrease the efficiency of DSS cells and should be suppressed.

As mentioned earlier, the first DSS cells contain a planar crystalline TiO_2 electrode covered with adsorbed dye. Because of the low surface area and low concentration of dye, the conversion efficiency was low. The situation was changed when not only the nanostructured metal oxide, mainly TiO_2, but also ZnO, Nb_2O_3, and SnO_2 were applied. The high porosity of mesoscopic semiconductor films enables the incorporation of dyes in large concentrations. Nowadays, working DSS cells contain photoelectrode made from TiO_2 (anatase) nanoparticles, deposited by hydrothermal method on flexible plastic support covered with a transparent conducting layer of fluorine-doped thin oxide or thin-doped indium oxide. The average size of the nanoparticles is 20 nm. They are covered and shelled by specially designed light-harvesting dye molecules. Mostly, Ru(II)bisbipyridyl complexes are used [10]. They are known as being very stable and possessing the light absorption coefficient α in the range 2–4·10^3 cm^{-1} so the absorption length of these sensitizers, $1/\alpha$, is about 2.5–5 μm. The light-harvesting efficiency (LHE) of these dyes is over 90% for the wavelength range near the absorption maximum for the thickness of the nanocrystalline film (*d*) equal to 6 μm. The LHE is determined by expression:

$$\mathrm{LHE}(\lambda) = 1 - 10^{-\alpha d} \tag{11.5}$$

As other PV cells, the quality of DSS cell is determined by fill factor ff, the power conversion efficiency (ε) of DSSC (see eqs. (11.1) and (11.2)), and by the external quantum efficiency sometimes referred to as incident photon to current conversion efficiency. This parameter depends upon the $LHE(\lambda)$, quantum yield $\varnothing$, and collection efficiency $\varepsilon_{\mathrm{coll}}$ (EQC). In this case, the quantum yield $\varnothing$ determined the electron injection from excited dye to the CB of SC:

$$\mathrm{EQC}(\lambda) = \mathrm{IPCE}(\lambda) = \mathrm{LHE}(\lambda)\varnothing\varepsilon_{\mathrm{coll}} \tag{11.6}$$

The collection efficiency of charge carriers (here electrons) depends on the relationship between the electron diffusion length L_n, the thickness of the nanocrystalline layer of SC (*d*), and the absorption coefficient (α). The good collection is obtained when $L_n > d > \frac{1}{\alpha}$. The best conversion efficiency is about 11–12%, so DSS cells can compete with conventional solar cells. One can find much more about DSS cells in articles [10–12] and references therein.

11.3.3 Quantum dot-sensitized solar cells

In the last decade, the huge development of nanotechnology is observed. It creates the possibility of the application of nanostructural materials in solar cell design. As mentioned earlier, in the DSS cells the nanocrystalline metal oxide semiconductors of high value of bandgap energy were used with the organic dyes as a sensitizer. In QDSSC, semiconductor oxide nanoparticles, mostly TiO_2, are also used. However, as sensitizer, the other semiconductors are applied (PbS, CdSe, and CdS) [12]. They have lower band gap energy than oxide support and are used in the form of quantum dots. During illumination, in quantum dots (QDs) the electron–hole pairs are also generated. The main problem here is the effective separation of charge carriers. For the efficient electron separation and injection, the bottom level of the CB of QDs should have higher energy than the bottom level of the CB of oxide SC. After separation, electrons from QDs are injected to CB of oxide and the holes are transferred to electrolyte oxidizing the reduced form of the redox couple. There are some advantages of the application of QDs as sensitizer when compared with dyes. Most of the dyes have narrow absorption band, while the QDs can absorb the energy of solar spectrum $h\nu \geq E_g$. Other advantages of QDs are the amount of light that they can absorb and the tunability of the bandgap energy. When QDs become smaller, their bandgap widens due to the quantum confinement effect. In consequence, it is possible to form the cascade SC photoelectrode containing the quantum dots with controlled sizes, covering all solar spectrum. There are some specific features of QDs, which can find application in solar cell but need special efforts. One feature is the generation of "hot" carriers when QDs absorbed light of $h\nu > E_g$, whereas the other is the carrier multiplication process. It means that one photon can excite two or more electrons. These two effects may increase the open-circuit photopotential and photocurrent; however, the lifetime of "hot" carriers (electrons) is very short so some methods should be developed to dissolve this problem. Up till now, the QDSSC show smaller fill factors, lower photocurrents, and lower power conversion efficiency (about 4%) in comparison with their dye analogs [12].

11.3.4 Photoelectrosynthetic cells

In these photocells the light energy is applied to initiate and carry out the reactions, resulting in the formation of the new products storing the chemical energy. The Gibbs free energy of overall reactions that take place in such cells is larger than 0 ($\Delta_r G > 0$). This energetic barrier can be overcome with the help of light energy. However, in many cases, additional energy of the electric field is required, supplied by the external source. The most popular reaction in which the rich energy products are formed is the splitting of water to hydrogen and oxygen. The cell where the splitting of water occurs is called photoelectrolytic cell. The first photoelectrolysis of water with the application of the SC electrode was carried out by Fujishima and Honda in 1972. They used n-TiO_2 photoanode and Pt counter electrode in a two-compartment cell separated with diaphragm. During illumination with ultraviolet (UV) radiation, they observed both O_2 and H_2 evolution at photoanode and counter electrode, respectively. As the energy of light was not sufficient, they had to apply the external voltage to facilitate the reaction. That work initiated the huge development of research in the field of the photoelectrolysis of water and is continued in our days [13, 14]. The overall reaction of water splitting is

$$2\,H_2O \rightarrow 2\,H_2 + O_2$$

The standard Gibbs free energy of this reaction ($\Delta_r G^0 = 237$ kJ/mol) is equal to 1.23 eV per electron. In photocells, this reaction is the sum of two half reaction taking part at semiconductor photoanode (n-SC) and at metal cathode or semiconductor photocathode (p-SC):

$$2\,H_2O + 4h^+ \rightarrow 4\,H^+ + O_2, \qquad E^0 = +1.23\ \text{V (NHE)}, \qquad \text{n} - \text{SC}$$

$$4\,H^+ + 4\,e \rightarrow H_2, \qquad E^0 = 0\ \text{V}, \qquad \text{Me (Pt) or p} - \text{SC}$$

Thus, the difference of standard potentials of the water-splitting cell

$$\Delta E^0 = E_C^0 - E_A^0 = -1.23\,\text{V}, \quad \text{at pH} = 0(\text{H}^+ = 1\ \text{M})$$

The water oxidation reaction is not spontaneous. To oxidize the water by applying solar energy the photoexcited electrons must have the energy higher than 1.23 eV. Such energy is also required for the photons; in the wavelength scale, the energy is equal to 1,200 nm. Therefore, all solar spectra up to this value can theoretically be used in the water splitting. In practice, these are the losses in energy resulting from losses in absorption of light, dissipation of energy in heat form, or some hindrances in the kinetics. These losses are estimated to be 0.8 V, so the minimal energy of photons required for the water photodecomposition is about 2.1 eV, that is, <620 nm.

Two types of photocells for water splitting can be distinguished: one is referred to as direct photoelectrolysis and the other as assisted photoelectrolysis [6, 7]. Their energy/potential diagrams under illumination are presented in Fig. 11.5(a) and (b). Both cells contain n-SC electrodes working as photoanodes and Pt cathode. The acid solution is used as electrolyte. During illumination, the electrons and holes are photogenerated and instantly separated in the electric field of the space charge layer. The electric field causes the flow of the holes to the n-SC/electrolyte interface, where H_2O molecules are oxidized to O_2. The electrons flow through the external circuit to the Pt, where they reduce H^+ ions. This is only possible when the value of flat-band potential E_{fb} versus NHE is more negative than the potential of hydrogen electrode (E_{eq} of H^+/H_2) (see Fig. 11.5(a)).

If the value of E_{fb} is more positive than E_{eq} of H^+/H_2 system, the potential for hydrogen evolution is not achieved. The additional external potential (U) has to be used to move the potential of Pt cathode to the more negative value than the potential of the hydrogen electrode (see Fig. 11.5(b)).

In this case, for the splitting of water, not only the light energy is used, but also the electrical energy. Such a process is called the assisted photoelectrolysis. The other

(a)

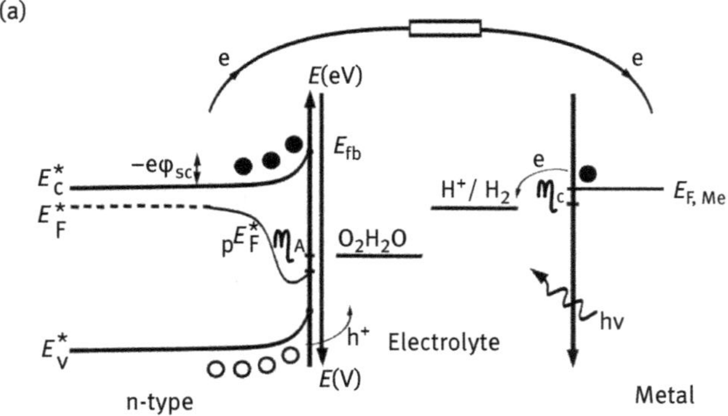

(b)

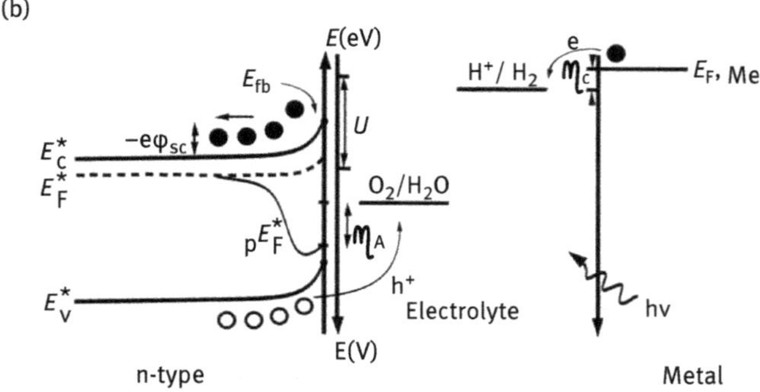

Fig. 11.5: (a) Schematic energy diagram of photocell under illumination for direct water splitting. The value of flat band potential (E_{fb}) in a potential scale should be more negative than the redox potential of H^+/H_2 couple. (b) Schematic energy diagram of photocell under illumination for assisted water splitting. The value of flat band potential (E_{fb}) in a potential scale should be more positive than the redox potential of H^+/H_2 couple. External potential U is applied.

cells that can be used for photoelectrolysis of water contain two semiconductor electrodes. In such a cell, Pt cathode is replaced by a p-type SC electrode working as photocathode. The applied n- and p-type semiconductors differ in their bandgap energies: band gap energy E_g of n-SC should be greater than E_g of p-SC. Under illumination, the holes generated in photoanode (n-SC) flow to SC/electrolyte interface oxidizing the water molecules, while the electrons flow to the photocathode, where they recombine with the holes of the valence band of p-SC. The minority carriers–electrons generated in photocathode (p-SC) flow to the interface reducing the H^+ ions, whereas holes recombine with electrons flowing from n-SC. The energy/potential diagram of such cells is shown in Fig. 11.6.

As described earlier, the photocells for splitting of water contain one or two semiconductor electrodes. The choice of semiconductor materials for such purpose is limited by several requirements. Let us consider these requirements to be fulfilled by the photoanodes (n-type SC):

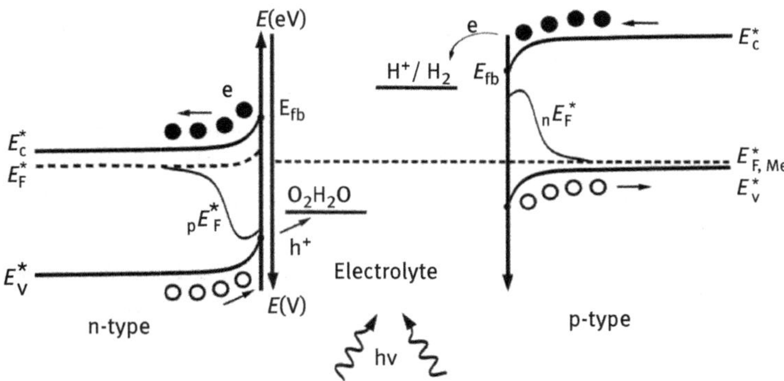

Fig. 11.6: Schematic energy diagram of a photocell under illumination for water splitting. Photocell contains two photoelectrodes: photoanode (n-SC) and photocathode (p-SC).

1) The bandgap energy of SC should be at optimal values, which means at the energy that allows for the efficient absorption of visible light. The range of visible light changes from about 400 to 700 nm, and the corresponding energy ranges from 1.8 to 3.1 eV. It is worth to mention that about 30–40% of solar energy reaching the Earth`s surface lays in the visible region.
2) The absorption coefficient α should be large, and α is large in the direct band semiconductors.
3) The doping of SC should be substantial, securing $L_{SC} > \alpha^{-1}$, then the separation of photocarriers is effective.
4) The diffusion length and lifetime of minority carriers should be large.
5) SC material should be chemically stable in water and in the acid solutions.
6) SC electrodes should be resistive against photocorrosion, so

$$E_{F,\,\text{Redox}} > E_{F,\,\text{dec},\,p} \text{ or } E_{F,\,\text{Redox}} < E^0_{\text{dec},\,p},\ \text{Redox} - O_2 \big| H_2O$$

7) For the direct photoelectrolysis, the flat band potential of semiconductor electrode U_{fb}, or the potential of the bottom of the CB should be more negative than the redox potential of H^+/H_2 (vs NHE). The potential of the top of the valence band should be more positive than the corresponding redox potential of O_2/H_2O.

Actually, such semiconductor materials that fulfill all these requirements do not exist. The semiconductors such as $SrTiO_3$, $KTiO_3$, $LaTi_2O_7$, and TiO_2 have E_{fb} potentials more negative than the potential of hydrogen electrode (at pH = 0). However, their bandgap energy is large: 3.3, 3.5, 4, and 3.2 eV, respectively, so they can absorb only UV part of solar spectra. Unfortunately, UV represents only a small fraction (about 4.5%) of the sunlight reaching the Earth. The other semiconductors materials such as metal chalcogenides have appropriate values of the energy gap, for instance, ZnTe (2.3 eV), CdS

(2.4 eV), and ZnSe (2.6 eV). All of them are direct band semiconductor and their flat band potential is much more negative than the redox potential of H^+/H_2 (vs NHE). However, they are not stable in the aqueous acid solution and undergo photocorrosion during illumination. For the water splitting, the semiconducting metal oxides are widely used as the photoanodes. These materials are chemically stable and avoid photocorrosion during illumination in acid solutions. Unfortunately, most of them have large bandgap energy and they do not absorb the light efficiently in the visible range. For instance, the $E_g$value of TiO_2 and ZnO, often applied in the PEC, is 3.2 eV. Better absorption ability has WO_3 and Fe_2O_3 with E_g equal to 2.7 and 2.2 eV. However, the flat band potentials of those oxides are about 200 mV or more positive than the potential of hydrogen electrode.

As mentioned earlier, for the water splitting, instead of Pt cathode we can use semiconductor photocathode (p-type SC). In this case, the flat band potential of photocathode should be more negative than the potential of hydrogen electrode E_{eq} of H^+/H_2, since in the reduction of H^+ ions participate the electrons photogenerated in p-SC (Fig. 11.6).

Additionally, the p-type of SC electrode should fulfill the requirements (1–4) described above and to be resistive against the photocorrosion, the decomposition potential of p-SC electrode should be lower than the potential of hydrogen electrode:

$$E_{\mathrm{F,\,Redox}} < E_{\mathrm{F,\,dec,\,n}} \quad \text{or} \quad E^0_{\mathrm{dec,\,n}} < E_{\mathrm{F,\,Redox}}, \quad \mathrm{Redox} - \mathrm{H^+}|\mathrm{H_2}$$

Such cells with two SC photoelectrodes do not require an external source of potential. The main problem there is the protection of photocathode against the photocorrosion, since the semiconductors with low E_g have to be applied to fulfill the condition E_g (n-SC) > E_g (p-SC). The p-Cu_2O (2.2 eV), p-GaP (2.3 eV), and p-WSe_2 (1.7 eV) were tested as photocathodes.

It was pointed out that presently available semiconducting materials cannot afford direct splitting of water in the photocell containing metal cathode. Therefore, the external source of potential is required. Instead of applying the electrical battery we can use solid or liquid PV cell, connected in series with PEC [15]. The STH (solar to hydrogen) efficiency in such tandem cells depends on the design. Up to date, the highest STH efficiency reached in the tandem PV-PEC system containing one semiconductor/electrolyte junction (PEC) is 12.4%.

Up till now, plenty of the semiconducting metal oxides have been reported to be active for the water splitting; however, most of them absorb UV light. Many efforts were made to improve the absorption properties of oxides and to shift the onset of absorption toward the visible range of light. One of the methods is the application of so-called Z-scheme photocatalysts [16]. A "Z-scheme" can be considered as a special tandem cell, in which two semiconductors are connected by an electrolyte containing redox system as mediator. Two photocatalysts are illuminated simultaneously, the electrons are generated in SC with lower CB, and holes from SC with higher VB react with redox couple, while the remaining electrons and holes participate in the reaction of water splitting. The correlation of energy bands of SC and the equilibrium potential of the redox system is shown in Fig. 11.7.

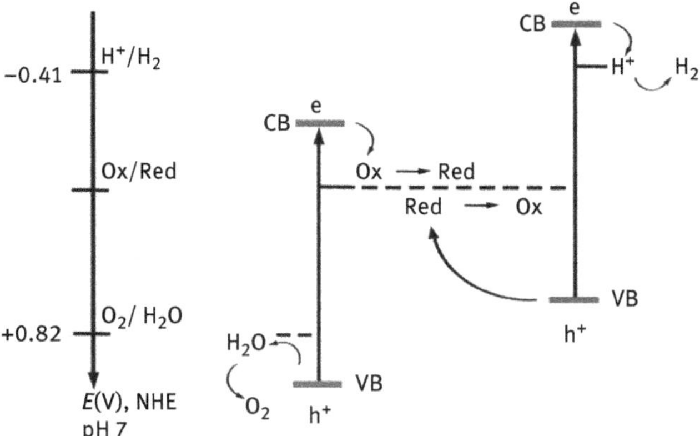

Fig. 11.7: Schematic energy diagram of the two-step photocatalytic water splitting system (photocatalytic Z-scheme).

Efforts are made to construct Z-scheme without a liquid redox system, replacing it by a solid electron mediator (metal nanoparticles). The other approach for improvement of the absorption properties is the doping of metal oxides with metal or nonmetal ions (V, Mo, Ru, Li, Si, or others). However, the doping induces the additional energy levels in the bandgap energy, which may work as recombination centers. There are several methods for the reduction of the recombination, such as the modification of SC materials with metal particles (Ag, Au, Pt), the modification of semiconductor metal oxide with other semiconductor particles, and the application of nanostructures for photoanodes formation. There is very rich literature concerning the splitting of water and modifications of applied semiconductors. More can be found in the reviews [16–19] and references therein.

The solar energy can be exploited in other important reaction – the reaction of CO_2 photoreduction. This reaction can be carried out in the PEC with semiconductor photocathode, or on the semiconductor particles.

The mechanism of CO_2 reduction is complicated and may be multielectron process, resulting in the formation of many carbonate compounds [20, 21]. For instance, eight or 12 electrons are involved during formation of methane or ethanol, respectively. Let us consider only two-electron processes of formation of CO and formic acid (HCOOH) and six-electron process of methanol (CH_3OH) formation. The last two compounds may be used as fuel in the fuel cells. The CO_2 reduction can be written as follows:

1) $CO_2 + 2\,H^+ + 2\,e \rightarrow CO + H_2O$, $\Delta_r G^0 = 19.9$ kJ/mol, $E^0 = -0.1$ V, $E_{eq}(\text{NHE, pH7}) = -\ 0.52$ V.

2) $CO_2 + 2\,H^+ + 2\,e \rightarrow HCOOH$, $\Delta_r G^0 = 33$ kJ/mol, $E^0 = -0.17$ V, $E_{eq}(\text{NHE, pH7}) = -0.58$ V.

3) $CO_2 + 6\,H^+ + 6\,e \rightarrow CH_3OH + H_2O$, $\Delta_r G^0 = 9$kJ/mol, $E^0 = -0.046$V, $E_{eq}(\text{NHE, pH7}) = -0.\ 37$V.

The values of $\Delta_r G^0$ were calculated using thermodynamic data from physical chemistry [22].

The values of redox potentials of reactions are close to each other, so from the thermodynamic point of view during the CO_2 reduction, the mixture of compounds may be formed. However, the main problem that should be solved is the competition between the reduction of CO_2 and the reduction of hydrogen ions that can occur at photocathode. The equilibrium potential of the hydrogen electrode at pH 7 is equal to −0.41 V (NHE). Since the potential of hydrogen electrode is more positive than the equilibrium potentials of reactions 1 and 2 and is only slightly more negative than for reaction 3, the minority carriers (electrons) photogenerated in photocathode may participate in the reduction of H^+ ions. The flat band potential, E_{fb}, for the reduction of CO_2 should be therefore more negative than the equilibrium potential of reactions (1, 2, 3) at a given pH. There are few semiconductors that fulfill such condition: p-Cu_2O, p-GaP, and p-CdTe. However, these semiconductor materials have low values of E_g, so they are susceptible to photocorrosion. Applying mixed SC oxides with small and large E_g, we can suppress the photocorrosion.

In the third type of cells (photocatalytic cells), the reactions can occur even in dark, since their $\Delta_r G < 0$. However, they proceed so slowly that light energy is utilized for acceleration. The suspensions of semiconductor particles may be considered as photo microcells working in the open-circuit conditions.

In the next chapter, we will consider briefly the properties of such systems and their applications.

Bibliography

[1] Crabtree GW, Lewis NS. Solar energy conversion. Phys. Today 2007, 37–42, www.physicstoday.org

[2] Skompska M. Hybrid conjugated polymer/semiconductor photovoltaic cells. Synth. Met. 2010, 160, 1–15.

[3] Saunders BR, Turner ML. Nanoparticle-polymer photovoltaic cells. Adv. Colloid. Interface Sci. 2008, 138, 1–23.

[4] Nozik AJ. Quantum dot solar cells. Physica E. 2002, 14, 115–120.

[5] Chen Z, Dinh HN, Miller E. Efficiency definitions in the field of PEC. In Photoelectrochemical water splitting: Standards, experimental methods and protocols. Berlin, FRG, Springer, 2013, 7–16.

[6] Pleskov YV, Gurevich YY. Semiconductor photoelectrochemistry. NY, USA, Consultant Bureau, Division of Plenum Publishing Corporation, 1986.

[7] Pleskov YV. Solar energy conversion. A photoelectrochemical approach. Berlin, FRG, Springer, 1990.

[8] Sharon M. The photoelectrochemistry of semiconductor/electrolyte solar cell. In : Licht S., ed. Semiconductor electrodes and photoelectrochemistry. Weinheim, FRG, Wiley-VCh. Verlag Gmbh, 2002, 287–316.

[9] Vlachopoulos N, Liska P, Augustynski J, Gratzel M. Very efficient visible light energy harvesting and conversion by spectral sensitization of high surface area polycrystalline titanium dioxide films. J. Am. Chem. Soc 1988, 110, 1216–1220.

[10] Gratzel M. Solar energy conversion by dye-sensitized photovoltaic cells. Inorg. Chem. 2005, 44, 6841–6851.

[11] Mc Evoy AJ, Gratzel M. Dye sensitized regenerative cells. In: Licht S. ed. Semiconductor electrodes and photoelectrochemistry. Weinheim, FRG, Wiley-VCh. Verlag Gmbh, 2002, 397–406.

[12] Kamat PV, Tvrdy K, Baker DR, Radich JG. Beyond photovoltaic: Semiconductor nanoarchitectures for liquid-junction solar cells. Chem. Rev. 2010, 110, 6664–6688.

[13] Licht S. Photoelectrochemical solar energy storage cells. In: Licht S. ed. Semiconductor electrodes and photoelectrochemistry. Weinheim, FRG, Wiley-VCh. Verlag Gmbh, 2002, 317–345.

[14] Sharon m, Licht S. Solar photoelectrochemical generation of hydrogen fuel. In: Licht S. ed. Semiconductor electrodes and photoelectrochemistry. Weinheim, FRG, Wiley-VCh. Verlag Gmbh, 2002, 345–357.

[15] Prevot MS, Sivula K. Photoelectrochemical tandem cells for solar water splitting. J. Phys. Chem. C 2013, 117, 17879–17893.

[16] Ismail AA, Bahneman DW, Photochemical splitting of water for hydrogen production by photocatalysis: A review. Sol. Energy. Mater. Sol. Cells. 2014, 128, 85–101.

[17] Kudo A, Miseki Y. Hetergeneous photocatalyst material for water splitting. Chem. Soc. Rev. 2009, 38, 253–278.

[18] Augustynski J, Aleksander BD, Solarska R. Metal oxide photoanodes for water splitting. Top Curr. Chem. 2011, 303, 1–38.

[19] Li J, Wu N. Semiconductor-based photocatalyst and photoelectrochemical cells for solar fuel generation: a review. Cat. Sci. Technol. 2015, 9, 1360–1384.

[20] Bockris J O M, Reddy AKN. Modern electrochemistry 2B. Electrodics in chemistry, engineering, biology and environmental science. New York, USA, Kluwer Academic/Plenum Publisher, 2000.

[21] Fujishima A, Tryk DA. Fundamentals of photocatalysis. In: Licht S. ed. Semiconductor electrodes and photoelectrochemistry. Weinheim, FRG, Wiley-VCh. Verlag Gmbh, 2002, 497–535.

[22] Atkins P. Physical chemistry. NY, USA, Freeman WH and Company, 1994.

12 Semiconductor particles in photocatalysis

12.1 General remarks

In the previous chapter, we have described how by using photoelectrochemical cells the solar energy can be converted in the electrical or chemical energy. The reactions that we considered in detail were the splitting of water and reduction of CO_2. There are many others important reactions for human being such as oxidation or reduction of inorganic/organic pollutions in water and air. Some of these reactions were investigated on illuminated, conventional semiconductor electrodes. Now, we will describe the applications of semiconductor particles in photocatalysis.

Photocatalysis is a kind of heterogeneous catalysis, where the light energy is applied for acceleration of reaction occurring on semiconductor surface. Depending on the place where photoexcitation occurs, we can distinguish two types of processes. If the initial photoexcitation of electrons takes place in the molecules adsorbed on SC surface, followed by injection of photogenerated electrons to the catalyst, the process is referred to "sensitized photoreaction." The example of such processes is the photoexcitation of dye molecules adsorbed on SC particles (Fig. 12.1(a)). In the second type, the initial photoexcitation occurs in the catalyst – semiconductor particles. The photogenerated charge carriers flow to the SC particle | solution interface, interface and react with acceptors and donors (Fig. 12.1(b)).

Such process is referred to as the "catalyzed photoreaction" and will be described briefly below.

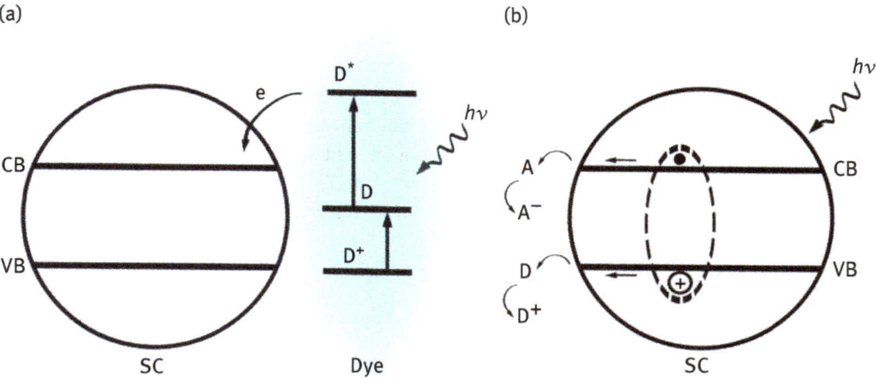

Fig. 12.1(a, b): Scheme illustrating photocatalytic processes: sensitized photoreactions (a) and catalyzed photoreactions (b).

https://doi.org/10.1515/9783111160986-013

In the photocatalytic studies, the colloid semiconducting particles or particle suspensions are often used. Their physicochemical properties depend on their nature and size. Their charge and potential drop across the particle–solution interface is influenced by the solution properties (ions, pH, kind of solvent). Much broader information about colloids, suspensions, micro-, and macroemulsions can be found in bibliography [1, 2].

12.2 Size effect and dimensionality

Since the properties of particles are size dependent [3], their behavior differs from that of bulk conventional semiconductor electrodes. Some important features of the reduction of particle size into nanometer scale can be described as follows:

1) Quantum size (QS) effect is observed when the size of particle is significantly reduced. It results in an increase in band gap energy. For instance, by changing the size of CdS particles its band gap energy can be tuned between 2.4 and 4.5 eV. However, there are some semiconductors with large effective electron mass, in which the QS effect occurs only at a very low value of particles radius. Such example is TiO_2. Only when particle size is < 3 nm the QS effect appears. The band widening causes the shifting of potentials of electrons (bottom edge of CB) to more negative and holes (upper edge of VB) to more positive values. In consequence, the reduction and oxidation ability of charge carriers increases.
2) The size is also important for the light absorption. In a suspension of isolated, individual nanoparticles, light is absorbed by spherical particles and also scattered on them. The relationship between optical penetration depth $x = \frac{1}{\alpha}$, the width of space charge layer L_{sc} (if exists), and the particle size should be appropriate. If not, the recombination may dominate in the vicinity of particles.
3) The development of band bending and the formation of space charge layer depend on the particle size. When the dimension of particle (d) is low $d \leq L_D$, ($L_D \geq L_{sc}, L_D$ means Debye length, Sections 10.2 and 10.4), the band bending does not exist. In this case the energy bands are flat. For $d > L_D$ the band bending is developed.
4) The reduction of particle size leads to on an increase of surface area and surface to volume ratio, causing the significant increase of the number of active catalytic centers.

Let us compare the processes taking place during the illumination of semiconductor particles and conventional (bulk) n-SC electrode (Fig. 12.2(a–c)) immersed in electrolyte solution containing redox couple. In dark, the state of bulk SC and particles depends on the relation between E_F and $E_{F,\text{Redox}}$.

If $E_F > E_{F,\text{Redox}}$, the space charge layer is formed in the depletion state (bulk n-SC) and proper band bending is developed (Fig. 12.2(a)). What is going on in the particles depends on their size. If the particles are small, the energy bands are flat (see point 3

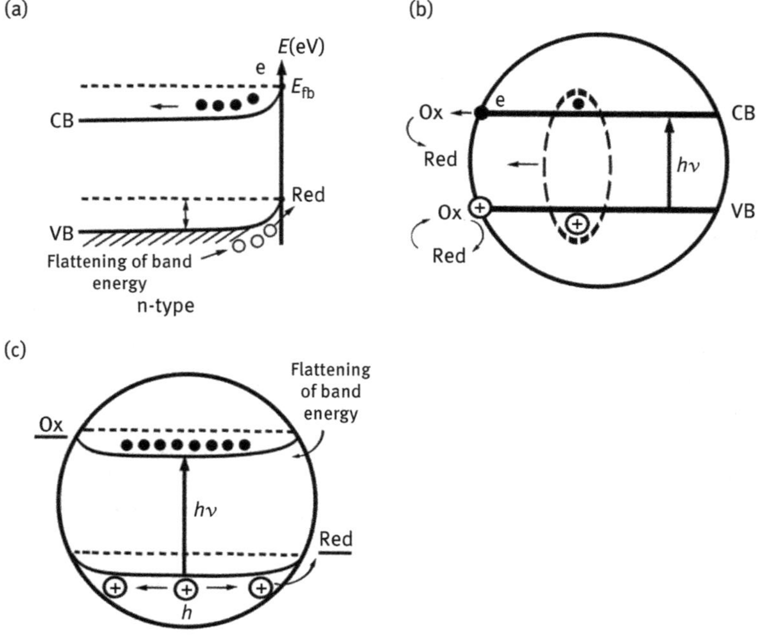

Fig. 12.2(a–c): Illustration of differences in behavior under illumination between n-SC electrode (a) and SC particles: small particle with dimension $d < L_D$ (b) and large particle with $d > L_D$ (c); CB, VB – conduction, valence band, respectively; Ox, Red – oxidized, reduced form of a redox couple in a solution.

above; Fig. 12.2(b)). If the particles are large and n type, they lose their electrons to the Ox form of a redox couple and exist in the depletion state (Fig. 12.2(c)). They can also exist in accumulation state ($E_{F,\text{Redox}} > E_F$), as a result of electron flow from Red form of a redox system to the conduction bands of n-SC particles (such case is not considered here). The equilibrium in dark is characterized, for SC electrode and particles, by equality of Fermi energy levels $E_F = E_{F,\text{Redox}}$ and equality of anodic and cathodic currents. Note the difference between the two systems: the potential of SC electrode can be changed using external potential source (potentiostat), while in the case of particles their potential is determined by redox systems in the solution.

In practice, the particles may be treated as microelectrodes at open circuit potential, with the reduction and oxidation reactions proceeding simultaneously at the same surface.

If we illuminate the n-SC electrode that possesses the space charge layer in the depletion state, with the photon energy $h\nu \geq E_g$, the generated excitons (bound electron–hole pairs) instantly dissociate in the electric field of space charge layer, forming free electrons and holes. The minority carriers, holes, flow to the SC | electrolyte interface reacting with Red form of a redox couple. The majority carriers flow to the back contact of SC electrode. It results in the unbending of energy bands (Fig. 12.2(a)) and generation of photopotential and photocurrent. The behavior of SC electrodes in dark and under illumination was described in details in Chapter 10 (Sections 2–6). Now, if we illuminate the small particle, the excitons are also generated; however, they cannot dissociate since the space charge layer does not exist. They diffuse to the interface and dissociate in the electric field of the interface SC Ps | electrolyte or recombine before that. Furthermore, electrons and holes take part in the reactions with Ox and Red forms of a redox couple (Fig. 12.2(b)).

During the illumination of large n-SC particles (n-SC Ps), which exist in the depletion state (Fig. 12.2(c)), the generated excitons dissociate in the electric field of space charge layer. The holes and electron migrate to the interface taking part in the reactions with Red and Ox forms of a redox couple. However, in the case of n-SC particles in depletion state, the reaction of holes (minority carriers) with the Red form will be more preferable. In consequence, the reaction of holes causes the negative charging of particles and flattening of the energy band. The particles are negatively charged, and the oxidation reactions take place under illumination, so such particles may be treated as micro-photoanodes.

Let us summarize the main processes which occur in/at particles:

$$\mathrm{SC} + h\nu \rightarrow \mathrm{e(CB)} + \mathrm{h^+(VB)}, \text{ photoexcitation, charge separation} \quad (1)$$

$$\mathrm{e(CB)} + \mathrm{A} \rightarrow \mathrm{A^-}, \qquad \mathrm{e(CB)} + \mathrm{Ox} \rightarrow \mathrm{Red} \quad (2)$$

$$\mathrm{h^+(VB)} + \mathrm{D} \rightarrow \mathrm{D^+}, \qquad \mathrm{h^+(VB)} + \mathrm{Red} \rightarrow \mathrm{Ox} \quad (3)$$

There are also other processes that can take place in the semiconductor particles after photoexcitation, such as trapping of charge carriers, relaxation of trapped carriers, and recombination:

$$\mathrm{e(CB)} \rightarrow \mathrm{e_{trap}}, \qquad \mathrm{h^+(VB)} \rightarrow \mathrm{h^+}_{\mathrm{trap}}, \qquad \text{trapping}$$

$$\mathrm{e_{trap}} \rightarrow \mathrm{e}, \qquad \mathrm{h^+}_{\mathrm{trap}} \rightarrow \mathrm{h^+}, \qquad \text{relaxation}$$

$$\mathrm{e{+}h^+}_{\mathrm{trap}}, \qquad \mathrm{e{+}h^+}\ (\text{heat, photon}), \qquad \text{recombination}$$

Such processes, as trapping and recombination, decrease the photocatalytic efficiency. All the above reactions have different time scale. Their time scale depends on the nature of particles, nature of trap centers, reagents, and so on. However, some generalization may be made. The separation of carriers and their trapping are the quickest processes; their time scale is femtoseconds. For relaxation, the time scale is in picoseconds. For the recombination, the time scale depends on its type. The radiative recombination is quick (nanosecond or less). If the trapped carriers take part in the recombination, the time scale can increase up to microseconds. The time scale for charge transfer reactions changes from nanosecond to microsecond depending on the type of reagents and charge carriers participating in the reaction, whether they are free or trapped. In the last case reaction (2) and (3) can be written:

$$e_{trap} + A \rightarrow A^-, \qquad e_{trap} + Ox \rightarrow Red$$

$$h^+{}_{trap} + D \rightarrow D^+, \qquad h^+{}_{trap} + Red \rightarrow Ox$$

Much more about the electron transfer dynamics you will find in the review [4] and references therein.

Up till now, we have considered the freestanding 0D (zero-dimensional) nanoparticles; they are typically applied as photocatalysts. The SC nanoparticles assembly can be used for the formation of photoelectrodes. This type of electrodes has a large specific area for reactions, but unfortunately are also characterized by slow diffusion and high, interparticles recombination of charge carriers. For the photoelectrode construction, the so-called 1D nanomaterials, such as nanowires, nanorods, nanotubes, vertically aligned on the transparent conducting substrates, as well as 2D thin films are more suitable. They are characterized by increased light path length, higher charge mobility, and lower charge recombination. Nanophotocatalysts, as well as nanostructured photoelectrodes, formed by an arrangement of 0D and 1D photocatalysts are studied intensively. For more broad information about the fundamentals of photocatalysis, strategy of development, and perspectives, their interested reader can find in the excellent book [4].

12.3 Photocatalytic degradation of organic/inorganic pollutants

In the last decade, the heterogeneous photocatalysis has been proved to be efficient method for environmental remediation. Among wide spectrum of semiconductors, TiO_2 (anatase and rutile) has attracted attention due to its excellent properties such as the sufficient band gap energy, photostability, nontoxicity, and low cost. TiO_2 was used in PEC for water splitting (Chapter 11, section 11.3), where photogenerated holes and electrons initiate the oxygen and hydrogen evolution. These charge carriers can also be involved in degradation of various organic pollutants, creating very reactive radicals according to the below reactions [5–8]. Light absorption and photogeneration of holes and electrons initiate the chain reactions involving charge carriers, water (moisture), and pollutants:

$$TiO_2 + h\nu \rightarrow e(CB) + h^+(VB)$$

holes:

$$h^+(VB) + H_2O_{ads} \rightarrow {}^\bullet OH_{ads} + H^+, \quad \text{radical may be also free in solution} \tag{1}$$

$$h^+(VB) + OH^-{}_{ads} \rightarrow {}^\bullet OH_{ads} \tag{2}$$

electrons:

$$e(CB) + O_2 \rightarrow O_2^{\bullet -} \tag{3}$$

$$O_2^{\bullet -} + H^+ \rightarrow HO_2^{\bullet} \tag{4}$$

$$2\,HO_2^{\bullet} \rightarrow H_2O_2 + O_2 \tag{5}$$

$$2\,O_2^{\bullet -} + 2\,H^+ \rightarrow H_2O_2 + O_2 \tag{6}$$

$$H_2O_2 + e(CB) \rightarrow {}^{\bullet}OH + OH^- \tag{7}$$

$$H_2O_2 + O_2^{\bullet -} \rightarrow {}^{\bullet}OH + OH^- + O_2 \tag{8}$$

$$\text{organic pollutants} + {}^{\bullet}OH \rightarrow \text{intermediate} \rightarrow CO_2 + H_2O \tag{9}$$

Reactions (1) and (2) generate highly reactive hydroxyl radicals $^{\bullet}OH$, which are mainly responsible for the degradation of organic pollutants (9). The reactions occur on the particle surface, where radicals and pollutants may be adsorbed.

The superoxide radical $O_2^{\bullet -}$ formed in reaction (3) is not so highly reactive; however it may be involved in the formation of hydroperoxy radical $HO_2^{\bullet}$ or H_2O_2 (4, 6), which can be further reduced to the adsorbed or free $^{\bullet}OH$ radical. The oxidative and reductive ability of radicals depend on their equilibrium redox potentials. Since H^+ or OH^- ions are involved in the reactions, the potential is pH dependent and decrease (is more negative) with an increase in pH. The same tendency is observed for SC metal oxide. The position of the edges of the conduction and valence band of SC is shifted to more negative potentials with an increase in pH. It results from acid/base equilibrium for the metal oxide surface groups. Therefore, the E_{fb} potential of semiconducting metal oxides is pH dependent: $E_{fb}(V) = E_{fb}(pH = 0) - 0.059pH$

Figure 12.3 presents the correlation between positions of conduction and valence bands of some SC oxides with the equilibrium redox potentials of important radicals $^{\bullet}OH$, $O_2^{\bullet -}$. From the thermodynamic point of view, in order to initiate the pollutant degradation, the bottom of CB must be located at more negative potentials than the redox potential of $O_2|O_2^{\bullet -}$ couple (−0.28 V vs NHE, pH 7). In addition, the top of VB should be located at potential more positive than the redox potential of $OH^-|\,^{\bullet}OH$ couple (1.6 V vs NHE, pH 7). Looking at Fig. 12.3 we see instantly that oxides such as TiO_2 and ZnO can be applied for pollutants degradation. Other oxides, such as SnO_2, ZrO_2, Fe_2O_3, Nb_2O_5, and so on, have also been intensively studied.

In practice, to choose the best photocatalyst for given reaction, we have to determine and compare the activity of photocatalysts. Quantum yield $\varnothing$ is often used for such comparison [9]. Quantum yield in the heterogeneous photocatalysis is defined as the ratio of reaction rate and the absorption rate of photon flux:

$$\varnothing = \frac{v}{I_a}, \quad I_a = I_0 F, \tag{12.1}$$

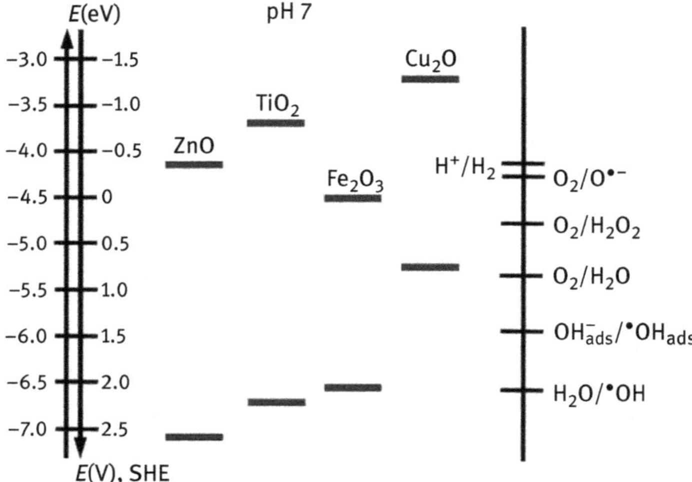

Fig. 12.3: Correlation between the conduction and valence bands of SC oxides (in the energy and potential scales) with the equilibrium potentials of oxygen radicals.

v is the photocatalytic reaction rate, I_a is absorbed photon flux, I_0 is intensity of incident photon flux, and F is the integrated absorption fraction.

It is difficult to assess the absorption rate of photon flux experimentally. Serpone et al. [10] had proposed a standardization protocol to evaluate relative photonic efficiency ($\Im$), using phenol as a standard organic reagent and TiO_2 (P-25 Degussa) as a standard photocatalyst.

$$\Im = \frac{\text{rate of organic compound degradation}}{\text{rate of phenol degradation}} \tag{12.2}$$

Both the initial rates in eq. (12.3) have to be determined under same experimental conditions. Such procedure eliminates the influence of light intensity, reactor geometry, and TiO_2 concentration on the photonic efficiency $\Im$.

The parameter that is used in chemistry for comparing the reaction rates is the rate constant. In the heterogeneous catalysis of organic compounds, the Langmuir–Hinshelwood model is used for kinetic description. This model is based on the assumption that only one reagent (in our case – the pollutant molecule) is adsorbed on the catalyst surface. The Langmuir isotherm is applied for description of surface concentration of reagent (this isotherm is described in Chapter 4). Rate of the first order reaction, v_r, is expressed by the equation:

$$v_r = -\frac{dc}{dt} = k\Theta \tag{12.3a}$$

$$v_r = -\frac{dc}{dt} = k\frac{K_{ads}c}{1 + K_{ads}c} \tag{12.3b}$$

k is the rate constant, Θ is degree of surface coverage, K_{ads} is equilibrium adsorption constant, and c is reagent concentration.

If the initial concentration is low, $K_{ads}c \ll 1$, the equation can be reduced to

$$v_r = -\frac{dc}{dt} = kK_{ads}c = k_{app}c \tag{12.4}$$

k_{app} is apparent rate constant and may be easily evaluated from the plot $\left(-\ln\frac{c}{c_0}\right)$ versus t. To compare the photoactivity of different catalysts, the apparent rate constant should be determined for a model reaction. Decomposition of phenols or dyes, such as methyl orange, rhodamine B, and methylene blue, belong to these reactions.

There are plenty of inorganic pollutants in air such as CO_2, CO, NO_2, SO_2, and so on. They are produced in the road traffic, in the domestic heating, or by industry. Important pollutant is carbon dioxide, whose concentration increases drastically, resulting from application of fossil fuel in the production of energy. There are few methods of reduction of CO_2 contaminations [11]. The most promising seems to be the photocatalytic reduction, in which solar energy is applied. The photoreduction of CO_2 in PE cell was described in Section 11.3, where the thermodynamic data were also presented for the formation of different CO_2 reduction products. In PE cell, the p-type of semiconductor has to be used as photocathode. There are at least two advantages in application of SC particles in CO_2 reduction. One is the increase of reaction efficiency, as the surface area of particles is higher than the surface of conventional SC electrodes. The second, very important, is the possibility of application of any kind (n or p type) of semiconductor particles. It results in the widening of the spectrum of semiconductor materials. It also enables better correlation between CB and VB edges of semiconductor and redox potentials of reaction of CO_2 reduction. Such scheme for few SC is shown in Fig. 12.4.

TiO_2, ZnO, CdS, GaP, and other SC particles were tested as photocatalysts for CO_2 reduction. More broad information on this topic can be found in review and references therein [11]. Up till now, the best photocatalyst for degradation of organic and inorganic pollutants is TiO_2. Its photocatalytic activity depends on physicochemical properties, including the primary particle size, degree of aggregation, crystalline structure, morphology, surface area, and on the external conditions, such as pH, kind of acceptors/donors in solution, and irradiation intensity. Commercially available TiO_2 Degussa P25 is regarded as model catalyst, widely used as a reference to other photocatalysts. The main disadvantage of TiO_2 and other SC oxides is their large E_g resulting in the absorption of light energy in UV range. This problem was discussed briefly in Section 11.3.

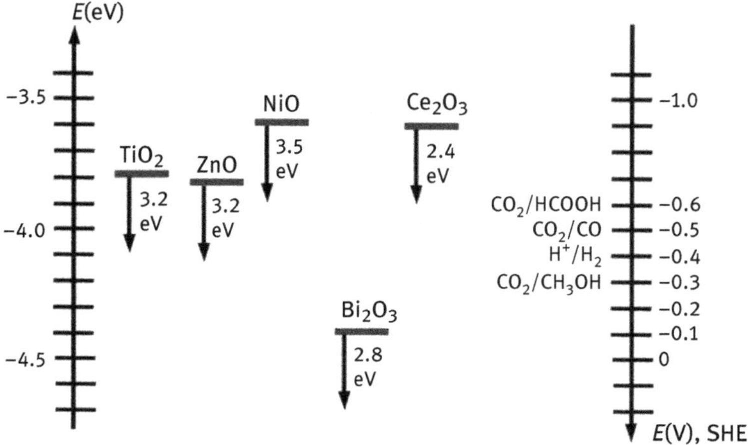

Fig. 12.4: Correlation between the conduction bands of some of SC oxides (in the energy and potential scales) with the redox potentials of CO_2 reduction reactions, pH-7. Arrows point to the position of valence bands, and the bandgap energy, E_g, is marked at arrows.

Bibliography

[1] Butt H-J, Graf K, Kappl M. Physics and chemistry of interfaces. Weinheim, FRG, Willey- VCH GmbH, 2003.

[2] Lyklema J. Fundamentals of colloids and surface science: Solid liquid interface. London, United Kingdom, Academic Press, 1995.

[3] Cao G. Nanostructures and nanomaterials. Synthesis, properties and applications. London, United Kingdom, Imperial College Press, 2006.

[4] Schneider J, Bahnemann D, Ye J, Li Puma G, Dionisiou (eds). Photocatalysis. Fundamentals and perspectives. Cambridge, United Kingdom, RSC Publishing, 2016.

[5] Li J, Wu N. Semiconductor-based photocatalyst and photoelectrochemical cells for solar fuel generation: A review. Catal. Sci. Technol. 2015, 5, 1360–1384.

[6] Schneider J, Matsuoka M, Takeuchi M, Zhang J, Horiuchi Y, Anpo M, Bahnemann DW. Understanding TiO_2 photocatalysis: Mechanisms and materials. Chem. Rev. 2014, 114, 9919–9986.

[7] Linsebigier AL, Lu G, Yates JT Jr.. Photocatalysis on TiO_2 surfaces: Principles, mechanisms and selected results. Chem. Rev. 1995, 95, 735–758.

[8] Fujishima A, Zhang X, Tryk DA. TiO2 photocatalysis and related surface phenomena. Surf. Sci. Rep. 2008, 63, 515–582.

[9] Zhang J, Chen G, Bahnemann DW. Photoelectrocatalytic materials for environmental application. J. Mater. Chem. 2009, 19, 5089–5121.

[10] Serpone N, Sauve G, Koch R, Tahiri H, Pichat P, Piccinini P, Pelizzeti E, Hidaka H. Standardization protocol of process efficiencies and activation parameters in heterogeneous photocatalysis: Relative photonic efficiencies ζr. J. Photochem. Photobiol. A: Chem. 1996, 94, 191–203.

[11] Wang W-N, Soulis J, Yang YJ, Biswas P. Comparison of CO2 photoreduction system: A review. Aerosol Air Qual. Res. 2014, 14, 533–549.

Part IV: Electrochemistry in environmental science – selected topics

13 Sensing the environment

13.1 General remarks

There exist major concerns for the safety of life from a number of harmful pollutants (anthropogenic or not) in the environment: atmosphere, ground (soil), and water. Moreover, on top of the chemical pollution, biological micropollutants (bacteria, viruses, and others), coming from sources such as many human and livestock wastes, enter the environment, groundwater in particular, causing many diseases. Some of such microorganisms acquired multi-drug resistance as the undesired effect of pharmaceutics abuse by humans. The large spectrum of pollutants present in the environment is usually categorized based on their mode of deteriorating effect, chemical structure, or both.

Apart from different initiatives, social concerns, and legislative actions, means of controlling and reporting such threats necessitate the development of sensitive, selective, and rapid detection techniques, screening the environment, and triggering the remediation processes [c.f. 1]. For these purposes, two different approaches are usually considered. The first one, more traditional, typically involves a multi-step approach, such as sampling, handling of a sample, collection of samples, and transportation to qualified laboratories. This time-consuming and costly procedure is nowadays being challenged with in-situ monitoring of environmental threats with the help of electrochemical sensors. Thus, electrochemistry becomes a sustainable way to monitor and resolve the analytical complexity of environmental pollution in a mild, green, and real-time way. Moreover, it can be used for the cost-effective prevention of the formation of toxic materials during industrial processes and their leakage into the environment. A very important from the point of view of "green" electrochemistry for improving the quality and safety of life, is the area of the food industry, where electrochemical sensors attract more and more interest to the industrial community. This is because the issue of food safety has become a significant public concern and awareness worldwide due to its relation to increased human diseases and economic burden. Therefore, there is a need for real-time monitoring of food safety and quality during its production, packaging, and storage, to screen and control at each manufacturing step the presence of contaminant residues, such as antibiotics, allergens, pesticides, toxins, pathogenic bacteria or the microbial breakdown of the food products. In all these areas the advantages of electrochemical sensors such as high selectivity and sensitivity, and minimal power requirements for the related instrumentation and equipment are obvious. Moreover, the use of electrons in electrochemical sensors for signal acquisition and transduction is considered a "***green***" model of detection and analysis of environmental pollution with no waste generation, low costs, fast analysis, and possibilities for miniaturization. Therefore, we will discuss some basic information about electrochemical sensors, their types, and applications for use in the detection of various environmental pollutants af-

https://doi.org/10.1515/9783111160986-014

fecting the everyday quality of life. It is worth keeping in mind, that the rapidly developing technologies and advances in microelectronics and microengineering coupled with the new concepts in nanoscience, allow for the design of miniaturized sensors with enhanced efficiency, sensitivity, selectivity, and durability for the real-time control of the environment. For in-depth information, the interested reader is directed to several publications [2–4]. All the needs mentioned above in environmental monitoring, clinical and food diagnostics, paralleled with rapid development in biotechnologies, also paved the way for the design of biosensors with biological sensing elements, such as enzyme, DNA, aptamer and protein sensors, as well as immunosensors. The use of biological molecules as sensing elements opened up a competitive area of research aimed at selecting the proper element, defining its sensing behavior in the composite matrices of the controlled environment (wastewater, soil, air), and its integration with signal transducer and analyzing setup. For the sensor activities, numerous enzymes from the group of oxidases are utilized, among them urease for metals, laccases, and tyrosinases for phenolic compounds, acetylcholine esterase or organophosphorus hydrolase (anodic oxidation of *p*-nitrophenol as a product of the enzymatic reaction of the latter enzyme). Some of the above-mentioned enzymatic biosensors for their proper performance are coupled to the competitive immune reaction (Chlorsulfuron, Simazine), thus forming the immunoassay-based biosensors. Formaldehyde and sulfur dioxide biosensors rely on formaldehyde dehydrogenase or sulfite oxidase/cytochrome *c*, respectively, for direct redox reaction generating the amperometric signal proportional to the pollutant concentration in the air at equilibrium with aqueous supporting electrolyte. However, as we discussed earlier in this book (Section 9.3), most biosensors are based on reactions catalyzed by macromolecules (e.g., enzymes) that are isolated from their original biological environment, the lifespan of these biosensors is limited, creating a major drawback and motivating permanent efforts to counteract such an issue [5].

13.2 Electrochemical sensors

Electrochemical sensors are by far most frequently employed for the detection of various pollutants, such as heavy metals, herbicides, pesticides, pharmaceutics, and dyes in complex environmental matrices such as air, water, and soil. Taking the definition by the IUPAC of a chemical sensor, the electrochemical sensors in the environment can be defined as devices that produce an electrical, analytically usable signal proportional to the concentration of pollutant as a result of pollutant(s) reaction with sensing electrode, transducing/converting such reaction into an electrical signal. If an integral part of the sensing electrode is of biological origin, then the sensor is called the biosensor and was discussed earlier in this book (Part II, Chapter 9 (Interfacing applied electrochemistry and biology)). Nevertheless, all of the electrochemical sensors can be divided into several groups, depending on the type of transduced signal, such as amperometric, potentiometric, conductometric (impedimetric), photoelectrochemical, and electrogenerated chemiluminescence. The most reliable from the point of view of environmental monitoring are the first three types of sensors, because of the ease of relaying the signal (changes of current, potential, or conductivity) into the control unit, sometimes far away from the sensor location. Therefore, below we will describe the behavior of these three types of sensing electrodes (electrochemical sensors). Of course, the design of those sensors varies, depending on the target pollutant and the environment that is to be monitored. With the rapid technological progress, the sensing processes – the "heart" of such sensors, were transferred from typical, aqueous solutions to less volatile solvents, such as ionic liquids, described in the previous chapter (Section 1.3.1) and to the all-solid-state sensors, utilizing various types of solid-state ion conducting, ion selective or semiconducting membranes capable of monitoring pollutants at high temperatures (e.g., the SO_x, NO_x, and CO_x monitoring sensors in the car exhaust systems). Since this is beyond the scope of this book to cover all existing solutions, we will focus mainly on the electrochemical reactions responsible for the sensing performance in aqueous phase-containing systems. However, there are several key parameters, describing the sensors' performance, regardless of their design. These are the sensor's linear range, its sensitivity, and selectivity. The linear range means the range of pollutant concentration for which the sensor's output/response scales linearly. The sensitivity is defined by the ability to detect minute changes in the sensor environment, whereas selectivity is the sensor's capacity to differentiate between various pollutants. With the sensor's sensitivity also another parameter is related, namely the limit of detection, or LOD (LoD). According to IUPAC, the limit of detection is the lowest sensor response (signal) that can be monitored with sufficient statistical significance. However, the exact threshold (level of decision) used to decide when a signal is significant above the background noise remains somehow arbitrary. The third important parameter defining the sensor's performance is its selectivity. It is quantified by the selectivity coefficient as:

$$log k_{A,Int} = log\left(\frac{S_{Int}}{S_A}\right) \tag{13.1}$$

where S_A is the selectivity against a pollutant A, and S_{Int} is the selectivity against a particular interferent. The more negative values of this coefficient, the more selective is the sensor.

13.2.1 Amperometric sensors

Amperometric sensors utilize controlled voltage applied between the sensing and reference electrode, measuring the resultant oxidation (or reduction) current, being in direct quantitative relation with the pollutant concentration, given by the Cottrell equation (see Chapter 2):

$$i_F(t) = nFADc^0 \frac{1}{(\pi Dt)^{1/2}} \tag{13.2}$$

Under semi-infinite linear diffusion control, if the applied potential is sufficiently negative (for reduction) or positive (oxidation), all pollutant species diffusing to the surface of a sensing electrode are immediately and totally reduced or oxidized (depending on the type of sensing reaction), this current is called the limiting current, i_L, scaling linearly with the pollutant concentration. This type of sensor is widely utilized to control/measurements of concentrations of different pollutants, such as gaseous pollutants (in the gas phase) or pollutants dissolved in a liquid (sewage, wastewater, groundwater, etc.). They provide good sensitivity and a relatively large linear range. They are simple to use and easy to manufacture by, for example, microfabrication technology. Apart from environmental monitoring, these sensors found their use in industrial safety and medical applications. Such sensors can be used to control a wide range of substances in different phases – gaseous and liquid). They are most frequently used to monitor the concentration levels of oxygen (O_2, both in gaseous and liquid phase), nitric oxide (NO, gas phase), sulfur dioxide (SO_2 both liquid and gas phase), ammonia (NH_3, gas phase), carbon monoxide (CO, gas phase), hydrogen sulfide (H_2S, gas) to name a few. The sensitivity range for amperometric sensors varies from part per million (ppm) level to micromolar concentration, depending on the pollutant. Since gas detection is essential to everyday life; the use of sensors for toxic and hazardous gases can set the difference between life and death. Therefore, below we will now consider some examples of amperometric sensors for some most important gas pollutants.

Nitric oxide (NO)

Nitric oxide together with nitrogen dioxide, among other toxic gases, are emitted to the environment in combustion processes that occur, for example, in fossil fuel electric power plants and transportation (diesel, gasoline) engines, affecting our health and the health of animals and greeneries on earth as well as in waters. Nitric oxide is a colorless gas, a radical (N=O), and should not be confused with brownish gas – nitrogen oxide, and a major pollutant. Nitric oxide in the atmosphere can react with ozone, forming nitrogen dioxide and oxygen, thus decreasing the level of ozone in the atmosphere. When exposed to oxygen, nitric oxide also generates nitrogen oxide, according to the reactions:

$$3NO + O_3 \longrightarrow 3NO_2 \text{ or } NO + O_3 \longrightarrow NO_2 + O_2$$

$$2NO + O_2 \longrightarrow 2NO_2,$$

However, in an aqueous or humid environment, these reactions proceed to form nitrous and nitric acids:

$$4\,NO + O_2 + 2\,H_2O \rightarrow 4\,HNO_2$$

$$4\,NO_2 + O_2 + 2\,H_2O \rightarrow 4\,HNO_3$$

Nitrous acid, nitric acid, and sulfuric acid contribute to acid rain precipitation, resulting in corrosive and otherwise deteriorating effects on the environment. However, one can acknowledge also that nitric acid in rainwater is an important source of nitrogen for plants.

Hence, the development of a NO gas sensor for measuring NO levels is very important to monitor and control the evolution of NO gas. Electrochemical detection of NO is typically based on its diffusion-controlled electrooxidation because it results in higher sensitivity and better performance in the presence of interfering dissolved oxygen. The electrooxidation proceeds according to the following reactions:

Sensing electrode (WE): $NO \rightarrow NO^+ + e$
$NO^+ + OH^- \rightarrow HNO_2$ (fast, highly irreversible chemical step)
$HNO_2 + H_2O \rightarrow NO_3^- + 3H^+ + 2e$

Under the neutral or physiological pH, the reaction proceeding at the counter electrode (the cathode) is:

$$O_2 + 4H^+ + 4e \rightarrow 2H_2O$$

The overall redox reaction will be:

$$4NO + 2H_2O + 3O_2 \rightarrow 4\,NO_3^- + 4H^+$$

Depending on the electrode material, nitric oxide oxidation must be performed at sufficiently positive potentials, ranging from +0.7 to 1.0 V vs. Ag | AgCl electrode. Different types of electrode materials have been reported over time, including platinum, gold, glassy carbon, graphite, and other electrodes [6]. Modification of the sensor with gas-permeable membranes significantly improved the sensor's selectivity and overall performance. These membranes, such as Nafion, PTFE, or cellulose acetate, not only facilitated the NO diffusion to the internal electrolyte but also prevented its leakage and evaporation from the sensor. They were also designed to prevent interference of carbon monoxide, CO, but in the case of sensors operating under "wet" conditions, such as wastewater or biological fluids, the interference of ascorbic acid, uric acid, and acetaminophen can be diminished, too. The amperometric sensors can monitor very low redox current (e.g., down to pA) produced by the oxidation of NO over time at a fixed (poise) voltage potential. The response time of this amperometric sensor is only a few seconds, and coupled with its high sensitivity, it provides a fast, quantitative measurement of very small changes in NO concentration.

For a critical review of recent advances of NO sensors, particularly those operating in physiological fluids, the interested reader is referred to [7].

Sulfur dioxide, SO_2

Monitoring and control of sulfur dioxide gas emission to the environment is a serious global problem in relation to the atmospheric pollution and poses a serious threat to population health. Its presence in the atmosphere and subsequently in groundwater mainly comes from industrial petrochemical fuel combustion and coal burning, refineries, and other chemical plants. In contact with humidity and rainwater, it produces acidity in rainwater (acid rains, acid fogs), being a major source of corrosion of buildings and other constructions. The major health concerns associated with exposure to high concentrations of SO_2 are respiratory illness, alterations in the lungs' defenses, and intensification of existing cardiovascular disease.

The reaction that occurred on the sensor's working electrode in the presence of sulfur dioxide in the environment, and in the case of a strong acid as an internal electrolyte, is the oxidation reaction:

Sensing electrode (WE): $SO_2 + 2H_2O \rightarrow SO_4^{2-} + 4H^+ + 2e$

At the same time, at the counter electrode, the cathodic reaction proceeds:

$$\text{CE:} \quad 2H^+ + 2e \rightarrow H_2$$

Thus, the overall reaction in the sensor is:

$$SO_2 + 2H_2O \rightarrow SO_4^{2-} + 2H^+ + H_2$$

Ammonia, NH_3

The most simple ammonia sensor relies on a straightforward oxidation reaction of NH_3 to nitrogen and hydrogen at the sensing electrode. As in the case of other amperometric sensors, the redox current output is used to determine ammonia concentration.

Thus, at the sensing electrode:

$$\text{(WE):} \quad 2NH_3 \rightarrow N_2 + 6\,H^+ + 6\,e$$

At the counter electrode:

$$\text{(CE):} \quad O_2 + 4H^+ + 4\,e \rightarrow 2H_2O$$

And the overall reaction:

$$4NH_3 + 3O_2 \rightarrow 2N_2 + 6H_2O$$

This sensor, even though simple, depends on the availability of oxygen in its electrolyte solution, as well as other electrolyte components. Once the diffusion of oxygen becomes the limiting rate of the reaction, the sensor is no longer capable of detecting ammonia. Therefore, electrochemical ammonia sensors should be used only when the normal ambient background concentration of ammonia is sufficiently low to allow a reasonable operational life. Similar limitations for amperometric sensors are imposed when ambient oxygen reaction proceeds on cathode.

Carbon monoxide, CO

Carbon monoxide is a product of an incomplete combustion of carbon fuel. It is a colorless, odorless, tasteless, and poisonous gas that even in small doses claims permanent health damage or death. In households, ca. 80% of CO-lethal poisoning comes from heating systems and engine-driven tools (including cars). Poisoning by carbon monoxide results from its binding to red blood cells (hemoglobin), blocking their capabilities of transporting oxygen to our body.

In industry, carbon monoxide is important in the large-scale production of numerous specialty compounds, including surfactants, drugs, and fragrances in the processes called hydroformylation. In the Fischer–Tropsch catalytic process, it is used to produce liquid fuels in a series of chemical reactions of the overall scheme: $(2n+1)H_2 + nCO \rightarrow C_nH_{2n+2} + nH_2O$, with n typically above 10. However, upon emission to the atmosphere, carbon monoxide can also contribute to climate change. It is also interesting to note, that CO at very low concentrations can serve as an endogenous neurotransmitter, being highly toxic at higher concentrations, as pointed out above. Because of its high toxicity, carbon monoxide sensors are widely used for process control and safety applications. CO sensing is based on the oxidation reaction on the working electrode, (WE), in the presence of water molecules (either from the supporting electrolyte or from the surrounding air, depending on the sensor's design), according to the scheme:

$$\text{(WE)}: \quad CO + H_2O \longrightarrow CO_2 + 2H^+ + 2e$$

And, at the counter electrode, similar to the NO and NH_3 sensors:

$$\text{(CE)}: \quad O_2 + 4H^+ + 4e \longrightarrow 2H_2O$$

Thus, the overall reaction is:

$$2CO + O_2 \longrightarrow 2CO_2$$

Oxygen, O_2

The necessity of monitoring industrial oxygen related to the safety and health of workers prompted the research on amperometric oxygen sensors. Perhaps the best-known pioneering work in this area was the invention of the Clark electrode used for monitoring oxygen levels in liquids, water, and blood in particular. Briefly, Leland C. Clark's oxygen-sensing concept relies on the use of a two-electrode cell with an oxygen-permeable (breathable) membrane separating the sensor electrode (cathode) from the electrolyte in the cell. Oxygen can diffuse through the membrane being subsequently reduced on the cathode, typically bare platinum, generating current scaling with the concentration of oxygen in the sample, according to the reaction similar to those at counter electrodes for NO, NH_3, and CO sensors: (please note the reaction for DMFC, Chapter 8):

$$(\mathrm{WE}):\quad O_2 + 4H^+ + 4e \rightarrow 2H_2O$$

$$(\mathrm{CE}):\quad 4Ag + 4Cl^- \rightarrow 4AgCl + 4e$$

As one can see, the counter electrode for this sensor can be made from silver, and both electrodes: Pt and Ag are immersed in aqueous saturated KCl solution. Thus, the overall reaction in this sensor is:

$$4Ag + O_2 + 4H^+ + 4Cl^- \rightarrow 4AgCl + 2H_2O$$

It is important to note, that the so-called oxygen concentration in the sample is in fact the partial pressure of oxygen (pO_2) dissolved in the sample at equilibrium, and under a given pH, as described in Section 3.3, eq. (3.10). If one uses these two electrodes only, the precise quantification of oxygen level is frequently inaccurate, because the electrochemical sensing reaction consumes high oxygen due to the rapid oxygen diffusion. Moreover, unintended reactions can interfere with the sensing reaction, altering the readouts of the sensor. Therefore, the introduction of oxygen-breathable, nonconductive membrane mentioned above, was a major development for this sensor. Such membranes can be structured to allow only oxygen to slowly diffuse through them, barring nontarget reactions to occur on the sensing electrode. Yet another problem can affect the readout of Clark's electrode, namely that in an acidic pH the sensing reaction can stop to yield the hydrogen peroxide:

$$O_2 + 2H^+ + 2e \rightarrow H_2O_2$$

Typically, this problem is solved by using a buffer at the Pt electrode that stabilizes the pH of the solution around the platinum electrode.

Apart from serving as a medical device, adaptations of Clark's electrode are used for measurements of the water quality in the oceans, seas, freshwater, underground, etc., with oxygen level as one of the indicators for the, for example, water treatment plants.

Careful readers will certainly recognize that many of the sensing reactions are similar to those discussed in the case of fuel and biofuel cells. However, in those cases, the reactions were carried out in reversed directions.

Some technical notes

A careful reader will immediately find that the electrochemical amperometric sensors are based on the same theory, but different redox reactions, and therefore use different materials for their construction and design, heavily depending also on the environment of their work. They have to perform their sensing tasks, frequently in harsh atmospheres, elevated temperatures, and pressure. It is impossible to cover all currently used solutions, nevertheless let us outline, even briefly, some key factors in the construction of the amperometric electrochemical sensors.

– Electrolyte. It has to be selected as a supporting electrolyte (both solvent and dissociated solutes) to carry out the redox sensing reaction (electrolysis), rapidly and effectively attaining the limiting currents (diffusion-limited, as described in Section 2.1.2) at the working electrode. Obviously, then, the supporting electrolyte should not undergo the redox processes within the potential window of sensing reactions. Moreover, it should not be too volatile to avoid signal interference due to solvent evaporation. This brings forth the issue of the sensor working at elevated temperatures when water as solvent will be unsuitable because of its relatively low boiling point. Therefore, the current focus is on the use of nonvolatile ionic liquids (ILs) as electrolytes. They are advantageous over aqueous electrolytes because they have wide electrochemical windows, high chemical and thermal stability, and intrinsic conductivity. ILs have also very good solvating properties necessary for the pollutant solution, including gases, as well as wetting the electrodes. They are also less leaky and corrosive, improving the lifespan of the sensor in dry or humid, and hot environments. Even more robust and flexible sensors are made with all-solid-state electrolytes (SSEs) placed between the sensing and counter electrodes. Even though the sensing redox reaction is the same as in the case of liquid electrolytes, the volumetric charge transfer mechanism and the mobile species transferring this charge within the solid-state electrolytes are different. The mobile charges (ions) movement occurs in the solid state. As already stated, the main advantage of such electrolytes is the complete removal of liquid components, greatly enhancing the robustness, temperature of the surrounding environment, and safety of the sensor device. However, there is some price that has to be paid, namely, the conductivity of SSEs is much lower compared to the liquid electrolytes. Depending on their chemical nature, all-solid-state electrolytes can be divided into inorganic solid electrolyte (ISE), solid polymer electrolyte (SPE), and composite polymer electrolyte (CPE). This is the most general classification and the interested reader can find more detailed descriptions in the references at the end of this section.

Inorganic solid electrolyte (ISE). An inorganic solid electrolyte (ISE) is constituted by an inorganic material in the crystalline or glassy state, and its ionic conductivity proceeds by diffusion through the crystalline or glassy lattice. The main advantages of this class of solid-state electrolytes are high ionic conductivity and thermal, and chemical stability. They are generally brittle and have relatively low compatibility with the electrode materials, leading to a rapidly increasing interfacial resistance in time, thus limiting their lifespan. The ISE materials can be based on oxides, sulfides, or phosphates, including their mixed crystalline structures like perovskites, garnets, and titanates, to name a few. Frequently applied examples of ISE include LISICON ($Li_{14}Zn(GeO_4)_4$, lithium superionic conductor, NASICON ($Na_3Zr_2Si_2PO_4$, sodium superionic conductor), LIPON (lithium phosphorous oxynitride, a family of materials with the general formula of $Li_xPO_3N_z$, etc.

Solid polymer electrolyte (SPE). Solid polymer electrolyte (SPE) contains salt solution in a polymer host network. It can conduct ions through the polymer chains. Compared to ISEs, SPEs are much easier to engineer by direct solution casting on the electrode material, therefore they can be preferable for manufacturing processes on a large scale. They seem to be more compatible with common electrode material, showing also high plasticity at the electrode surface. Salts that are dissolved in the polymer matrix are typically Li salts due to their high concentration being achieved in the polymer. However, SPEs suffer from generally lower ionic conductivity compared to the ISEs. Some polymeric matrices used for SPE are poly(ethylene oxide) – based (PEO) and its polycarbonates, polyesters, polyamines, polysiloxanes, as well as some eco-friendly biopolymers like chitosan, cellulose or lignin or their blends.

Composite polymer electrolyte (CPE). When SPE is modified with nanoparticles inert to ion conduction, such as alumina, titania, or silica nanoparticles (extrinsic passive fillers), mixed nanoparticulate oxides of lithium or other metals used in, for example, supercapacitors or fuel cells (extrinsic active fillers), then it is called a composite polymer electrolyte. CPE can also be obtained by copolymerization, blending, and crosslinking with other polymers and all these procedures aim at tailoring the properties of host SPE for better dissolution of ionic salts, its ion-conducting properties (ionic mobilities) or only increasing its amorphous state by reducing crystallinity and compatibility with the electrode surface.

It is hard for one material to fulfill the criteria for a good all-solid-state electrolyte to be used in solid-state gas sensors. Below, Tab. 13.1 exemplifies some materials applied for detecting three already discussed pollutant gases.

Tab. 13.1: Examples of solid electrolytes used for sensors of selected pollutant gases.

Pollutant gas	Examples of solid electrolytes
NO	$Ba(NO_3)_2$, Na-β/β″-alumina/$NaNO_3$ (or $Ba(NO_3)_2$, NASICON/$NaNO_2$ + Li_2CO_3
SO_2	K_2SO_4, lithium aluminum titanium phosphate (LATP, $Li_{1+x}Al_xTi_{2-x}(PO_4)_3$) modified with Li_2SO_4
CO_2	NASICON/Li_2CO_3-$BaCO_3$
CO, CO_2	Yttria-stabilized zirconia (ZrO_2/Y_2O_3)

– Electrode materials, catalysts. The electrodes used for gas sensors (as well as for any other type of sensors) should fulfill the following conditions:

- They should be chemically and mechanically stable in the solution or on the substrates
- They must provide excellent contact with the solid electrolyte
- They must be compatible with the breathable membrane/filter
- They must have a geometry that is suitable for sensor construction
- Ideally, they should possess catalytic properties toward the sensing reaction.

Therefore, several metals, such as Ti, Pd, Pt, and Au, carbon-based materials, such as GCE, carbon nanotubes, and conducting polymers, for example, polyaniline, polyacetylene, polypyrrole, etc. are used for the construction of sensors. To improve the sensor performance, catalytic metallic nanostructures are also utilized. Table 13.1 exemplifies some metal nanostructured catalysts used for pollutant detection.

Tab. 13.2: Examples of metal catalysts used for selected pollutant gases.

Pollutant gas	Electrocatalyst
NO, NO_x	Au
SO_2	Au, Au/C
CO	Pt
O_2	Au, Ag, Pt

– Breathable membrane. Typically it separates/covers the sensing electrode but allows the target gas to diffuse to the electrode surface (see Fig. 13.1A below). It should also filter out interfering species and unwanted particles (such as dust). In the case of liquid electrolytes, it should also protect them from drying or leaking. Since condensed water present in harsh environments or high-humidity human breath samples can result in false detection, the breathable membrane should also be hydrophobic. To achieve such properties, hydrophobic polyethylene (PE), polyurethane (PU), and polytetrafluoroethylene (PTFE)-based membranes, among other polymeric structures, are widely used. In particular, PTFE is of high usefulness because it is a highly hydrophobic, thermoplastic with excellent thermal stability and chemical resistance properties. Moreover, it can operate within a wide range of temperatures, from −150 °C to 250 °C. PTFE film is also an ideal host material for further modifications tailored for the specific sensor design improving also selectivity characteristics of gas sensors.

The technology described in this chapter can be modified according to the target gas, working environment, pressure, and temperature, as well as other factors, such as geometry. Apart from applying different materials for the sensors' design, their specificity and selectivity can be further improved by the application of bias potentials specific to the redox reaction of pollutants of interest, and chemically selective filters to block potentially interfering species.

Fig. 13 presents some possible configurations of amperometric gas sensors with liquid or SSE, exemplified for CO and oxygen gas sensors.

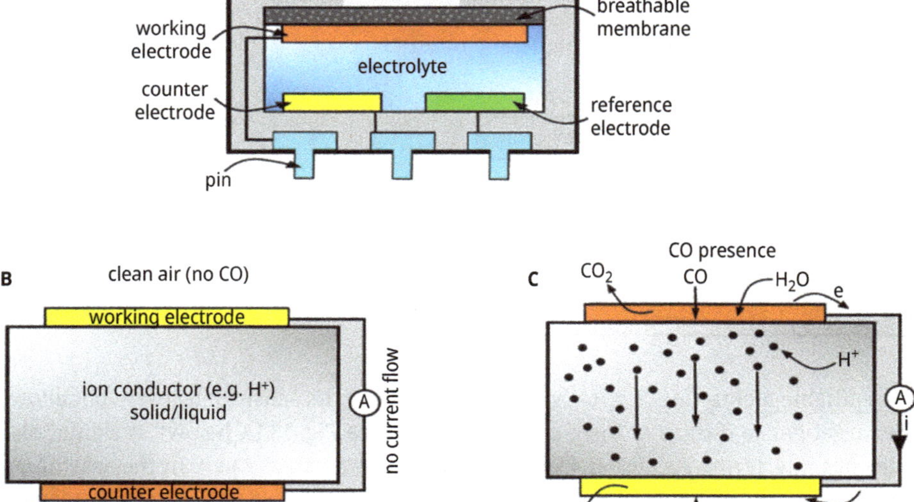

Fig. 13.1: General scheme of a 3-electrode amperometric gas sensor with liquid electrolyte (**A**); scheme of sensing mechanism for the amperometric gas sensor in sensing of CO (**B**), and oxygen (**C**).

13.2.2 Potentiometric sensors

A potentiometric sensor operates under conditions of near-zero current flow and measures the difference in potential between the working electrode (sensing) and a reference electrode. Essentially, when these electrodes are immersed in the solution containing pollutant molecules showing redox properties, even this simple configuration can provide potential signal scaling linearly with the natural logarithm of the concentration of this pollutant according to the Nernst equation (compare Chapter 1, eq. (1.53)):

$$E_N = E^f + \frac{RT}{nF} ln \frac{c_{p,ox}}{c_{p,red}} \tag{13.3}$$

Here, E^f is the formal potential, and $c_{p,ox}$, $c_{p,red}$ are the pollutant's concentrations in its oxidized and reduced forms, respectively. Please note that to apply this equation, one needs a redox-active pollutant and its liquid solution. Moreover, such a design, as in the case of amperometric sensors using liquid or aqueous solution, lacks the necessary robustness and compatibility with in-situ or in-field monitoring. Therefore, the solid-state potentiometric sensors were designed, in which liquid solutions are replaced by ion exchange membranes working as transducers. Ion-selective electrodes (ISEs), of which the pH electrode (Section 1.4) is the best-known example, are the most important of this class of transducers, particularly in control of hazardous ions and compounds in aquatic systems, such as groundwaters, seawaters, sewage, agricultural, industrial and pharmaceutic wastes. Upon immersion in the pollutant solution, across such ion-selective membrane a certain value of potential difference can develop, called the open circuit potential (OCP), or if the quasi-equilibrium conditions are reached, the potential at the sensing electrode can attain the Nernstian slope, according to the equation:

$$E = const. + \frac{2.303RT}{nF} logc_{Me^{n+}} \tag{13.4}$$

Referring to the scheme shown in Section 1.3, the origin of this equation can be explained by the following Scheme 13.1:

Scheme 13.1: Simplified scheme of potentiometric sensor arrangement.

Here, *E(I)* and *E(II)* are the half-cell potentials of reference electrode I and reference electrode II in their solutions, respectively, while $\Delta\varphi_M$ is the membrane potential. Thus, the following potential difference will develop in the sensor at equilibrium:

$$E = E(I) - E(II) + \Delta\varphi_M = const.' + \Delta\varphi_M \tag{13.5}$$

Taking advantage of eq. (1.51), and keeping *E(I)* and *E(II)* half-cell potentials and concentration of the internal solution (II) constant, we can easily derive that the membrane potential in such a case can be described as dependent only on the concentration of pollutant ion Me^{n+} species in the solution (I).

Similar considerations based upon the equilibrium chemical potentials and partial pressures of pollutant gas can lead to the analogous equation for gas sensors:

$$E = const. + \frac{2.303RT}{nF} logp_{pol} \tag{13.6}$$

where p_{pol} is the partial pressure of pollutant gas to be measured, however not all gaseous pollutants can be addressed this way, mainly due to the difficulties in defining the processes at the reference side of a membrane.

Specificity is imposed by the structure and properties of selective membranes, which may be formed from metal salts, oxides, mixed compounds, or polymer membranes containing ion exchangers or neutral carriers. Depending on its physical state and chemistry, generally, there are three ISE membrane types and according to this criterion the sensing electrodes are classified:

– Solid state/crystalline and mixed chemistry membrane electrodes. pH electrodes can belong to this class. Other membranes are typically composed of an insoluble inorganic salt, doped with ions to be detected. For instance, $Ag_2S/AgCl$ can be used for Cl^- detection, or Ag_2S/CuS for the detection of Cu(II). The same Ag_2S polycrystalline matrix can be doped with "insoluble" sulfide of other cations to be detected. Ag_2S has a very low solubility product constant, $pK_{sp} \sim 50$, therefore it is very stable in aqueous solutions. Moreover, it is a relatively good ionic conductor, transducing the equilibration of ion exchange between the pollutant ion in the solution with the dopant ion in the silver sulfide matrix. There are many other structures, such as fluorides of several of the rare-earth metals (e.g., LaF_3 with unusually low resistivities for F^- flow) of ion-selective solid-state membranes, depending on their target ions or even neutral molecules that can be found in the relevant literature.

– Liquid membrane electrodes. This is an organic membrane material, very viscous, and immiscible with water, inaccessible to hydrophilic ions from an aqueous solution. Therefore, an ion-exchanger or ionophore molecules are dissolved in this membrane. Mostly this membrane is not a free-standing one but is supported by a porous structure separating the analyte solution from the internal reference solution. In the case of divalent alkaline earth metal cations, such as Ca^{2+} or Mg^{2+}, the ion exchanger usually is an aliphatic diester of phosphoric acid which has a high affinity to this cation. The ion exchange can be described as:

$$Ca^{2+}{}_{aq} + 2(RO)_2PO^-_{2\,mebrane} \leftrightarrow [(RO_2)PO_2]_2Ca_{membrane}$$

where R is an aliphatic hydrocarbon chain containing 8 to 16 carbon atoms.

As in the case of all ISEs, equilibrium is established at both membrane | solution interfaces. Other ion exchangers may include $R–S–CH_2–COO^-$ for Cu^{2+} and Pb^{2+}, substituted phenantroline ligands for NO_3^- and BF_4^-. Ionophores that can be used for such membranes, depending on analyte ions, include crown ethers, or valinomycin (for K^+). In contrast to charged ion exchangers, the important feature of ionophores is that they are neutral carrier molecules whose structural cavities fit exactly the dimensions of the analyte molecule or ion. An excellent example is the antibiotic – valinomycin which effectively chelates K^+ ions into its electron-rich cavity presenting the outer lipophilic part to the organic hydrophobic membrane. However, from the practical point of view of monitoring the environment, such liquid membrane ISEs are now quite obsolete due to the fast development of solid-state membranes as well as polymer membranes, which provide much more robust sensors showing better selectivities than liquid systems.

– Polymer membrane electrodes. These electrodes constitute more robust and better-performing systems compared to the "wet" membranes discussed above. In these membranes, polymer works as a host matrix for ion exchanger or carrier. To improve the exchange process and ionic mobility, a plasticizer is usually added. The performance of these electrodes is highly selective and they have been used for environmental and health control purposes to determine ions such as K^+, Ca^{2+}, Cl^-, and NO_3^-. Particularly, measurements of blood potassium levels are carried out with polymer membrane electrodes containing valinomycin. To date, there is a range of commercially available ISEs that can detect specific ions (i.e., calcium, nitrate, potassium, copper, barium, chloride, etc.). The environmental control applications of ISEs include the monitoring of concentration levels of trace metals such as iron(III), mercury(II), cadmium(II), copper(II), lead(II), and chromium(VI), as well as important anions such as fluoride, phosphate, sulfate, nitrate, nitrite, chloride and cyanide, mostly in natural waters, but also in tap water and agricultural and industrial drainage and sewage.

The elimination of liquid components in Scheme 13.1 above resulted in the development of solid-contact material for ISE, where the reference, internal solution, and electrode (II) were replaced by metal contacts, intermediated by a layer of conducting polymer. The conducting polymers placed between an electronically conducting metal contact and ionically conducting ion-selective membrane work as the ion-to-electron transducers, completing the charge flow and thus connecting the device to the external electronic circuits reporting the sensor response. Moreover, such configuration simplifies the sensor design and offers unique possibilities for manufacturing miniaturized, personal sensors also for health control and clinical analysis. The performance of such solid contact is dictated by its redox state and ionic equilibrium which are affected by the polymerization conditions and by the composition of the pollutant solution. This performance has been significantly improved down to the nanomolar level, particularly in terms of the detection limits of solid-contact ISE sensors. Conducting polymers applied for sensing devices include polyaniline (PANI, pH sensing, used also in the nonaqueous environment), polypyrrole (ISE for nitrates in waters), poly(3,4-ethylenedioxythiophene), (PEDOT) or polypyrrole (PPy, e.g., for K^+ ISE) and numerous other polymers, co-polymers and their mixtures, some of them biocompatible, that can be found in the literature.

All these conducting polymer (CP) materials have to possess highly reversible redox behavior and ease of penetration of the analyte ion to compensate for the charge in the polymer. It is also obvious, that the polymer layer must form a stable mechanical contact between the metal electrode (Me) and ISE. Taking an example of PPy doped with X^- anion, the ion-to-electron conversion process, in the case of, for example, K^+ analyte ions, can be illustrated as follows:

$$\mathrm{PPy^+X^-(CP) + K^+{}_{aq} + e(Me) \leftrightarrow PPy(CP) + X^-(ISE) + K^+(ISE)}$$

where $PPy^+X^-(CP)$ and PPy(CP) are the oxidation and neutral (reduced) states of PPy in the CP phase, respectively, whereas K^+_{aq} and $K^+(ISE)$ are the concentrations of the analyte ion (K^+) in aqueous and ISE phase, respectively; $X^-(CP)$, $X^-(ISE)$ are the concentrations of X^- anion in CP (polypyrrole) and ISE, respectively, whereas e(ME) are the electrons being transferred between the metal and CP. Please note, however, that this is the overall ion-to-electron reaction scheme, because, as the careful reader should certainly notice, this reaction involves three charge transfers at equilibrium at three-phase boundaries: (i) electron transfer at Me | CP interface, (ii) ion transfer of X^- at the

CP | ISE interface and (iii) analyte ion transfer at the ISE | solution interface. All these equilibria can be described by the Nernstian-type equations.

The potential of the solid-contact CP ISE is the sum of all three interfacial potentials, as discussed above:

$$E = E_{Me|CP} + E_{CP|ISE} + E_{ISE|sol} = const. + \frac{RT}{F}\ln[K^+]_{aq}, \tag{13.7}$$

if the first two potential drops are fixed. Thus, this equation explains the sensitivity to K^+ analyte ions taken as an example. Of course, such an approach can describe the ion-to-electron conversion process of other systems with ISEs and CPs appropriately coupled to fulfill the demands of sensor performance in a given environment, pressure, and temperature.

Some technical notes

All potentiometric sensors are based on the same principle of establishing a potential difference between the sensing and reference electrodes at equilibrium or quasi-equilibrium. The majority of technical requirements described above in the case of amperometric sensors have to be met in the construction of potentiometric sensors. In both types, the elimination of liquid components enables the sensors to work at elevated temperatures. In many industrial processes, such as metallurgy, power plants, or the petrochemical industry, high temperatures are a must. In these areas, the sensors are utilized to analyze and control the emissions to reduce the released hazardous gases by optimizing the combustion processes. If the sensor operates at elevated temperatures, it usually uses solid ceramic electrolytes or ISEs, such as doped zirconium oxide, that can work both in amperometric and potentiometric modes. Such oxygen sensors combining both modes have been recently developed for automotive applications. They are applied to obtain the maximum power from gasoline combustion by controlling an optimum air-to-fuel ratio. However, the traditional design of potentiometric zirconia-based oxygen sensors is based on a thimble-type shape and these sensors are often called the lambda sensors (see Fig. 13.2 below). The sensors are based on yttria- or scandia-stabilized zirconia (YSZ or SSZ) placed between two porous Pt electrodes. The thimble shape and sensor mounting in the exhaust system provides separation of air (reference) and exhaust gas (measured) oxygen partial pressures. At a sufficiently high temperature, the oxygen gas, the mobile oxygen ions in the zirconia lattice, and the electrons in the electrodes are in thermodynamic equilibrium. To achieve it, electronic charge transfer occurs between Pt and the zirconia solid electrolyte as oxygen is introduced or removed from the zirconia lattice. Moreover, the electrochemical potential of oxygen ions has to be constant throughout the whole system. Therefore, the whole system can be represented as an electrochemical cell:

$$O_2(p_{O_2}(\text{gas})),\ \text{Pt}|\ \text{YSZ (or SSZ) matrix with mobile } O^{2-} \text{ions}|\text{Pt},\ O_2(p_{O_2}(\text{reference}))$$

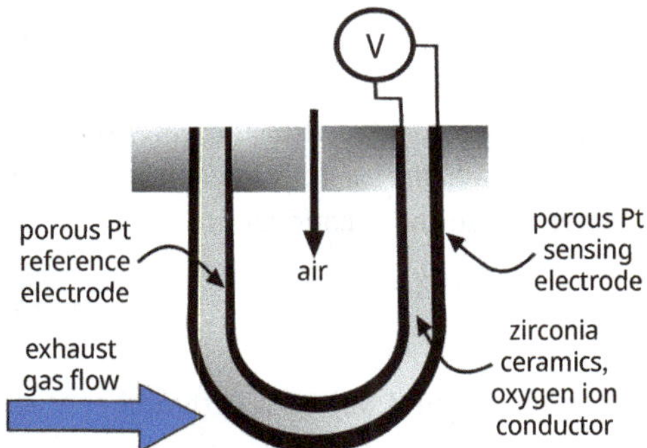

Fig. 13.2: A cross-section of a lambda sensor monitoring the oxygen content in an exhaust gas.

If the sensing electrode and reference electrode are exposed to different oxygen partial pressures, at equilibrium, the electromotive force (EMF) of such sensor (electrochemical cell) can be described by the Nernst equation:

$$E = EMF = t_i \frac{RT}{2nF} \ln \frac{p_{O_2(gas)}}{p_{O_2(reference)}} \tag{13.8}$$

where t_i is the average ionic transference number of mobile ions in the zirconia matrix (for only one ionic species it can be approximated as 1), all other symbols have their usual meanings. Since this equation contains only thermodynamic quantities, this sensor is very robust and stable. Nevertheless, it is important to keep in mind that the above equation implies that only oxygen is involved in the potential determining reaction. However, the exhaust gases contain also NO_x, CO, CO_2, SO_x, as well as volatile organic compounds (VOCs), that can undergo a series of electrode reactions interfering with the sensor readout. Usually, these reactions are nonequilibrium, so to facilitate and improve the sensor performance; the sensing electrode also contains catalytically active materials and is operated at temperatures above 600 °C. At the end of this section it is also worth noting the similarity of the design of the potentiometric lambda sensor with doped zirconia in between the Pt electrodes with the amperometric sensor design shown in Fig. 13.1C, where the ion-conducting ISE (can be zirconia) was placed between the anode and cathode polarized from the external source and generating the flow of electric current as a sensor output signal.

13.2.3 Conductometric sensors

Conductometric sensors rely on modulating their electrical conductivity in response to the environment. Conductivity, σ, or specific conductance (reciprocal of resistivity, ρ) is a material property normalized with respect to area, potential gradient, and time:

$$\sigma = \frac{j}{E} \tag{13.9}$$

where j is the current density passing through the sensor placed in the potential gradient E. The conductometric sensors are also named *chemiresistors* to clearly indicate their functions. Just to remind that $\rho = 1/\sigma$ is defined in the second Ohm's law:

$$R = \rho \frac{l}{A} \tag{13.10}$$

where R is the resistance of the conducting sample, l and A are its length and contact area, respectively, whereas ρ is the sample's intrinsic property directly related to the number of charge carriers within the sample volume. The inherent dependence between current and resistance/conductance (first Ohm's law) sometimes causes problems in separating the amperometric and conductometric sensors. Both types of sensors are electrochemical cells and as such they should be treated with general rules of thermodynamics and electrochemistry. In considering the sensor response, for the sake of simplicity, let us take into account only the pollutant/analyte influence on the overall conductivity of the sensor material. A more detailed analysis can be found in [8]. Let us consider further bipolar arrangements in the sensor construction, in which the chemiresistive material separates two electrodes as in Fig. 13.3.

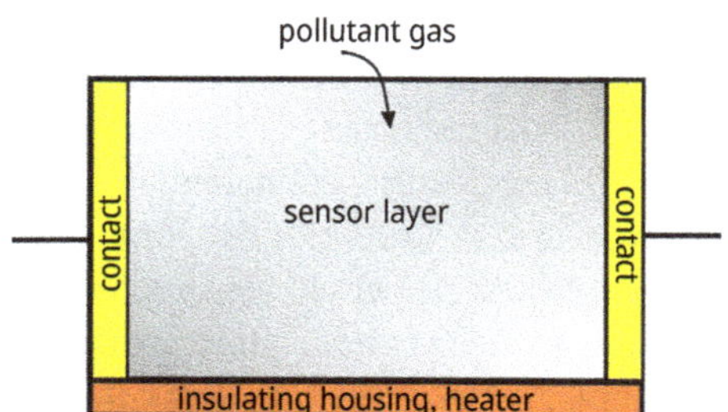

Fig. 13.3: A chemiresisting sensor layer separates two contact electrodes. The sensor is placed on a housing/heater layer to improve its performance.

This arrangement is suitable for controlling the gaseous environment in nonconducting liquids because the conducting samples would short out the chemiresistor. The conductivity of sensing layer for all types of conductors generally depends on the following parameters:

1. Charge carrier mobility (u_i).
2. Number of charge carriers (n_i).
3. Elementary charge (e).
4. The valency of the mobile charges (z_i).

All these parameters are related to conductivity, σ, by the following equation:

$$\sigma = e \sum_i u_i n_i z_i \tag{13.11}$$

This equation is exactly the same as eq. (1.76), however, it is related to single charges, not to the molar concentration of charges, as in the case of eq. (1.74).

Since the vast majority of conductometric sensors are employed for the detection of hazardous gases at high temperatures and aggressive environments, the heater as shown in the schematic graph above is not necessary, however, at ambient temperatures it is generally required to improve the sensors' performance.

The best sensing layers are usually made of semiconducting solids and therefore, the overall conductivity of the sensor depends on the net concentration of available free charge carriers in the semiconducting material:

$$\sigma = e\left(c_n \cdot \mu_n + c_p \cdot \mu_p\right) \tag{13.12}$$

where e is the elemental charge, μ_n, and μ_p are the mobilities of electrons and holes, respectively, whereas c_n, c_p are the concentrations of electrons and holes, respectively.

The vast majority of conductometric gas sensors are based on oxide-type semiconductors, such as ZrO_2, ZnO, or SnO_2 (n-type), or NiO, CoO (p-type). Although the exact mechanism of conductometric changes of the semiconductor gas sensors under the influence of pollutants is still under thorough examination, it is accepted that they are caused by surface adsorption and catalysis. Thus, the principle of operation is based on the presence of surface states forming the space charge layer (discussed earlier in Section 10.2.2), that can exchange electrons with the bulk of the semiconductor. The surface states can be formed by the adsorption of ambient (air) gas molecules (e.g., oxygen – an electron acceptor or hydrogen – donor). In the case of an n-type semiconductor, the adsorbed oxygen species capture electrons from the inner of the semiconductor forming a carrier depletion layer at the surface, reducing the semiconductor conductivity. If the sensing reaction involves oxidation of a pollutant, the electrons delivered to the semiconductor increase its conduction, whereas in the reduction process, involving the electrons from the pollutant, the conduction decreases. For p-type semiconductors, the situation will be the opposite. A more detailed description of the operation principles of semiconducting sensors can be found in [8].

This simple case discussed above assumes a monocrystalline lattice of a semiconducting sensing element, however, the real semiconducting sensors utilized for environmental control are polycrystalline or even amorphous. Then, the morphology, the grain contacts, and other parameters that are beyond the scope of this book become also important.

Some technical notes

The tunable conducting behavior and high sensitivity of their surface states to changes in the composition of the environment led to the great success of semiconductor conductometric gas sensors. These devices can be simple, easy to manufacture on a large scale, compact, and durable, therefore can be designed at reasonable production costs. The semiconducting gas sensors have also some drawbacks. The most important one, particularly for metal oxide semiconductors, is their typically high operation temperature, usually above 500 °C. The elevated temperature is required to mitigate their high sensitivity to moisture and improve their response time and sensitivity. On the other hand, such temperatures may cause poor stability and therefore induce poor energy consumption efficiency. Nevertheless, since they are relatively simple and robust, they are widely put to practical use when the sensing processes are necessary in demanding environments and at elevated temperatures, (e.g., combustible hazardous gases from metallurgy, refineries, or exhaust from transportation), where the carrier mobilities and reaction kinetics are advantageous for sensors' performance. In order to alleviate the need to heat the sensing element and thus decrease the level of power consumption without compromising the sensor performance, the development of chemiresistive gas sensors, new gas sensor materials are being sought. For this purpose nonoxide semiconductors, such as cadmium sulfide, cadmium selenide, and cadmium telluride, and their composites have been explored. It was reported that these materials offer an additional advantage of enhanced pollutant gas response, that is, the electrical conductivity change ratio (gas sensitivity), under illumination compared to those in the dark. This phenomenon is explained by the effect of light on gas adsorption/desorption and excitation of the electrons in the semiconductor. Also, their operating temperatures are much closer to the room temperature, decreasing the power consumption and diminishing the operational safety issues.

Other possible directions that are based on rapidly developing nanotechnologies, include nanostructurization of the sensor design (nanowires, nanobelts) [9], combining multiple types of semiconducting sensing materials [10, 11].

13.3 Future perspectives and challenges

Permanent heterogeneity in time and space of the pollutant in various environments, such as aquatic, earth and air, complicates monitoring and appropriate, fast responses. This is particularly complicated in a largely unpredictable environment – atmosphere. Air pollution is one of the main challenges to health and environmental sustainability. According to the World Health Organization, the vast majority of the world's population lives in places where air quality exceeds the limits recognized as safe. The concentration of air pollutants depends on localization, population and industrial densities as well as time. Taking an urban environment as an example, its local atmosphere is affected mainly by anthropogenic and natural processes and varies in cycles related to urban activities and related emissions. Thus, better sensing and local pollution assessment is a must. Therefore, the static, traditional monitoring at fixed locations becomes frequently insufficient. In response to these needs, backed by the new capabilities in microelectronics, microfabrication, and nanomaterials science, new research in portable sensing device design and development is rapidly growing. This may finally lead to personalized air pollutant sensors and monitoring systems, assessing both the health and pollutant exposures of individuals. However, proper assessment of the relation between the exposure and negative health effects on humans (to say nothing on other organisms) is limited by the availability of reliable data, collection and computational/modeling capabilities. One of the pathways to overcome these difficulties is the development of miniaturized sensor arrays that can work in liquid or atmospheric conditions. If each sensor in the array has different sensitivity and selectivity, and the array is coupled to signal processing and pattern recognition engine, then such system is called electronic tongue (e-tongue for liquids) or electronic nose (e-nose for air). This is roughly shown by the scheme below (Scheme 13.2):

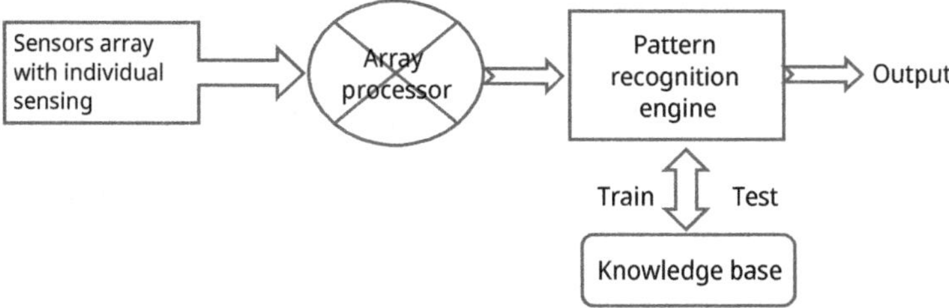

Scheme 13.2: Diagram illustrating the functioning of e-tongue or e-nose. Adapted after [10].

Thus, the combination of modern electrochemical techniques with breakthroughs in microelectronics and miniaturization allows the introduction of powerful analytical devices for effective pollution control. This in turn, coupled with improvement of the sensor response, development of machine learning and deep learning can pave the way for pollution monitoring even at a personal, miniaturized level. However, regardless of the type of output signal being generated, the final decision-making for a greener environment is in our hands.

One final word is worth adding: where environmental sensing and power supply start to synergize, new opportunities for multifunctional devices based on similar chemistry to store energy or detect environmental hazards (or both), are emerging simply by the choice of electrodes and electrochemistry. One of such example can be self-powered sensors with the ability to scavenge energy from the environment to self-drive the sensor's operation. They can harvest and convert into electric energy, solar, mechanical, thermal energies, etc., that are available in the environment for long-term sustainable, "green" operation [12].

Bibliography

[1] Kim M-Y, Lee KH. Electrochemical sensors for sustainable precision agriculture – a review. Front. Chem. Sec. Electrochem. 2022, 10, 1–14.

[2] Simoes FR, Xavier MG. Electrochemical sensors. In: Nanoscience and its applications. Elsevier, 2017, 155–178.

[3] Hussain CM, Keçili R. Electrochemical techniques for environmental analysis. In: Modern environmental analysis techniques for pollutants. Elsevier, 2020, 199–222.

[4] Baranawal J, Barse B, Gatto G, Broncova G, Kumar A. Electrochemical sensors and their applications: A review. Chemosensors 2022, 10, 363.

[5] Electrochemical biosensors. In: Cosnier S, ed. Pan Stanford series on the high-tech of biotechnology, Vol. 3. Pan Stanford Publishing Pte. Ltd, 2015.

[6] Xu T, Scafa N, Xu L-P, Su L, Li C, Zhou S, Liu Y, Zhang X. Electrochemical sensors for nitric oxide detection in biological applications. Electroanalysis 2014, 26, 449–468. doi: 10.1002/elan.201300564.

[7] Brown MD, Schoenfisch MH. Electrochemical nitric oxide sensors: Principles of design and characterization. Chem. Rev. 2019, 119, 11551–11575.

[8] Janata J. Principles of chemical sensors, Chapter 8, Conductometric sensors. Springer US, 2009. doi: 10.1007/978-0-387-69931-88.

[9] Neri G. Thin 2D: The new dimensionality in gas sensing. Chemosensors 2017, 5, 21. doi: 10.3390/chemosensors5030021.

[10] Korotcenkov G, Cho BK. Engineering approaches for the improvement of conductometric gas sensor parameters Part 1. Improvement of sensor sensitivity and selectivity (short survey). Sens. Actuators B 2013, 188, 709–728.

[11] Ando M, Kawasaki H, Tamura S, Haramoto Y, Shigeri Y. Recent advances in gas sensing technology using non-oxide II–VI semiconductors CdS, CdSe, and CdTe. Chemosensors 2022, 10, 482.

[12] Alagumalai A, Shou W, Mahian O, Aghbashlo M, Tabatabaei M, Wongwises S, Liu Y, Zhan J, Torralba A, Chen J, Wang ZL, Matusik W. Joule 2022, 6, 1475–1500. doi: doi.org/10.1016/j.joule.2022.06.001.

14 Green fuel: hydrogen production

14.1 General remarks

As global warming and pollution of the environment are increasing, the investigation and development of renewable energy sources are crucial. The most promising source of clean energy, considered as the fuel of the future, is hydrogen.

Hydrogen plays an essential role in many industrial processes such as the production of ammonia (Haber process), the production of methanol, oil refining, and others. As a fuel, hydrogen can directly react with oxygen (air) in the fuel cells to provide electrical energy or can be burned in combustion engines instead of fossil fuel. So it is not surprising that the production of hydrogen increases every year and the global hydrogen market is expected to grow to about US$220 billion by 2030. Hydrogen can be produced from various nonrenewable and renewable energy sources. Currently, the main sources for commercial production of hydrogen are fossil fuels: natural gas, coal, and oil. Application of fossil fuels accounts for about 95% of total H_2 production worldwide. Plenty of methods are used for hydrogen generation from fossil fuels such as steam reforming of natural gas, partial oxidation of heavy hydrocarbons, or coal gasification. Other processes and technologies currently developed include direct methanol reforming, reforming and pyrolysis of biomass, and fermentation of biomass. However, in many methods besides hydrogen, carbon dioxide is formed, which is emitted to the atmosphere, causing temperature increase and climate changes. To reduce the amount of CO_2 released into the environment, some technologies called carbon capture and storage (CCS) have been developed. Currently, in industry, hydrogen is mainly produced from natural gas (methane) by using the steam methane reforming (SMR) method. In this way, about 50% of industrial H_2 is generated. In the first stage of the SMR method, methane (natural gas) is heated to the temperature of 700–1,000 °C at a pressure of 3–25 bar in the presence of steam and a proper catalyst (nickel). In this step, except hydrogen, carbon monoxide is produced. In the second stage, carbon monoxide and steam react at a lower temperature of about 360 °C, forming carbon dioxide and hydrogen in a so-called water-gas shift reaction:

$$CH_4 + H_2O \rightarrow CO + 3H_2 \quad \text{first stage, endothermic reaction}$$

$$CO + H_2O \rightarrow CO_2 + H_2 \quad \text{second stage, exothermic reaction}$$

Finally, the carbon dioxide and other impurities have to be removed from the gas stream. When CO_2 is released into the atmosphere, the obtained H_2 is often termed "gray hydrogen," and when it is captured (separated) by CCS processes, it is termed "blue hydrogen." Comparing hydrogen as a fuel with common fossil fuels used, we can find some advantages and disadvantages. The gravimetric energy density of hydrogen (141.8 MJ/kg) is about 2–3 times higher than that of some fossil fuels, which means that hydrogen can generate several times more energy than conventional fuels. For instance, the gravimet-

https://doi.org/10.1515/9783111160986-015

ric density energies of methane and diesel are 55.6 MJ/kg and 45.4 MJ/kg, respectively. Hydrogen has the lowest flash point (−231 °C), so even at very low temperatures enough hydrogen evaporates for ignition. The flammability range of hydrogen is the widest (4–75%) among other fuels. It is its disadvantage, but on the other hand, as water vapor is produced during the combustion of hydrogen, flames are not toxic. An alternative technology applied for the production of hydrogen from fossil fuel is water electrolysis. Using these technologies, very pure hydrogen can be obtained (up to 99.99%), suitable for application in fuel cells. More information about other methods of hydrogen production, hydrogen economy, and technologies of production and storage can be found in [1].

14.2 Water electrolysis

Water electrolysis is the simplest method of hydrogen production. In the process, the reactant, renewable H_2O, is decomposed to pure H_2 and O_2 by utilization of the electrical energy. Currently, only about 4% of total H_2 world production is generated by using water electrolysis. To date, the cost of hydrogen production through water electrolysis is too high to be economically competitive with established technologies of hydrogen production from natural fossil fuels (about 5 times). This is due to the high electrical energy demand for electrolysis and low efficiency, resulting from energy losses during the reaction of water splitting. Plenty of efforts were and are made for the development of existing technologies of water electrolysis, to make them more effective and less expensive. Much research has been done related to the improvement of the efficiency of electrolytic reactions, the finding of new low-cost catalysts, or the application of alternative, renewable sources of electrical energy, such as wind energy or solar energy. Hydrogen obtained in this way is termed "green hydrogen."

A few types of methods and electrolytic cells (electrolyzers) can be used for water splitting. Two of them, alkaline water electrolysis (AWE) and polymer electrolyte membrane electrolysis (PEM), are applied in the industry, however, on different levels. As the technology of AWE is very well established, it is widely used for H_2 production on a big scale, while PEM is applied in small factories. Other methods such as solid oxide electrolysis (SOE) and microbial electrolysis cell (MEC) are still developed. All these methods are described briefly in Section 14.2.2. More comprehensive information regarding water electrolysis is given in reference [2], and others are cited in the text.

14.2.1 Thermodynamics, energy losses, and efficiency

On a laboratory scale, the electrolysis of water can be carried out in simple electrochemical two-chamber cells containing two planar electrodes: a cathode and anode made from inert metal (mostly Pt), and an aqueous solution of electrolyte, by using a DC electrical power source. Electrolysis of pure water is also possible, but as the conductivity resulting from hydronium (H^+) and hydroxide (OH^-) ions (5.7×10^{-6} S/m, $K_w = 1 \times 10^{-14}$) is low, the process requires more electrical energy than in the case of aqueous electrolyte solution. Different electrolytes can be applied, but their ions cannot undergo any competitive electrode reactions. Because of that, the standard redox potentials of electrolyte cations should be lower than the standard redox potential of H^+ and the standard redox potential of electrolyte anions should be higher than the standard redox potential of OH^- ions. The chambers containing the cathode and anode are separated by a diaphragm or an ion-conducting membrane. The reactions that take place at the cathode and anode depend on the pH of the solution.

In acidic media, they are as follows:

$$\text{Cathode:}\quad 4H^+(aq) + 4e^- \longrightarrow 2H_2(g),\ E^0 = 0\,\text{V} \tag{14.1a}$$

$$\text{Anode:}\quad 2H_2O(l) \longrightarrow O_2(g) + 4H^+(aq) + 4e^-,\ E^0 = 1.229\,\text{V} \tag{14.1b}$$

In neutral or basic solution, hydroxide ions participate in an oxidation reaction:

$$\text{Cathode:}\quad 4H_2O(l) + 4e \longrightarrow 2H_2(g) + 4OH^-(aq),\ E^0 = -0.828\,\text{V} \tag{14.2a}$$

$$\text{Anode:}\quad 4OH^-(aq) \longrightarrow O_2(g) + 2H_2O\,(l) + 4e^-,\ E^0 = 0.401\,\text{V} \tag{14.2b}$$

E^0 means the standard electrode potential of an electrode in equilibrium, related to SHE (Pt/H_2, $T = 298.15$ K, H_2 ($p = 1$ bar, 1×10^5 Pa), H^+ ($a = 1$)):

$$\Delta E^0 = E^0_A - E^0_c = 1.229\,\text{V} \sim 1.23\,\text{V}$$

$$\Delta E_{eq} = \Delta E^0 - \frac{RT}{4F}\ln\frac{p^2_{H_2}p^2_{O_2}}{a^2_{H_2O}} \tag{14.3}$$

In (14.1) and (14.2), the overall reactions are the same, gaseous hydrogen and oxygen are generated at the cathode and anode, respectively:

$$2H_2O(l) \longrightarrow 2H_2(g) + O_2(g)$$

The reaction is the inverse of the hydrogen–oxygen fuel cell reaction, where the chemical energy of the reaction is transformed into electrical and thermal energy. As a matter of fact, some fuel cells can operate also as water electrolysis cells (electrolyzers), depending on the current direction. The formation of H_2 and O_2 from water requires energy. This energy is equal to the enthalpy reaction $\Delta_r H$ of water decomposition:

$$\Delta_r H = \Delta_r G + T\Delta_r S \tag{14.4}$$

where $\Delta_r G$ is the Gibbs free energy of the reaction and $\Delta_r S$ is the entropy of the reaction, and can be supplied to electrolytic cells in the form of electrical and thermal (heat) energy or only as electrical energy. The thermal energy compensates for the losses connected with changes in the entropy of the reaction and is equal to product $T\Delta_r S$. The minimal electrical energy required for water decomposition is described by the Gibbs free energy of reaction $\Delta_r G$. When energy is delivered to the electrolytic cell only in the form of electrical energy, the amount of electrical energy should be higher as the entropy loss has to be compensated.

In standard conditions (T = 298.15 K, p = 1 bar), the total energy (work + heat) of decomposition of 1 mol of water is described by the standard enthalpy of reaction $\Delta_r H^0$ equal to 285.83 kJ/mol. The contributions of the electrical energy (work) $\Delta_r G^0$ and thermal energy $T\Delta_r S^0$ (heat) are 237.13 kJ/mol and 48.7 kJ/mol, respectively.

Let us consider two cases. First is when the heat energy 48.7 kJ/mol is supplied to the electrolysis cell (for instance, by thermostat) together with electrical energy, and the second is when only electrical energy is used. In the first case, assuming that all the thermal energy is consumed during the reaction, we can calculate the equilibrium potential of electrolysis cell, ΔE^0, by using the equation $w = \Delta_r G^0 = nF\Delta E^0$ (compare equation (8.3) and note the positive sign of work, because it is introduced into the cell), where $n = 2$, $F = 9.648 \times 10^4$ C/mol, $\Delta_r G^0 = 237.13$ kJ/mol. ΔE^0 is equal to 1.229 V and is also the standard minimal potential of water splitting referred to the low heating value (LHV) of hydrogen, equal to 241.8 kJ/mol or 120 MJ/kg. LHV is described as the thermal energy of hydrogen combustion and formation of water in a vapor state and can be determined by subtraction of the heat of water vaporization ($\Delta_v H^0 = 44.0$ kJ/mol) from the enthalpy of water formation in a liquid state. When electric energy is only applied for water electrolysis, without heat contribution, we used for potential calculation the value of total energy, which means the enthalpy of reaction. In this case, $\Delta_r H^0 = nF\Delta E^0_{\mathrm{tn}}$, where $n = 2$, $F = 9.648 \times 10^4$ C/mol, $\Delta_r H^0 = 285.83$ kJ/mol, ΔE^0_{tn} is termed thermoneutral potential of electrolysis cell and is equal to 1.481 V. The thermoneutral potential is also the minimal potential that should be used for the electrolysis of liquid water in standard conditions. This potential is referred to as the high heating value (HHV) of hydrogen equal to 285.83 kJ/mol or 142 MJ/kg. HHV is equal to the heat of hydrogen combustion and formation of water in its liquid state (enthalpy of water formation in liquid state at standard conditions).

Comparing the standard values of free Gibbs energy and enthalpy of reaction, it is clearly seen that about 20% more energy is required for electrolysis when only electrical power is applied. The increase in temperature causes the decrease in $\Delta_r G$ and influences the minimal potential of water splitting. For instance, at 150 °C, $\Delta_r G^*$ is equal to 221.49 kJ/mol and E^* to 1.15 V (asterisk means not standard temperature). The influence of T on water splitting is more significant at higher temperatures. The increase of temperature up to 1,500 °C reduces the value of the reversible thermodynamic potential of water decomposition E^* to ~0.7 V [3]. Considering only thermodynamic factors, we assumed that reactions occur ideally and all energy losses in operating electrolytic cells can be neglected. In practice, to overcome the energy losses and to maintain the electrolysis of water, we have to apply higher values of potentials (higher than 1.23 V or 1.48 V). This difference in potentials called overpotential is described by the equation

$$\eta = \Delta E - \Delta E_{\mathrm{eq}} \tag{14.5}$$

where ΔE is the potential applied to the electrolysis cell and ΔE_{eq} is the difference of equilibrium potentials of electrodes in the cell described in equation (14.3), when current does not flow ($I = 0$). If we consider the reaction at one electrode (cathode or anode), this difference is the measure of polarization of electrode $\eta = E - E_{eq}$.

A few kinds of overpotentials can be considered in the operating electrolytic cell: (i) ohmic overpotentials, η_{ohm}, often termed Ohmic losses; (ii) activation overpotential η_a originated from hindrances in the kinetics of electrode reaction at cathode and anode; (iii) concentration overpotentials, η_c, resulting from problems with transportation of reactants or products to/from electrodes.

Therefore, the potential that has to be applied for water electrolytic cell is described by the equation:

$$\Delta E = \Delta E_{eq} + \sum |\eta| = \Delta E_{eq} + \eta_{ohm} + \left(\eta_{a,C} + \eta_{a,A}\right) + \left(\eta_{c,C} + \eta_{c,A}\right) \tag{14.6}$$

C and A represent cathode and anode, respectively.

In practice, the two overpotentials, η_{ohm} and η_a, significantly increase the minimal potential required for water electrolysis. The concentration overpotential is governed by the diffusion process, depending on the concentration gradient of reactants, and can be eliminated to a greater extent by a continuous flow of reactant (H_2O) to electrodes and efficient removal of products H_2 and O_2. For these purposes, special systems for water pumping and circulation, gas separators, and collectors are designed for the different types of electrolyzers. Much more about the origin of concentration overpotential, you can find in Section 1.1.3.2.

To carry out the electrolysis of water with good efficiency, we should minimize the losses of energy caused by ohmic and activation overpotentials. The ohmic overpotential depends on the internal resistance of the cell R_{in}:

$$\eta_{ohm} = IR_{in} \tag{14.7}$$

where I is the current flowing through the cell.

There are a few components of internal resistance: (i) resistance of the electrolyte solution depending on the conductivity of the electrolyte and on the volume of the electrolyte between electrodes, and (ii) resistance of electrodes depending on the conductivity of electrode materials, on the area of the electrode, and the special design of electrodes with dispersed electrocatalysts. Note that the working, active surface of porous electrodes loaded with dispersed catalyst is difficult to determine. If planar, hemispherical, or disk electrodes are applied, the geometrical area of the electrode is usually used for density current calculation. In other cases, special methods should be used to determine the real, active electrode area. However, the main factor determining the internal resistance of the cell is the design of separators of the cathodic and anodic parts of the electrolytic cell. Depending on the type of the electrolytic cell, the polymer ion conducting membranes or inorganic diaphragms are used as separators. Their resistance depends on many factors; for instance, the resistance of proton conducting membrane depends on hydration, the ionic state of the membrane, history of membrane preparation, and treatment. In electrolyzer operating at high temperatures, in the range 800–1,000 °C, the solid oxide (SO) electrolytes are used, their conductivity depends on composition and temperature, and may be increased by special treatment. By good engineering, the ohmic losses can be reduced to ~0.1 V.

The activation overpotential η_a has its origin in the kinetics of reactions, their slowness, their irreversibility, and the formation of new phases and is minimized by the application of proper electrocatalysts. The relation between η_a and current density (j) can be evaluated from the Butler–Volmer equation (see Section 1.1.3.3). For higher values of applied potential or current, this relation is known as the Tafel equation:

$$\eta_a = a + b\ln|j| \tag{14.8}$$

where j is the current density

$$a = \frac{RT}{\alpha nF}\ln j_0, \qquad b = -\frac{RT}{\alpha nF} \quad \text{for } E \ll E_{eq}$$

$$a = -\frac{RT}{\beta nF}\ln j_0, \qquad b = \frac{RT}{\beta nF} \quad \text{for } E \gg E_{eq}$$

R, T, F, and n have the usual meaning, α, β are transfer coefficients, j_0 is the exchange current density, and coefficient b is often called the Tafel slope.

Note that these relations are accomplished, when the reaction is going at one working electrode and also when we consider the operating cell with different reactions on the cathode and anode.

In this case

$$\eta_{a,C} = -\frac{RT}{\alpha nF}\ln\frac{j}{j_{0,C}} \tag{14.9a}$$

and

$$\eta_{a,A} = \frac{RT}{\beta nF}\ln\frac{I}{j_{0,A}} \tag{14.9b}$$

The exchange current describes the oxidation and reduction currents, which flow through the electrodes at equilibrium potential and is proportional to the standard rate constant of the reaction (1.104). As the standard rate constant depends on the standard Gibbs free energy of activation of the reaction (1.92), all factors that decrease the activation energy of the reaction cause the increase of the rate constant and the exchange current. In consequence, the activation overpotential decreases. Another parameter, the Tafel slope (b), gives us insight into the reaction mechanism providing information on elementary steps and rate-determining steps (rds). To increase the reaction rate, catalysts play a crucial role (see Chapter 4).

In the case of water electrolysis, the HER (hydrogen evolution reaction) at the cathode and OER (oxygen evolution reaction) at the anode are multistep reactions with intermediate formation, influencing the rate of reactions and efficiency. Pathways of HER are much simpler than OER. In the first step of HER, in an alkaline solution, the water is split with the formation of hydrogen adsorbed (Volmer step) on the free active site of electrode/catalyst (M):

$$H_2O + e \leftrightarrow H_{ads}(M) + OH^-$$

This is followed by an electrochemical step of hydrogen formation (Heyrovsky step) or by chemical recombination of adsorbed hydrogen atoms (Tafel step):

$$H_2O + H_{ads}(M) + e \leftrightarrow H_2 + OH^-$$

$$2H_{ads}(M) \leftrightarrow H_2$$

The overall HER is a two-electron reaction. The crucial factor, which determined the rate of HER and its activation overpotential is Gibbs free energy of H adsorption, $\Delta G^0_{ads,H}$ – a measure of the binding energy of H with electrode surface. If the binding energy is too low or too high, the exchange current densities of HER are low, in the range of 10^{-9}–10^{-7}A/cm^2. The moderate values of binding energy facilitate the HER. This is the case of Pt and metals of Pt group – Ir, Rh, and also of Co and Ni. Their H binding energy is in the range of 260–280 kJ/mol, and the values of exchange current densities lay in the range of 10^{-4}–10^{-2} A/cm^2, depending on the catalyst and solution used. Pt is considered the best catalyst for HER in acidic and alkaline media, but as the first step in alkali media is the sluggish reaction of water dissociation, the rate of HER is much lower than in acidic media (about 2 orders of magnitude). Generally, for good HER electrocatalysts, the overpotential should be lower than 0.1 V.

The OER is a much more complex reaction than HER and requires much higher overpotentials. To date, the mechanism of OER is discussed and further evaluated. Generally, the adsorption evolution mechanism is accepted assuming the four electron transfer steps with the formation of oxygen-containing intermediates, their adsorption, and binding with the active side (M) of the electrode/catalyst. The formation of oxygen-containing radicals such as $^{\bullet}OH$, $O^{\bullet}$, and $HOO^{\bullet}$ ($HO_2^{\bullet}$ notation is also used) is postulated. In alkali media, the first step is the charge transfer step with the formation of adsorbed $^{\bullet}OH$:

$$OH^-{}_{ads}(M) \leftrightarrow {}^{\bullet}OH_{ads}(M) + e$$

It is followed by the further charge transfer and formation of adsorbed $O^{\bullet}$:

$$OH^- + {}^{\bullet}OH_{ads}(M) \leftrightarrow O^{\bullet}{}_{ads}(M) + e + H_2O$$

Thereafter, two pathways may be considered: one similar to the Tafel reaction of hydrogen formation:

$$2O^{\bullet}{}_{ads}(M) \leftrightarrow O_2 + 2e$$

And the other with the formation of $HOO^{\bullet}$ ($HO_2^{\bullet}$) radicals and O_2 in the reactions

$$O^{\bullet}{}_{ads}(M) + OH^- \leftrightarrow HOO^{\bullet}{}_{ads}(M) + e$$

$$HOO^{\bullet}{}_{ads}(M) + OH^- \leftrightarrow O_2 + H_2O + e$$

The oxygen-containing intermediates (radicals $^{\bullet}OH$, $O^{\bullet}$, and $HOO^{\bullet}$) are adsorbed on the catalyst surface through O atoms forming one ($^{\bullet}OH$, $HOO^{\bullet}$) or two bonds ($O^{\bullet}$). The binding energy of these species with catalysts decides about the values of overpotentials of OER and rds. If the oxygen of the intermediate binds with the catalyst surface too strongly, the rds is the formation of $HOO^{\bullet}$, if too weakly – the rds is the oxidation of adsorbed OH^- ions. As in the hydrogen case, the best catalysts for OER have moderate values of binding energy. At present, the best catalysts applied for OER are IrO_2, RuO_2, or mixed oxides; however, even the benchmark IrO_2 catalyst has an overpotential of about 0.3–0.4 V, at a current density of 10 mA/cm^2. Plenty of efforts are made to improve the efficiency of electrocatalysts in OER. Much more about the development and progress in the field of electrocatalysts for HER and OER can be found in reviews [4–7] and references therein.

Theoretically, in standard conditions, the minimal potential that should be used for water splitting is 1.229 V (or 1.481 V), but as a result of energy losses, the potentials demanded are higher. In practice, most water electrolyzers operate in the potential range of 1.6–2.0 V depending on the type of electrolyzer, its construction, and operating condition.

If we want to estimate the usefulness of the water electrolytic cells for hydrogen production we have to compare their efficiency. Many formulas are applied, and there is controversy about what value HHV or LHV should be used in calculations.

For electrochemical processes, the Faradaic efficiency is a basic dependence. In the case of hydrogen production, it can be defined as the ratio of the amount of hydrogen (in mole, volume) evolved during electrolysis and the theoretical amount of hydrogen that should be produced in the same electrolytic condition. The theoretical amount of H_2 is calculated by using Faraday's law, assuming 100% of Faradaic efficiency; the amount of evolved H_2 during electrolysis may be determined by using gas chromatography or other methods:

$$\text{Faradaic efficiency} = \frac{\text{amount of } H_2 \text{ produced}}{\text{amount of } H_2 \text{ calculated}}$$

It is accepted in Europe and recommended that the water electrolysis efficiency should be calculated by using HHV of hydrogen. The standard potential corresponding to the HHV, termed thermoneutral potential, is equal to 1.481 V (see above). The formulas for the calculation of the efficiency of an electrolytic cell are as follows:

$$\text{Potential efficiency (HHV)} = \frac{\text{Thermoneutral potential}}{\text{Operating potential}}$$

The operating potential means the external applied potential for the cell:

$$\text{Electrical efficiency (HHV)} = \frac{\text{HHV of } H_2 \text{ production}}{\text{Electrical input}},$$

The electrical input means the total electrical energy applied for electrolysis.

The same formulas are used when the LHV value is applied instead of HHV. Therefore, it should be pointed out which value is used for calculation. The formulas described above apply to the electrolytic cells (electrolyzers) operating at temperatures below the boiling point of water, termed low-temperature electrolysis (LTE) cells . At present, the high-temperature steam electrolysis (HTE) cells are developed, operating in the range of temperature 700–1,000 °C. At such temperatures, the Gibbs free energy of water splitting ($\Delta_r G^*$) and enthalpy of reaction ($\Delta_r H^*$) are much lower than at standard temperature, resulting in a significant decrease of thermodynamic potentials ΔE^* and $\Delta E_{tn}{}^*$ calculated on their basis. In the case of HTE, the following formulas are used:

$$\text{Potential efficiency} = \frac{\text{Thermodynamic potential } (\Delta E^*)}{\text{Operating potential}}$$

$$\text{Electrical efficiency} = \frac{\text{HHV of } H_2 \text{ production}}{\text{Electrical input} + \text{heat supplied}}$$

14.2.2 Methods and technologies

Depending on the operating temperature, two types of electrolysis method and electrolytic cells (electrolyzers) can be distinguished: low-temperature electrolysis (LTE) and high-temperature electrolysis (HTE). To the first group belongs the alkaline water electrolysis (AWE) and polymer membrane electrolysis (PEM), to the second the solid oxide electrolysis (SOE). At present, low-temperature technologies such as AWE is used for hydrogen production on a large industry scale, while PEM electrolyzers are applied on a lower scale, in small factories. SOE technology is still under development. In the last years, the other low-temperature method of electrolysis, microbial electrolysis (MEC), is investigated and developed at the laboratory level.

14.2.2.1 Alkaline water electrolysis (AWE)

The technology of AWE was developed for many years (about a hundred) and is very well established. The principles of AWE are very simple. The electrolyzer contains an aqueous solution of 25–30% KOH or NaOH and two electrodes – cathode and anode – separated by the porous diaphragm. On the cathode, the reduction of water takes place with the formation of hydrogen atoms that recombine forming gaseous H_2. In solution, the hydroxide ions are charge carriers. They move across the diaphragm in the electric field to the anode where they are oxidized to water and O_2 (see reactions (14.2a) and (14.2b)). Both H_2 and O_2 are removed from the vicinity of electrodes by collectors. Figure 14.1 illustrates the principle of alkaline electrolyzer operation.

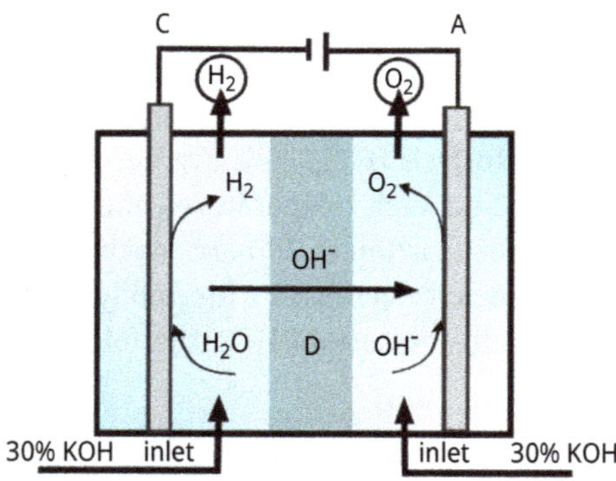

Fig. 14.1: Schematic illustration of alkaline water electrolyzer. A, anode; C, cathode; D, diaphragm.

As pointed out in Section 14.2.1, the HER and OER are complicated multistep reactions with the formation of adsorbed, intermediate products, which determine the rate of evolution of H_2 and O_2.

This rate can be increased by using a proper catalyst as an electrode material. For HER, the best catalysts are Pt and the metal of its group, and for OER, iridium oxide and ruthenium oxide are often used. However, as the cost of these materials is high, the other materials are applied in the commercial electrolyzers. Since in alkali media the steel and Ni are stable and resistant against corrosion, steel plates plated with Ni are commonly used as electrodes. However, at high operating temperatures, Ni can react with hydroxide ions forming nickel hydroxides. To eliminate such reactions, some metals (Fe, V, and Co) are added to electrode materials. The other Ni materials with high surface area were also tested such as Ni–Zn and Ni–Al alloys, or hot galvanized Ni meshes. A very important part of the electrolyzer is the diaphragm in the form of porous foil with a thickness between 0.05 and 0.5 mm. The diaphragm should avoid mixing of produced gases and be permeable for hydroxide ions. In alkaline electrolyzers, asbestos porous diaphragms are commonly used. However, as asbestos is toxic and undergoes corrosion at higher temperatures, other materials such as Zirfon, a composite material of zirconia and polysulfone, or the composite of potassium titanate (K_2TiO_3) fibers with polytetrafluoroethylene are tested. The new approach in AWE is the application of anion-exchange membranes (AEM) made up of polymer with anionic (hydroxide ions) conductivity. Conventionally working alkaline electrolyzers operate typically at temperatures and pressures in the range between 60 and 90 °C, 10 and 30 bars, and in the cell potential and current density range of 1.8–2.4 V and 0.2–0.6 A/cm^2, respectively. Energy efficiency is 70–80%. Hydrogen produced in this way has a purity of 99% [2, 8, 9].

14.2.2.2 Polymer electrolyte membrane electrolysis (PEM)

PEM water electrolysis technologies are similar to those applied in PEM fuel cells (Section 8.1.3.4). The PEM electrolyzer contains a membrane electrode assembly (MEA) together with current collector and plate separators. MEA consists of a polymer-exchange proton membrane, covered on both sides by proper catalysts for anode and cathode reactions. The ionomer solution is added to the vicinity of catalytic layers increasing the proton transport from catalysts to the membrane. Figure 14.2 illustrates the operating principle of the PEM electrolyzer.

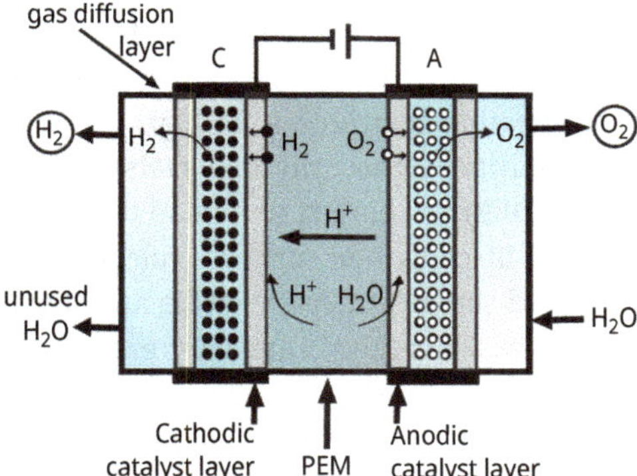

Fig. 14.2: Schematic illustration of PEM electrolyzer. A, anode; C, cathode; PEM, proton-exchange membrane.

The water is pumped to the anode reaching the catalyst layer, where it is oxidized to protons and oxygen. Protons go from the anode through the proton conducting membrane to the catalyst layer of the cathode, where hydrogen is generated (see reactions (14.1a) and (14.1b)). The gases left the cell via anodic and cathodic current collectors and separator plates made from materials with porous structure providing facilities for water transport to electrodes and elimination of gases from the cell. At present, porous titanium materials are used as collectors and separators. They are characterized with good electric and thermal conductivity and mechanical stability. Titanium materials are resistive against corrosion in acidic media; however, they corrode in O_2, so anode current collectors and separators should be protected by anticorrosion layers. The heart of PEM electrolyzer is a membrane with catalyst layers. Membranes that are commonly used in PEM electrolyzers are perfluorosulfonic polymers such as Nafion $(CH_2)_{18-28}SO_3H$, referred to as Nafion 115, 117, 212.

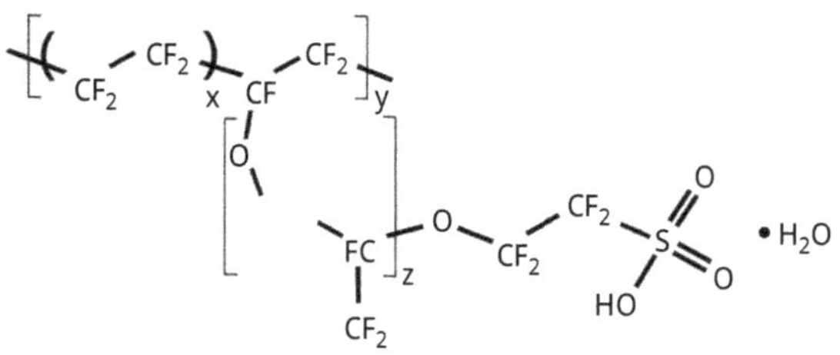

Fig. 14.3: Scheme of Nafion. See: https://en.wikipedia.org/wiki/Nafion.

The membranes are made from perfluorinated and sulfonated polyalkene chains. Their sulfonic groups dissociate in the presence of water causing high proton conductivity (0.1 ± 0.02 S/cm). Nafion membranes work as solid polymer electrolytes permeable only for proton ions. They are characterized with good mechanical stability and low resistance, and can work at higher current densities (2 A/cm). After the dissociation of membrane sulfonic groups, the concentration of proton is so high that it is equivalent to approximately 0.5–1 M H_2SO_4 solution. Hence, the materials used as electrocatalysts of electrode reactions (HER and OER) are limited to a relatively small group. Typically Pt, Pt-black, and Pt dispersed on black carbon support (Vulcan) have been used as cathode electrocatalysts, having good stability in acidic media and showing good electrocatalytic properties to HER. Pt-based catalysts are very expensive; therefore, the investigations are carried out in several directions such as (i) reduction of the Pt loading in the support materials, (ii) increase in the surface of catalyst particles by using highly dispersed carbon materials and carbon nanostructures (graphite nanofibers and carbon nanotubes doped with N, P, or S) as catalyst support, and (iii) investigations of materials being alternative to Pt. Materials such as NiC, Mo_2C, Ni_2P, WO_2, and Mo_2S on different carbon substrates are extensively studied.

Metal oxides, mainly IrO_2, RuO_2, or their mixture, are used as electrocatalysts for OER at the anode. Among these oxides, RuO_2 has the highest activity for OER. However, in oxygen, ruthenium oxide is not stable and undergoes further oxidation to RuO_4. IrO_2 is stable, but its activity toward OER is lower; therefore, mixed oxides of Ir and Ru are used. To minimize the catalyst cost, the research is focused on decreasing Ir and Ru amount by using transition metal oxides such as TiO_2, Nb_2O_5, and Ta_2O_5.

The technology of PEM electrolysis is partially established, and PEM electrolyzers are commercially used in small factories, but still plenty of work should be done to minimize the cost of electrocatalysts and hydrogen production. PEM electrolyzers operate typically in the range of temperature and pressure from 50 to 80 °C and from 20 to 50 bars, in the potential range of 1.6–1.8 V and current density of 1.0–2.0 A/cm^2. The energy efficiency of PEM electrolyzer is 80–90%. The obtained hydrogen has a purity of 99.99%. Much more about PEM electrolysis and catalysts applied can be found in [2, 9, 10] and references therein.

14.2.2.3 Solid oxide electrolysis (SOE)

Solid oxide electrolysis (SOE) of water is the high-temperature process carried out typically in the range between 700 and 1,000 °C, at a pressure of up to 15 bars, and current density ranging from 0.3 to 1.0 A/cm^2. The important part of the device is the electrolyte–SO ion conductor. In such electrolyte, the current flow occurs by the movement of thermally activated hopping of oxide ions through the crystal lattice directed by the electric field. As the movement of oxide ions is thermally activated, the conductivity of SO is highly temperature-dependent. Much more about solid electrolytes, their structure, properties, and applications can be found in Section 1.3.3, and in [11]. The principles of operating SO electrolyzers are shown in Fig. 14.4.

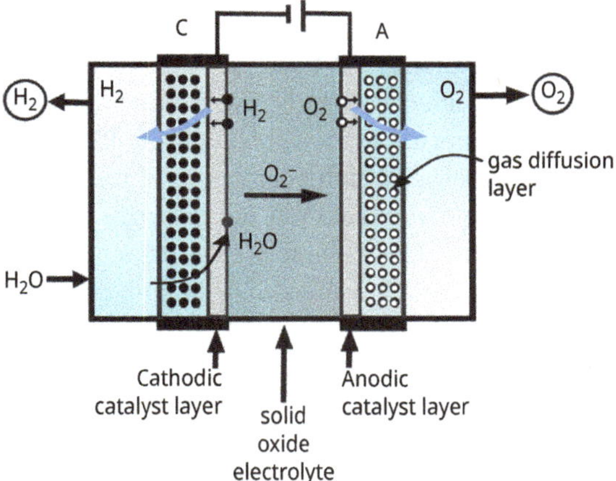

Fig. 14.4: Schematic illustration of solid oxide electrolyzers. A, anode; C, cathode.

The water steam is pumped in a porous cathode, where under an electric field it diffuses to the electrode surface. At the cathode interface, the water is reduced to oxygen ions (O^{2-}) and molecular hydrogen (H_2). Oxygen ions migrate across the SO electrolyte to the anode, where they are oxidized to molecular oxygen (O_2). Both gases diffuse back through the cathode and anode, respectively, and are collected. The reactions that take place during electrolysis are as follows:

$$\text{Cathode:} \quad H_2O + 2e \rightarrow H_2 + O^{2-}$$

$$\text{Anode:} \quad O^{2-} \rightarrow 1/2 \quad O_2 + 2e$$

The most common material used as an electrolyte in SO electrolyzer is an oxygen ion O^{2-} conductor ZrO_2 doped with 8 mol% of Y_2O_3 known as YSZ. YSZ is chemically stable up to 1,000 °C and has a high ionic conductivity of about 0.02 S/cm at 900 °C. Other materials were also tested as electrolytes, such as scandium-stabilized zirconia (ScSZ), ceria oxides doped with Gd_2O_3 or Sm_2O_3, and $LaGaO_3$ doped with Sr and Mg. In recent years, some of the ceramic proton conducting materials have been investigated and developed, such as yttrium-doped barium cerate, yttrium-doped barium zirconate, and yttrium-doped strontium zirconate. For the cathode, the commonly used material is porous cermet composed of Ni and YSZ, but also the perovskite-type lanthanum strontium manganese (LSM) and LSM doped with scandium were tested. For the anode, the composite containing LSM and YSZ is applied. The predicted electrical efficiency of hydrogen production in SOE is the highest among other water electrolyzers (90–100%), and also the purity of generated hydrogen is high. However, the main problem that should be solved before the technology is commercially available relates to the degradation of Ni/YSZ cathodes decreasing the lifetime of SO electrolyzers [2, 9, 12].

All these briefly described technologies have some advantages and disadvantages. For instance, the advantages of AWE are well-established technology, low-cost catalysts, and high durability. As disadvantages, we can consider the low current densities and low purity of gases which need further purification. The advantages of PME cells are compact system design, high current densities, and high purity of produced gases, but their technology is not fully established, the cost of catalysts is very high, and durability is low. SOE cells operate with low-cost catalysts, and their efficiency is the best in comparison with other water electrolyzers, but durability is low, design is large, and they are mostly at the laboratory stage.

14.2.2.4 Seawater electrolysis

In the water electrolyzers, briefly described above, fresh water is applied and enriched with proper electrolytes. However, such water constitutes about 1% of Earth's water and is unevenly distributed. Fresh water is a scarce resource in many parts of the Earth, and about 80% of the human population faces a risk of water shortage. On the other hand, about 71% of the Earth is covered by oceans and seas, so seawater is the richest source of water. Plenty of efforts are made to develop the technology for seawater electrolysis but researches are still at the laboratory level. Unfortunately, seawaters are salty waters, containing plenty of ions such as Na^+, K^+, Ca^{2+}, Mg^{2+}, Cl^-, Br^-, and SO_4^{2-}. Moreover, they contain not only different salts but also microorganisms and pollution, the products of human activity. Therefore, to remove sediments, plastics, and microorganisms, the seawater needs initial purification. After that, two different routes can be considered: (i) two-step indirect electrolysis with desalination of seawater by using reverse osmosis, and application of such purified water in commercially working electrolyzers (mainly AWE); and (ii) one-step electrolysis carried out by applying seawater directly to electrolytic cells, without initial desalination (direct seawater electrolysis). In electrolyzers filled with fresh water, H_2 and O_2 are evolved at the cathode and anode, respectively. If seawater is applied, the H_2 is still evolved at the cathode, but at the anode, instead of O_2, Cl_2 is formed:

$$\text{Cathode} \quad 2H_2O\,(l) + 2e \rightarrow H_2(g) + 2OH^-(aq), \quad E^0 = -0.828\,\text{V vs. SHE}$$

$$\text{Anode} \quad 2Cl^-(aq) \rightarrow Cl_2(g) + 2e, \quad E^0 = 1.358\,\text{V vs. SHE}$$

$$\text{Overall reaction:} \quad 2H_2O\,(l) + 2Cl^-(aq) \rightarrow H_2(g) + Cl_2(g) + 2OH^-(aq)$$

$$\Delta E^0 = E^0_A - E^0_C = 2.186\,\text{V} \sim 2.19\,\text{V}$$

$\Delta E^0 = 2.19$ V means the standard minimal potential required for seawater splitting and ΔE_{eq} is the equilibrium potential of the cell which depends on partial pressures of Cl_2 and H_2, concentration of Cl^-, and pH of the solution.

The overall reaction is very well known in the chloralkali industry, based on the electrolysis of brine (a concentrated solution of sodium chloride), and is applied for the production of chlorine. The primary products of chloralkali processes, besides chlorine gas, are hydrogen gas and sodium hydroxide solution (caustic soda). The concentration of sodium chloride in such type of electrolysis is about 7 M. Because of the very high corrosive properties of chlorine toward metals, some other materials for the anode are applied. For many years, graphite electrodes were applied in industry; at present, titanium–ruthenium oxide is commonly used. The applied potential for brine electrolysis is in the range of 2.9–4 V depending on the type of the electrolyzer. You can find much more on brine electrolysis in the *Electrochemistry Encyclopedia* online. It is beyond the scope of this book.

The concentration of chloride ions in seawater is much lower than in brine (about 0.5 M), also pH is near neutral, so electrolysis of seawater can be carried out in more friendly conditions, but still chlorine evolved at the anode is aggressive and should be eliminated. Therefore, the main task for researchers is to carry out seawater electrolysis in such a way that at the anode instead of Cl_2, the O_2 is evolved, as during electrolysis of fresh water.

Let us look more closely at the problems of anodic reactions. Let us suppose that both reactions, Cl_2 and O_2 evolutions, are competitive and depend on the electrolysis conditions. If we compare the standard potentials of the below reactions, we find out that they do not differ too much; however, only the potential of O_2 formation is pH-dependent:

$$2\,H_2O\,(l) \rightarrow O_2(g) + 4H^+(aq) + 4e\,, \; E^0 = 1.229\,\text{V vs. SHE}$$

$$2\,Cl^-(aq) \rightarrow Cl_2(g) + 2e, \quad E^0 = 1.358\,\text{V vs. SHE}$$

Evolved chlorine is not stable in water solution and undergoes disproportionation. This reaction creates hypochlorous acid (HClO) in acidic solutions and hypochlorite ions (ClO^-) in alkaline media (basic solutions):

$$Cl_2(g) + H_2O\,(l) \leftrightarrow HClO + H^+ + Cl^-, \quad K_{ac} = 4.2 \times 10^{-4}\,mol^2/(dm^3)^2$$

$$Cl_2(g) + 2OH^-(aq) \leftrightarrow OCl^- + H_2O\,(l) + Cl^-(aq), \quad K_b = 7.5 \times 10^{15}\,mol/(dm^3)$$

K_{ac} and K_b are equilibrium constants in acidic and basic solutions, respectively.

Hypochlorous acid and hypochlorite ions can also be generated at the anode as a result of the side reaction involving Cl^- ions:

$$Cl^-\,(aq) + 2H_2O\,(l) \rightarrow HClO\,(aq) + H^+\,(aq) + 2e, \quad E^0 = 1.484\,V\,\text{vs. SHE}, \; a_{H^+} = 1$$

$$Cl^- + 2OH^- \rightarrow ClO^- + H_2O + 2e, \quad E^0 = 0.89\,\text{vs. SHE}, \; a_{OH^-} = 1$$

In solutions with pH > 7.5, the equilibrium between $HClO/ClO^-$ is fully shifted to ClO^- ion formation. These ions can also be oxidized at the anode to ClO_2^- ions (chlorite ions) and further to ClO_3^- ions (chlorate ions) but as the amount of ClO^- ions is small in comparison with Cl^- ions, reactions that involve chloride ions are dominant.

Comparing values of potentials of O_2 evolution with values of potentials obtained for side reactions at different pH, we can find that in all pH range the potential of O_2 evolution is lower (see Pourbaix diagram, Fig. 3.5). This means that the OER is more favorable from the thermodynamic point of view than chlorine or hypochlorite ion formation. The factor that inhibits the evolution of O_2 is the high activation overpotential of this reaction described in Section 14.2.1. To illustrate the differences between both evolution reactions (Cl_2 and O_2), we can compare their exchange currents. The ratio of exchange current densities for chlorine and oxygen evolution on most anodes tested during seawater electrolysis is in the range of 10^3–10^7 [3]. It showed differences in the kinetics of both reactions and pointed out how easy chlorine evolution is. Therefore, the development of suitable OER electrocatalysts is crucial for the industrial application of direct seawater electrolysis. These electrocatalysts should not only provide high selectivity for OER but also should exhibit anticorrosive properties against aggressive chlorine and chloride ions. More about seawater electrolysis can be found in [2, 13].

The excess availability of seawater and wastewater, as well as the reduction in energy resources, has initiated a new thought of green process "waste to energy." Apart from the classic green technology like solar and wind systems, the utilization of seawater and wastewater as a source of energy depends on the conversion of the chemical energy trapped in waste to green energy.

For this purpose, a new bioelectrochemical system has been identified for sustainable recovery/production of valuable resources such as hydrogen and oxygen that utilizes wastewater and seawater electrolysis. Here, we will focus on a general description of such a system involved in the production of hydrogen on the cathode and a model reaction for microbial anode.

14.2.2.5 Microbial electrolysis (MEC)

At present, a new method of water electrolysis called MEC is being investigated and developed. This method is related to the microbial fuel cell (MFC, see Section 9.4) where the chemical energy is converted into electrical energy as a result of microorganism action. On the contrary, in microbial electrolysis cells (MECs), electrochemically active microorganisms (bacteria and microbes) participate in the production of chemical energy by oxidizing the organic compounds at the anode, producing CO_2, protons, and electrons. At the cathode, protons are reduced and H_2 is evolved. In microbial cells, hydrogen can be produced from pure organic compounds such as acetic acid, lactic acid, and glucose as well as from biomass and wastewater. Moreover, the demand for electrical energy is much lower than for other electrolytic cells, as a part of energy is derived by the microorganism activities. Such cells can operate in the potential range from 0.25 to 0.8 V – much lower, compared with typical water electrolytic cells operating at potentials higher than 1.6 V. The efficiency of H_2 production depends on the kind of organic substances used. In the case of lactic or acetic acids, the efficiency achieved is 87%; for not pretreated cellulose, or glucose it is about 63%.

A microbial electrolytic cell is a one- or two-compartment cell. In a two-compartment cell (Fig. 14.5), the cathodic and anodic compartments are separated by proton-conducting membranes.

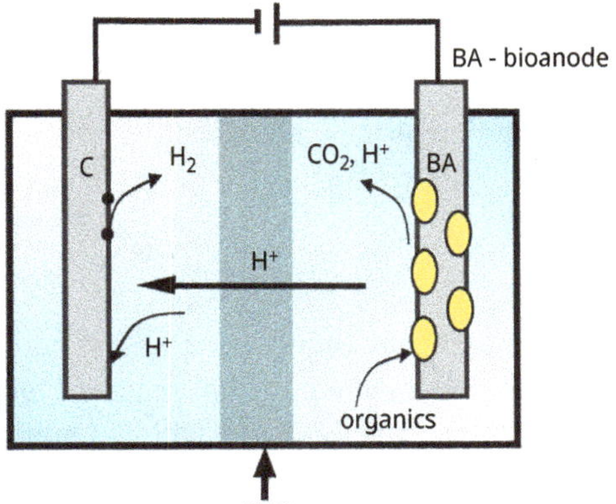

Fig. 14.5: Schematic illustration of microbial cell. BA, bioanode; C, cathode.

The membrane is an important part of a cell. It is not only the conductor for proton ions released at the anode but also the membrane reduces the crossover of organic compounds from anode to cathode, improving purity of the produced hydrogen and preventing microbial consumption of hydrogen. The most common membrane used in MECs is the proton-exchange membrane, but other membranes have also been tested such as the AEMs, and the bipolar membranes. Much more information about membranes and membrane potentials can be found in Section 1.4. The cathodic and anodic parts of the cell contain electrodes: conventional cathode and microbial anode. Anode is placed in the electrolyte containing microorganisms and the medium proper for their existence and growth. Some microorganisms can spontaneously adsorb on an anode forming electroactive biofilms, and catalyzing the oxidation of organic compounds. In this case, the anode works as a support for biofilm formation and electron collector.

The oxidation of acetic acid is considered a model reaction for microbial anode:

$$C_2H_4O_2 + 2\,H_2O \rightarrow 2CO_2 + 8H^+ + 8e, \quad E^0 = 0.071\,\text{V (vs. SHE)}$$

The equilibrium potential of the anode depends on pH:

$$E_A = E^0 - \frac{2.3\,RT}{F}\text{pH} + \frac{2.3RT}{8F}\log\frac{p^2_{CO_2}}{a_{C_2H_4O_2}} \tag{14.10a}$$

$$\text{if } C_{\text{acid}} - 1\,\text{M},\ \text{pCO}_2 - 1\,\text{bar},\ a_{H_2O} - 1,\ \text{formal potential } E_A{}^f = E^0 - \frac{2.3\,RT}{F}\text{pH} \tag{14.10b}$$

at pH = 7, it is equal to –0.291 V (vs. SHE).

Electrons released in this reaction participate in water reduction at the cathode and in the production of hydrogen:

$$8\,H^+ + 8e \rightarrow 4H_2$$

$$E_C = E^0 - \frac{2.3\,RT}{F}\text{pH} - \frac{RT}{8F}\log p^4_{H_2},\ E^0 = 0\,\text{V (vs. SHE)} \tag{14.11a}$$

$$\text{Formal potential } E_C{}^f = -\frac{2.3\,RT}{F}\text{pH} \tag{14.11b}$$

and at pH 7, it is equal to –0.413 V:

$$\text{Overall reaction:}\quad C_2H_4O_2 + 2H_2O \rightarrow 2CO_2 + 4H_2$$

At pH 7, the equilibrium potential $\Delta E = E_A{}^f - E_C{}^f$ of MEC containing acetic acid as substrate is equal to 0.122 V. This is the minimal thermodynamic potential required for hydrogen generation in MECs in such conditions. Comparing this value with the minimal thermodynamic value 1.229 V, required for conventional water electrolysis, we see the benefits and attraction of MEC. In practice, the potential that has to be used in MECs is higher (about 0.25 V and more up to 0.8 V), but still much lower than in the typical water electrolytic cells. In the case of conventional cells, the values of applied potentials are dependent on the energy losses. They result not only from activation overpotentials of reaction but also from ohmic losses. Due to the potential applications of MECs in "green chemistry," polluted electrolytes such as sewage media or wastewater are tested as a fuel. Their ionic conductivity is low, in the order of 0.002 S/cm; therefore, some salts have to be added to improve the conductivity. However, in microbial cells, the important anodic reactions are enzyme-catalyzed reactions, very sensitive to pH, salt concentrations, and temperature. Therefore, the microbial anode can operate efficiently in the limited temperature range of 25–40 °C, and in neutral, not too concentrated solutions.

Plenty of materials were tested as anodes or biofilm supports in MECs, such as stainless steel, carbon papers, graphite brushes, and graphite granules. As cathodes, stainless steel, nickel, and cobalt, and some alloys were applied. It seems that stainless steel after special treatment is a good material for both electrodes. The alternative to the abiotic cathode is the application of a biocathode. Microorganisms that are involved in the formation of such bioelectrode contain the enzyme hydrogenase, catalyzing the reaction $2H^+ + 2e^- \rightarrow H_2$.

Microbial cells, like other water electrolytic cells, have some advantages such as (i) low operating potential, usually in the range of 0.25–0.8 V; (ii) low cost of electrode materials, since special treatment of anode is not required as biofilm is formed spontaneously; and (iii) application of polluted media as a source of hydrogen. The main disadvantages of MECs are (i) low hydrogen production rate, (ii) low purity of hydrogen, and (iii) microbial anode is very sensitive to pH, T, and concentration of substrates. MECs are in the phase of investigations, and many efforts are ahead as their operations depend upon the type of microorganisms and organic compound applied. Much more about MEC can be found in [14, 15] and references therein.

14.3 Solar energy in photo water splitting

Hydrogen is the fuel of the future. The best way to produce hydrogen without pollution is the electrolysis of water. However, if fossil fuels are used as a source of electrical energy, the problem of CO_2 emission and other environmental pollution remains unsolved. Therefore, many efforts are made for the application of renewable energy sources. The Sun is the cleanest and inexhaustible energy source and its energy produces heat and can be converted to electrical energy.

Two approaches to the application of solar energy can be distinguished in water splitting. The first can be realized by the installation of complex devices containing PV (photovoltaic) panels and water electrolysis units. PV panels can convert solar energy into electrical one and this energy can be further used for feeding electrolyzers. The other way is the direct application of solar energy for the photoelectrolysis of water by using photoactive materials – semiconductors (SCs). In the future, photo one-step technology of water splitting may be economically more profitable than PV production, followed by water electrolysis. The advantages lay in the application of a single plant and the mitigation of efficiency losses inevitable in the case of two plants that are required in a two-step technology. There is a huge amount of research carried out on the improvement of efficiency of water photoelectrolysis and hydrogen production; however, we are still far away from commercial application of this technology, and it is song of future. The current approaches are focused on the evaluation of new light-sensitive electrode materials and multijunction configurations of the photocells.

Two terms should be distinguished for a better understanding of this section: photoelectrochemical electrolysis of water and photocatalytic water splitting. In both methods, solar energy can be applied. The term "photoelectrochemical electrolysis" is correctly used when photoelectrochemical reactions take place in electrochemical cells at SC electrode/electrodes (SCE). The form of SC photoelectrodes may be different. The solid plates, thin films, and SC particles (SCPs) distributed on different substrates are applied. The photocatalytic water splitting occurs on suspensions of SCPs. There are no essential differences between both methods, in the description of the photo mechanism and the applied SC materials. The crucial difference between SCEs and SCPs is the place of reactions. In a photoelectrochemical cell (PEC), the oxidation of water occurs at the n-SC anode (O_2 evolution), while the evolution of H_2 takes place at the cathode (Pt, or p-SC). Hence, the separation of O_2 and H_2 is much easier than in the case of direct photocatalytic splitting of water, where both reactions take place at the same particles. The process of separation increases the cost of production of photocatalytic H_2; therefore, some approaches described further are developed. The research of photoelectrolysis of water is carried out in the typical PECs, while for photocatalytic water splitting, chemical reactors are used. The SCPs where photoreactions occur are often regarded as short-circuited microelectrochemical cells with a cathode and anode placed together. It is worth mentioning that hydrogen may be obtained by the half-reaction of water splitting in the presence of sacrificial agents – donors. Then, oxygen is not evolved at the SCEs or SCPs. Such reactions cannot be named "water splitting." The term is reserved strictly for situation when water splitting and photo water splitting of H_2 and O_2 is produced in stoichiometric amount. In other cases, we suggest using the term photo-sacrificial water splitting or photo water splitting in the presence of a sacrificial agent.

Part III of this book was devoted to the electrochemistry and photoelectrochemistry of SC materials. In Chapter 10, we described the behavior of the SC/electrolyte interface in dark and under illumination. In Chapter 11, we discussed the types of photocells, their operation, and the application of SC materials. Special attention was devoted to the photosynthetic cell and photoelectrolysis of water, to the energetic diagrams of direct photoelectrolysis and assisted water photoelectrolysis. Some brief, basic information about SCPs, their photoreactions, and their applications can be found in Chapter 12. We encourage all readers interested in a deeper understanding of photoelectrochemical and photocatalytic processes, and their applications in the resolution of environmental problems to read carefully these chapters. Below we concentrate on comparison between photoelectrochemical and photocatalytic water splitting, looking for some similarities and differences and on the new developments and trends.

14.3.1 Thermodynamics, energy losses, and efficiency

The overall reaction of water photo splitting may be expressed, independently from the reaction location (SCE and SCP), in the form:

$$4\,h\nu + 2\,H_2O\,(l) \rightarrow 2H_2\,(g) + O_2\,(g)$$

where h is Planck's constant and ν is the frequency.

This reaction takes place if the energy of photons E_{ph} is equal to or larger than

$$E_{ph} = h\nu \geq \frac{\Delta_r G^0}{2N_A} \tag{14.12}$$

Taking into account the values of $\Delta_r G^0$ – the standard Gibbs free energy of reaction equal to 237.13 kJ/mol, and N_A – Avogadro's number equal to 6.022×10^{23} particles/mol, $1\ eV = 1.602 \times 10^{-19}$ J, we obtained the minimal energy of photons required for photo water splitting equal to 1.229 eV. This value is equivalent to the value of $\Delta E^0 = 1.229$ V, the standard minimal potential required for water splitting when electrical energy is used. As in the case of water splitting, also in photo water splitting, there are losses of energy, and some of them are common and some are different as SC materials and photons are applied. Taking into account all losses, we can roughly estimate the optimal value of bandgap energy of SC, and optimal photon energy as 2.0–2.4 eV. This range corresponds to the wavelength of light 620–517 nm, respectively, and lies in the visible range of the solar spectrum (see Fig. 11.2). Before going further, we should briefly consider the energy losses during photo water splitting at SCEs and SCPs. We can distinguish three main steps in photo water splitting.

A. The first step that takes place during illumination (irradiation) of SC materials (SCEs and SCPs) is the absorption of photons. Only photons with energy $E_{ph} = h\nu \geq E_g$ (E_g – bandgap energy of SC) are absorbed and lead to a generation of excitons: bonded pairs electron–hole, electrons (e) at the bottom of conduction band (CB), and holes (h^+) at the top of the valence band (VB) of SC:

$$h\nu + \mathrm{SCE\,(or\,SCP)} \rightarrow \mathrm{e\,(CB)} + \mathrm{h^+\,(VB)}$$

Note the differences between the direct and indirect SC bandgap energy (Section 10.1.1). In the first case, photons with energy $h\nu = E_g$ are absorbed; in the second case, photons should have energy higher than E_g. In both cases, the excess of energy is dissipated as thermal energy. What about photons with energy lower than E_g? The SC materials are transparent to them and their energy is lost. The other factor, decreasing available solar energy, is the reflectivity of applied materials. All these energy losses connected with sunlight reflection and threshold absorption are named optical losses, and they determine the threshold efficiency of solar energy conversion.

The efficiency of solar energy conversion is substantially decreased by recombination processes. The generation of electron–hole pairs during illumination perturbs the equilibrium of the charge carriers' concentration. The processes of recombination restore the equilibrium. The excited electrons in CB might fall back to VB and then recombine with holes. It is a so-called direct recombination with a one-step transition. Sometimes, the recombination occurs via impurities, defects acting as the bulk, or surface traps. Recombination annihilates the photogenerated electrons and holes, so they cannot participate in any redox processes. In consequence, the recombination decreases the efficiency of reactions that occur at SCEs and SCPs, and the efficiency of solar energy conversion. Therefore, the excited electron–hole pairs should be effectively separated before annihilation. Much more about the processes of direct and indirect recombination, surface recombination, and trapping can be found in Sections 10.1.2 and 10.5.

B. The second step is the separation of excited pairs e–h$^+$ and the motion of free charge carriers to the reaction sites. There are differences in separation processes occurring in solid SCEs and SCPs. In both cases, separation is caused by an electric field, but the origin of this field is different. Figure 14.6(a, b) illustrates the differences and the pathways of recombination.

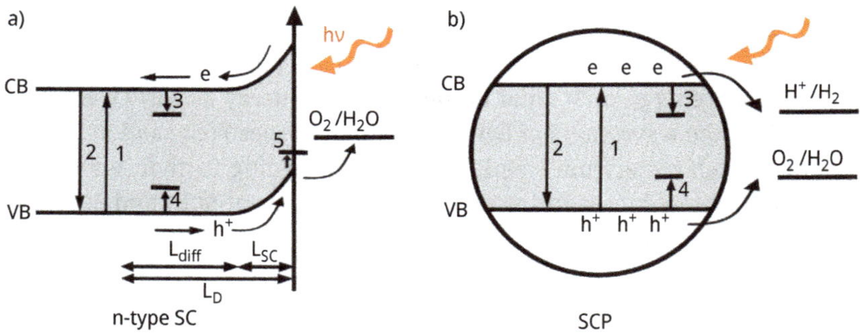

Fig. 14.6: Schematic illustration of generation and recombination of charge carriers (electrons and holes) in (a) n-type of SC and (b) SC particles. L_{SC}, thickness of space charge layer; L_{diff}, diffusion length; $L_D = L_{SC} + L_{diff}$. Arrows 1, 2 – generation, direct recombination; 3,4 – bulk recombination; 5 – surface recombination.

In the beginning, let us consider the formation of an electric field in a SCE. The concentration of electrons in SC materials is much lower than in metals; therefore, the compensation of the charge of electrolyte ions takes place in the thin layer of the SC. This charge region of SC is named the space charge layer, and its existence is crucial for SCE behavior in dark and under illumination. Much more about the space charge layer and its significance in electrochemistry and photoelectrochemistry of SCs can be found in Chapters 10 and 11. This space charge layer is characterized by the charge Q_{SC}, potential drop φ_{SC}, and thickness L_{SC}, and can exist in three states: depletion, inversion, and accumulation (see Fig. 10.7). The best charge separation is obtainable when the space charge layer exists in a depletion state. In this state, in n-type SC, the Q_{sc} is positive, created by ionized donor atoms. On the contrary, in p-type SC, the Q_{SC}

is created by ionized acceptor atoms and is negative. In Fig. 14.6(a), the energy diagram for n-type SC in a depletion state is shown. There, besides the space charge layer thickness L_{SC}, the diffusion length L_{diff} is marked. Note that the sum of L_{SC} and L_{diff} means the Debye's length L_D. What processes occur during illumination depends on the depth of photon penetration characterized by the reciprocal of the absorption coefficient ($x = \alpha^{-1}$). If photons are absorbed at $x > L_D$, the photogenerated e–h$^+$ pairs recombine (arrows 1 and 2 marked generation and recombination processes). If photons are absorbed in the vicinity of L_{SC}, the photogenerated e–h$^+$ pairs dissociate in the electric field and free minority carriers – holes migrate to the n-SC/electrolyte interface where they can participate in the oxidation reaction, or are trapped by surface states (arrow 5). The e–h$^+$ pairs photogenerated in the range of L_{diff} diffuse to the space charge region, where they are separated by the electric field. They can also be trapped by defects of the crystal lattice (deep energy levels, arrows 3 and 4) and recombine. The free majority carriers – electrons, obtained during dissociation, diffuse in the bulk of SC to the back contact or to the other electrode where they participate in reduction. They can also be trapped and recombined on their way.

A different situation exists when SCPs are illuminated, and then separation starts to depend on the particle size. The large particles (dimension higher than L_D) behave similarly to the bulk SCs. The separation of electrons and holes occurs in the electric field of space charge layers, and free charge carriers migrate to the interface of SCP/electrolyte taking part in redox reactions. The situation is different when small particles are illuminated (Fig. 14.6(b)). Then photogenerated e–h$^+$ pairs diffuse to the SCP/electrolyte interface, where they dissociate in the electric field of an electric double layer (Section 1.1.1), and there they participate in redox reactions. They can also be trapped by surface states created by particle inhomogeneity or species adsorbed from solution. Earlier, other processes can also occur in SCPs such as recombination and bulk trapping. All these processes have different timescales. It is worth mentioning that trapping processes at particle surface can also play a positive function, as they compete with surface recombination of charges. Trapped electrons or holes participate in a redox reaction at the particle surface after release. Size effect, other processes, and their timescale are briefly described in Section 12.2. In the case of SCPs, the main problem is the improvement of the separation of photogenerated e–h$^+$ pairs and the decrease of recombination processes.

C. In the third step, the photogenerated, separated free charge carriers (electrons and holes) participate in the redox reaction with substrates in the solution. The evolution of O_2 occurs at the n-SC photoanode, while H_2 is generated at the cathode (Pt or p-SC). In the SCP case, both H_2 and O_2 are evolved at the same particles. The half-reactions are as follows:

$$\text{Anodic side:}\quad 2\,H_2O\,(l) + 4h^+(SC) \rightarrow 4\,H^+\,(aq) + O_2\,(g)$$

$$\text{Cathodic side::}\quad 4H^+(aq) + 4e\,(SC) \rightarrow 2H_2\,(g)$$

These reactions will take place only when the relation between energy levels of CB (E_C) and VB (E_V), and energy levels of reactions H^+/H_2 and O_2/H_2O will be appropriate. Figure 14.7(a, b) presents the simplified energy diagrams for n-SCE photoanode and SCPs, where proper positions of energy levels for SC materials and redox systems (H^+/H_2 and O_2/H_2O) are marked.

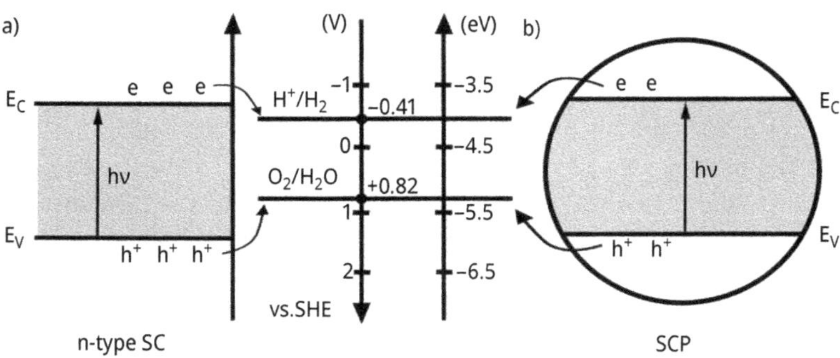

Fig. 14.7: Schematic energy diagram showing requirements for direct photo water splitting: (a) n-type SC electrode and (b) SC particles. pH is 7.

In electrochemistry, the potential scale is mainly applied, so two scales are used in this figure. The correlation of both scales is simple and is based on the value of the Fermi energy level (electrochemical potential of electrons) for the SHE electrode equal to −4.5 ± 0.01 eV. It means that 0 V vs. SHE is equivalent to −4.5 eV. The photogenerated electrons and holes should have sufficient energy to reduce protons and to oxidize water. Generally, the energy of the CB edge (E_c) for n-SCE and SCP has to be higher than the energy level of the H^+/H_2 redox couple, that is, it is located at a more negative potential relative to the H^+/H_2 redox potential in the solution. In the case of water oxidation, the energy of the VB edge (E_V) should be lower than the energy level of the O_2/H_2O redox couple, that is, it is located at more positive potential than the O_2/H_2O redox potential in solution. In the energy diagram of the SC electrode (Fig. 14.7(a)), the energy bands are flat. It means that the charge of the space charge layer, Q_{SC}, was compensated by an external source, for instance, the external potential applied to the SC electrode. This potential at which the energy bands are flat is referred to as the flat band potential, E_{fb}, and has to be more negative than the potential of the H^+/H_2 redox couple. Note that the redox potential of H^+/H_2 and O_2/H_2O couples depends on the pH of the solution; therefore, the position of their energy levels may be changed. In Fig. 14.7(a, b), they are marked at values corresponding to the redox potential of H^+/H_2 and O_2/H_2O couples at pH 7. If we applied in PEC the p-type SC photocathode, the photogenerated minority charge carriers, electrons, participate in the reduction of hydrogen ions. Also then, the flat band potential has to be more negative than the redox potential of the H^+/H_2 couple. If these conditions are not fulfilled for n-SC and p-SC electrodes, we have to use the additional external potential (bias voltage) to overcome the energy demand for photo water splitting. This type of photoelectrolysis is referred to as assisted water photoelectrolysis, contrary to direct photoelectrolysis. The detailed requirements, which have to be fulfilled by the photoanodes, photocathodes, and more advanced schematic energy diagrams for water photo splitting, can be found in Section 11.3.4. Therein, on the energy scheme, the anodic and cathodic overpotentials (η_A and η_C) are also marked (Fig. 11.5). As in the case of water splitting, they cause energy losses influencing the efficiency of solar energy conversion. Their formulas differ from those applied for the reaction at metal electrodes (see Section 10.5). The energy diagram for SC particles is shown in Fig. 14.7b. The requirements for the relation between E_C, E_V, and redox potentials of H^+/H_2 and O_2/H_2O are the same as for SC electrodes.

There are many definitions and equations describing the efficiency of solar energy conversion. They depend on whether the solar energy is converted into electrical energy (regenerative photocells), or is stored as chemical energy in a fuel (photosynthetic cells). Some of these definitions are common, as the basic processes in SC materials are the same. We will concentrate here on the efficiency of conversion of solar energy during photoelectrolysis of water and on the efficiency of photocatalytic water splitting. Some general information about solar energy, conversion, and efficiency of regenerative photocells can be found in Sections 11.1 and 11.2.

Generally, the efficiency of solar energy conversion in photodevices can be determined as the ratio of power output (P_{out}) of any devices and systems, and power input (P_{in}) of solar irradiation $\varepsilon = P_{out}/P_{in}$. The solar-to-hydrogen (STH) conversion efficiency is described for direct water photoelectrolysis by equation (14.13a). Sometimes, instead of Gibbs free energy of reaction, the enthalpy of reaction is used (14.13b). In this case, it is assumed that the hydrogen will be burnt and the stored energy of this reaction will be recovered as heat:

$$\varepsilon_{STH} = \frac{\Delta_r G^0 R}{P_{in} A} \tag{14.13a}$$

$$\varepsilon_{STH} = \frac{\Delta_r H^0 R}{P_{in} A} \tag{14.13b}$$

$\Delta_r G^0$ is the standard Gibbs free energy of water splitting (kJ/mol), $\Delta_r H^0$ is the standard enthalpy of water splitting (kJ/mol), R is the rate of hydrogen generation (mol/s), P_{in} is solar power (W/m^2), A is the illuminated (irradiated) area (m^2).

If we determined the solar conversion efficiency to hydrogen during the assisted water photoelectrolysis, we should also take into account the additional electrical energy, which is applied to fulfill energy demand for photo water splitting equal to UI (see Fig. 11.5). In this case, equations (14.13) are modified. In these equations, the standard potential $\Delta E^0 = 1.229$ V $\approx$ 1.23 V or the standard thermoneutral potential $\Delta E^0_{tn} = 1.481$ V $\approx$ 1.48 V of water splitting are used:

$$\varepsilon_{\mathrm{STH}} = \frac{\Delta_r G^0 R - UI}{P_{\mathrm{in}} A} \tag{14.14a}$$

$$\varepsilon_{\mathrm{STH}} = \frac{(1.23 - U)I}{P_{\mathrm{in}} A} \tag{14.14b}$$

$$\varepsilon_{\mathrm{STH}} = \frac{\Delta_r H^0 R - UI}{P_{\mathrm{in}} A} \tag{14.15a}$$

$$\varepsilon_{\mathrm{STH}} = \frac{(1.48 - U)I}{P_{\mathrm{in}} A} \tag{14.15b}$$

where U is the applied potential and I is the current flowing through the cell.

The efficiency of solar energy conversion is decreased by the processes described above. Their influence is taken into account in the equation:

$$\varepsilon = \varepsilon_{\mathrm{thresh}} \varnothing\, \varepsilon_{\mathrm{stor}}^{\mathrm{ch}} \varepsilon_{\mathrm{ohm}} \tag{14.16}$$

(i) $\varepsilon_{\mathrm{thresh}}$ – the threshold efficiency pointing out that only part of a photon is absorbed by SCE or SCP (see A above, Section 14.3.1)

$$\varepsilon_{\mathrm{thresh}} = \frac{E_g \int_{E_g}^{\infty} N(E)(1-R)dE}{\int_0^{\infty} EN(E)dE} \tag{14.17}$$

$N(E)$ is the number of photons striking the SC surface in unit time and R is the reflection coefficient

(ii) $\varnothing$ – the quantum yield is defined as a ratio of effective photon incidents N_{eff}, leading to the generation of photo e–h$^+$ pairs, to the total number of absorbed photons:

$$\varnothing = \frac{N_{\mathrm{eff}}}{N_{\mathrm{total}}} \tag{14.18}$$

$\varnothing$ is connected with the efficiency of separation of photogenerated electron–hole pairs (see B above, Section 14.3.1) and manifests itself in the value of $j_{\mathrm{sc,c}}^{\mathrm{ph}}$ short circuit photocurrent (see Section 11.2).

If the light energy is converted into chemical energy in direct photoelectrolysis, the efficiency of energy storage is determined by the ratio of free Gibbs energy of the overall reaction occurring in the photocell to the energy of absorbed photons $h\nu = E_g$. In the case of assisted photoelectrolysis, the power input (UI) should be taken into account:

$$\varepsilon_{\text{stor}}^{\text{ch}} = \frac{\Delta_r G}{E_g} \tag{14.19a}$$

$$\varepsilon_{\text{stor}}^{\text{ch}} = \frac{\Delta_r G - UI}{E_g} \tag{14.19b}$$

The ohmic losses in the photocell are manifested by changes in the fill factor, ff (see Section 11.2, Fig. 11.5). Note that the fill factor determines the quality of PE cells. The ohmic losses are mostly caused by the lower conductivity of SC materials, the problems with the transport of charge carriers to the reaction place, and the transport of electrons from photoanode to cathode where H_2 is evolved. The efficiency of PE cells may also be evaluated by IPCE, the incident photon to current conversion efficiency. IPCE is defined as the ratio of electron numbers generated by light to the number of incident photons:

$$\text{IPCE} = \frac{1,240 \times \text{photocurrent density}\,[\text{A/cm}^2]}{\text{wavelength}\,[\text{nm}] \times \text{photon flux}\,[\text{W/cm}^2]} \times 100\%$$

where 1,240 is the unit conversion coefficient.

All these equations may be used when solar energy is converted to chemical energy in a photoelectrochemical cell. To determine the efficiency of solar energy conversion at SCPs, we may use equations (14.13a) and (14.13b). In practice, the activities of SCPs applied in photo water splitting are usually assessed by the ratio of evolved gases H_2 and O_2 (mol/h) per SCP amount (g) under specified illumination conditions. Having the amount of H_2 or O_2, the apparent ∅ may be determined:

$$\text{Quantum yield}\,(H_2) = \frac{2 \times \text{number of evolved hydrogen molecules}}{\text{number of incident photons}} \times 100\%$$

$$\text{Quantum yield}\,(O_2) = \frac{4 \times \text{number of evolved hydrogen molecules}}{\text{number of incident photons}} \times 100\%$$

Here, numbers 2 and 4 point out that two or four photons are required to generate one H_2 and O_2 molecule, respectively.

To compare the efficiencies of solar energy conversion between different PECs and also at different SCPs, we should carry out the experiments under the same photo conditions. It is accepted to apply the photodevices generating irradiance of power 100 mW/cm^2, which also mimics the air mass 1.5 spectra (AM 1.5). The STH conversion efficiency in PECs depends on their configuration. Theoretically predicted STH conversion efficiencies for ideal photosystems containing single photoabsorber (S) or dual photoabsorber (D) are about 30% and 41%, respectively. In reality, taking into account all energy losses (including electrochemical ones), they do not exceed about 10% and 16% [16]. Note that a single photoabsorber in PEC means one SCE (photoanode or photocathode), while a dual photoabsorber means two n-SC and p-SC individual electrodes operating in PEC, or a special design of two SC materials coupled together. Such photocells are referred to as one-bandgap or two-bandgap devices. In the literature concerning PEC and STH, you can find notations: S2 and D4. These notations mean that two or four photons are required for the generation of one H_2 molecule. Data concerning activities and ∅ (quantum yield) of many SCPs applied in the photocatalytic water splitting can be found in review [17] and references therein. The obtained values depend on the modification of particles, applied solutions, and illumination conditions, but in comparison with the efficiency of SCE, they are much lower. To date, the efficiencies of conversion of solar energy to chemical energy are too low for practical, commercial application of photo water splitting. At present, to improve the efficiency of conversion, the investigations are carried out in a few directions: the evolution of materials applied for photoelectrodes and photocatalysis, the more sophisticated design of the PECs, and special arrays of SCPs in photocatalytic splitting of water. We very briefly describe these trends below.

14.3.2 Photoelectrolysis of water: trends

In photo water electrolysis, the typically studied cells are single photoelectrode and two photoelectrode cells. Single photoelectrode cells contain n-SC photoanode and metal cathode which is not light sensitive. In dual photocell, two electrodes are light sensitive, n-type SC photoanode, and p-type SC photocathode. The "hearts" of these cells are photoelectrodes. To be efficient in the photo water splitting, they have to fulfill several specific requirements described above. From a practical point of view, the photoelectrodes should be stable and the cost of their manufacturing should be low. The stability of photoelectrodes depends on corrosion and photocorrosion of SC materials. The problems of corrosion and photocorrosion of SC materials are described in Section 10.6. Let us concentrate firstly on the photoanodes applied in the photoelectrolysis of water. Most of them are fabricated from metal oxides or materials with oxide structure; they are n-type SCs. We can distinguish two main groups: the binary oxides and the perovskite-structured oxide materials (ABO_3). The first group is represented by TiO_2, ZnO, WO_3, Fe_2O_3, and In_2O_3. Perovskite-structured SCs have a general formula ABO_3. They are represented by titanates ($SrTiO_3$, $BaTiO_3$, and $FeTiO_3$), tantalates ($NaTO_3$), and niobates ($KNbO_3$). The energetic diagram for selected oxides is shown in Fig. 14.8(a). The bandgap energy (E_g) and positions of the energy band edges E_C and E_V are marked there in correlation to (E^0) standard potential of H^+/H_2 O_2/H_2O couples in the energy and potential scale.

The family of the perovskite-structured oxides is very large, but they are mostly applied in photocatalytic water splitting. The group of oxides studied for possible application as photoanodes in photoelectrolysis of water is much higher. The interesting materials tested lately are vanadates $BiVO_4$ (2.5 eV), $InVO_4$ (2.0 eV), and oxynitrides such as TaON (2.5 eV). Let us divide the studied oxides into two groups: one having the bandgap energy $E_g > 3$ eV and the other with $E_g < 3$ eV. Generally, oxides with $E_g > 3$ eV are chemically stable even in acidic solutions and avoid photocorrosion. Unfortunately, they absorb

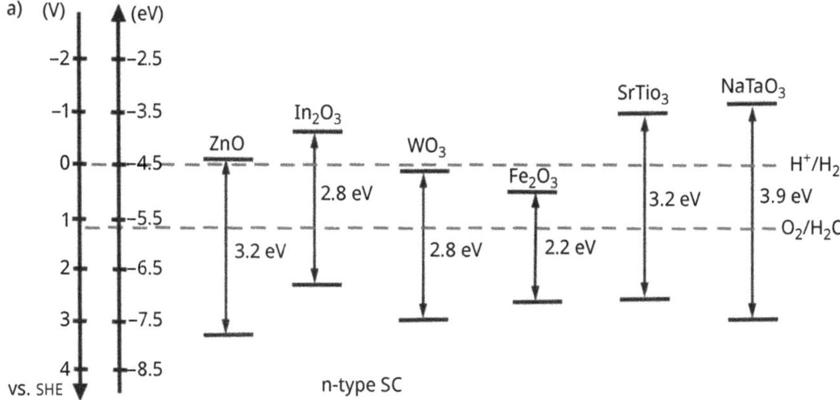

Fig. 14.8: Schematic energy diagram showing correlation between the energy of conduction and valence band edges (E_C, E_V) and redox potentials of H^+/H_2 and O_2/H_2O couples: (a) for n-type SC and (b) p-type SC materials (adapted from [17, 19]).

solar energy in the UV range of the solar spectrum. The better absorption ability in the visible range of the spectrum has oxides with $E_g < 3$ eV. However, they are not stable and undergo photocorrosion under illumination. Moreover, for WO_3 and hematite (Fe_2O_3), the CB edges lay at too positive potentials with respect to hydrogen evolution potential. Therefore, they cannot be used in direct photoelectrolysis of water. Others such as In_2O_3 are not direct SCs and absorb photons with energy higher than E_g.

In the photoelectrolysis of water, except photoanodes, the light-sensitive photocathodes, p-type SCs are used. In earlier studies, III–V and II–VI semiconductors such as GaAs (1.42 eV), GaP (2.25 eV), and CdTe (1.56 eV) were mostly tested. Great popularity enjoyed the cupric oxides: CuO (1.57 eV) and CuO_2 (2.47 eV). All these compounds can absorb photons from the visible range of the solar spectrum; however, they are unstable and undergo photocorrosion in the electrolyte solutions. In recent years, the attention of researchers has been devoted to other materials, called ternary oxides. They are represented by oxides with spinel structure (AB_2O_4) or the layered delafossite-type oxides (ABO_2). Among them are $CaFe_2O_4$, $CuBi_2O_4$, $CuFe_2O_4$, and $CuRhO_2$ (Fig. 14.8(b)). Note that they have the appropriate E_g values and can absorb photons in the visible range of solar spectrum. They have also the potentials of band edges more negative than the redox potential of the H^+/H_2 couple and more positive than the redox potential of water oxidation O_2/H_2O except $CaFe_2O_4$. Therefore, they may simultaneously evolve H_2 and O_2 and can be applied in the photocatalysis of water splitting.

Plenty of studies have been carried out to improve the photon absorption efficiency of SCEs possessing broad forbidden bands. Among them are: (i) modifications of electronic structure through doping of SCs with metals (Au, Pt, and Ag), nonmetals (N, C, S, and P), and ions (Ru^{4+}, Pt^{4+}, and Cr^{3+}); (ii) nanostructuring and formation of ordered arrays of SC (nanotubes, nanowires, nanorods, and nanosheets); (iii) modifi-

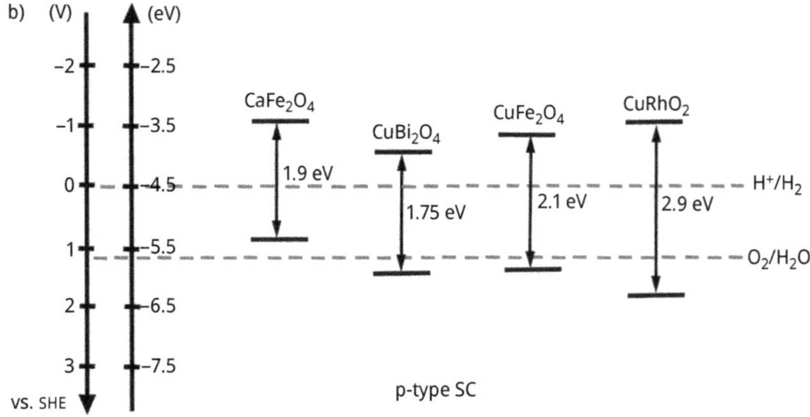

Fig. 14.8 (continued)

cations of photoelectrode surface with sensitizers such as organic dyes, SC quantum dots, or metal nanoparticles (Ag, Au, and Pt). Dye and quantum dot-sensitized solar cells are described in Sections 11.3.2 and 11.3.3. All these approaches have advantages and disadvantages. For instance, doping can create additional energetic levels in the vicinity of the energy gap, which can change the photon absorption threshold, moving it to the visible range of light. However, these additional energy levels can also facilitate the recombination of charge carriers decreasing their amount. Nanostructuring of photoelectrode materials increases the active surface area of light absorption and decreases the path of majority carriers limiting their recombination. The nanostructures also improved the light absorption by photon scattering. On the other hand, it was shown that nanostructuring reduces the open-circuit photopotential when the surface area of the SC/electrolyte interface increases. For in-depth information about oxides applied as photoanodes, photocathodes, their modification, and their influence on the efficiency of water photoelectrolysis, an interested reader is directed to the reviews [16, 18, 19] and references therein.

Many SC materials suitable energetically for the photoelectrolysis of water are unstable. To cope with this problem, the surface of such photoelectrodes was coated by a thin layer of metals (nanometer thickness layers of Au, Pt, and Pd) or a thin layer of stable semiconductor (TiO_2). The compact metal films protect the SCEs against photocorrosion. Additionally, they create the photosensitive metal/SC junction, and the photopotential of which can help to drive the reactions at the interface. From an electrochemical point of view, such electrodes work as metal electrodes, and the kinetics of reactions depends on the metal used. When the metal layers are porous, they do not protect the photoelectrodes; however, they can catalyze the reactions. The application of the thin films of a more stable semiconductor (SC_{outer}), with much higher E_g than SC substrate (SC_{inter}), has another advantage: this bilayer photoelectrode absorbs much more solar energy. The UV (short wavelength) part of the spectrum is absorbed in the outer layer, while the photons with lower energy (longer waves) penetrate across the outer layer and are absorbed in the substrate. The future of minority carriers generated in the interlayer depends on their energy, whether they have enough energy to overcome the barrier at the junction SC_{inter}/SC_{outer} and cross the outer layer. The studies of bilayer electrodes containing a thin layer of n-TiO_2 and p-type substrates GaP, GaAs, or Si confirmed that TiO_2 effectively protects those SCs against photocorrosion, but photoelectrochemical behavior is entirely determined by TiO_2. The new approaches in protections against photocorrosion focus on the application of catalytically active coatings: oxides (NiO_x and CoO_x), nitrides (GaN and TaN_x), carbides (TiC and ZrC), for photoanodes, and the application of oxides such as Al_2O_3, TiO_2, AZO (Al-doped ZnO/TiO_2) for photocathodes. Up to date, plenty of different coatings for photoelectrode protection have been tested, and much more can be found in review [20] and references therein.

The efficiency of H_2 and O_2 production is strongly dependent on the surface properties of photoelectrodes. The surface recombination, the low density of active sites, and the sluggish kinetics of charge transfer decrease the efficiency of photo water splitting. These problems can be overcome by the deposition of catalysts (cocatalysts) on the electrode surface. They may be deposited directly on the SC surface in the form of particles, nanoparticles, or embedded in the protective layer of the electrode, if such is used. The typical catalysts enhancing HER are noble metals such as Pt, Pd, Ru, and Au, as in the case of water electrolysis. The nonnoble metal catalysts NiS, CuS, and CoP, and binary systems Ni–Mo and Ni–Co have been explored as alternatives to them. To improve the efficiency of OERs, the oxides such as IrO_2, RuO_2, ZrO_2. NiO_2, and SnO_2 were tested. More detailed information about catalysts studied in the photoelectrolysis of water can be found in reviews [16, 21]. All these approaches described above improve the efficiency of STH at the individual electrodes. STH efficiency can also be increased by the appropriate design of photoelectrolysis cells.

The configurations of the most studied and developed photocells applied in water splitting are shown in Fig. 14.9(a–d). For many years, the investigations of photo water splitting have been concentrated on the photoelectrolysis cells containing a single photoanode or photocathode. Cells such as n-type photoanode PEC and p-type photocathode PEC are referred to as single bandgap devices (S2, Fig. 14.9(a)). In the last years, the huge development of dual-bandgap devices (D4) has been observed. The simplest dual bandgap photocells studied earlier contained two photoelectrodes: photoanode and photocathode immersed in electrolytic solution independently. The other configuration of PEC, referred to as "photochemical diode," also contains two photoelectrodes, and they are pressed together and connected in series by ohmic contact (Fig. 14.9(b)). In both types of cells, named p/n-PECs, the minority carriers (h^+) photogenerated in n-SC (photoanode) and (e) photogenerated in p-SC (photocathode) participate in the water splitting. Such dual bandgap configuration has some advantages in comparison with S2 photocells. They absorb a higher amount of solar spectra, as the

E_g of the photocathode is usually much lower than the E_g of the photoanode. The reaction of water splitting is divided between two SC/liquid electrolyte interfaces and is harvested by their properties. The open-circuit photopotential of the cell is higher than the photopotential of individual photoelectrodes and, in some cases, is higher than the potential required for direct water splitting. The efficiency of STH predicted theoretically for p/n-PECs, including the losses associated with reactions (activation and concentration overpotentials), is 16%.

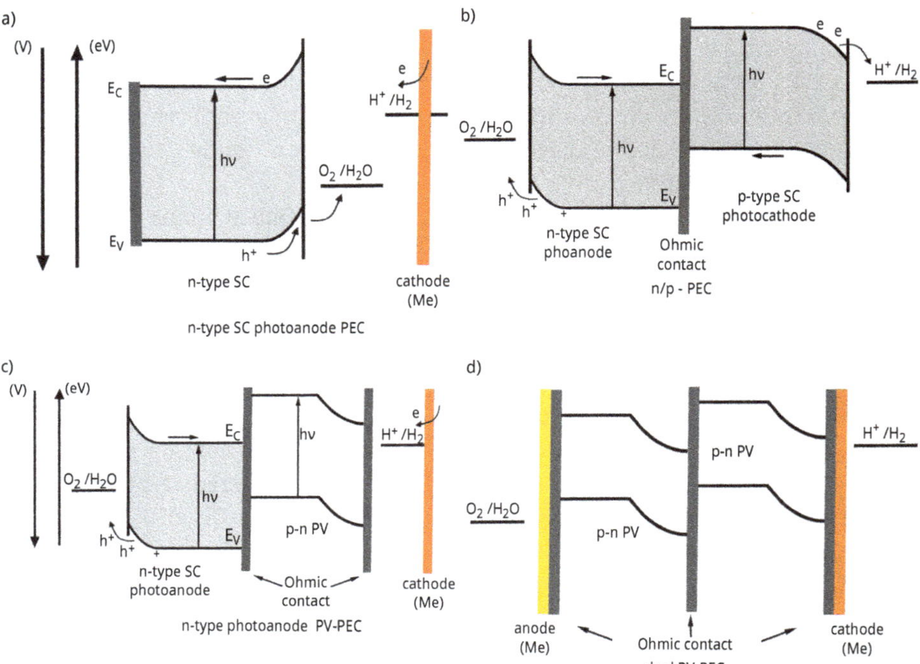

Fig. 14.9: Schematic illustration of different configurations of photoelectrochemical cell (PEC): (a) n-type photoanode PEC; (b) n/p PEC; (c) n-type photoanode PV-PEC; and (d) dual PV-PEC. PV, photovoltaic cell (adapted from [16]).

In practice, many PEC systems containing individual photoelectrodes were tested such as p-GaP/n-TiO_2, p-CdTe/n-TiO_2, and p-GaP/n-$SrTiO_3$. They show the STH values lower than 1%. Much better results were obtained when the photoelectrodes were pressed together and modified. For the p/n-PEC containing p-InP/n-GaAs system with the surface of p-InP covered with Pt and n-GaAs with MnO_2 layer, the efficiency of STH equal to 8.2% was reported. It should be mentioned that configurations applied in PECs "two in one" are referred to as "back to back" wireless configurations and are widely studied and applied. The back-to-back configurations are realized in the n-type photoanode-PV–PV and p-type photocathode-PV–PV cells. In these PECs, the photoelectrode is connected directly with the p–n PV layer cell (Fig. 14.9(c)). The photogenerated minority carriers participate in reactions of water oxidation (photoanode) or reduction of H^+ ions (photocathode), while the majority charge carriers photogenerated in the PV cell are provided to the counter metal electrodes reducing protons at the cathode or oxidizing water at the anode. The system containing p–n Si cell (PV) covered on the top by an n-type TiO_2 layer (photoanode) is an example. In this case, only the n-TiO_2 layer is exposed to the electrolyte solution. Such a design is more effective in the conversion of solar energy. TiO_2 absorbed the photon of $E_g \geq 3.2$ eV, and the remaining part of the solar spectrum is absorbed by the p–n Si part of the system (E_g of Si = 1.2 eV). The additional potential generated by the p–n Si cell is about 0.7 V and may be helpful in water splitting. Other PV cells can also be used such as dye-sensitized solar cells. The efficiency of STH obtained in such a dual system was 3.1%. Much higher efficiency of STH, equal to 12.4%, was reported for the dual system containing p-type photocathode $GaInP_2$ coupled together with n–p GaAs PV cell.

Another interesting approach is the application of devices containing two n–p PV cells connected together back to back. Their outer surfaces are covered with metal, forming a cathode and anode (Fig. 14.9(d)). In this case, the splitting of water takes place at the metal/electrolyte solution interface, such as in the direct water electrolysis (Section 14.2.2). The electrical energy required for water electrolysis is delivered by SC junctions. The energetics of these junctions are not directly related to the redox potentials of the H^+/H_2 and O_2/H_2O couples, and the relation between these potentials and the position of E_C and E_V edges is not important. Such designs are referred to as PV-PECs. They are similar to the system containing a water electrolyzer fed by PV panels, where electrical energy is delivered by wires, but PV-PEC systems are wireless. Note that ohmic contact should be formed between p–n PV junctions (Fig. 14.9(b–d)). Efficiency of STH, equal to 16%, was reported for such PV-PEC systems containing n/p $GaInP_2$/GaAs covered on both sides with Pt. Much more detailed information about the configuration of photodevices applied in water splitting and their efficiency can be found in reviews [16, 22] and references therein.

14.3.3 Photocatalysis of water: trends

In the photocatalysis of water, powdered SC materials are applied in the form of suspensions, SCPs embedded in a matrix, or spatially organized nanostructures. All these SC materials operate as photoabsorber of solar energy catalyzing photosplitting of water. As in photoelectrolysis, the water is split into H_2 and O_2. However, during photocatalysis, H_2 and O_2 are not separated between electrodes and evolved simultaneously at the same particles. The requirements that should be fulfilled by SC materials applied in these processes are common and described in Section 14.3.1. Here also the equations for efficiency of STH and apparent quantum yields are given. In practice, the efficiency of H_2 formation is determined by the amount of H_2 evolved (mol/h) in unit time (h) per unit mass (g) of SCPs. It was pointed out that STH conversion efficiency of 5–10% obtained in photocatalysis of water would render this method economically acceptable for H_2 production. However, the STH values published in articles are much lower. On the laboratory levels, they can reach values of 1–3%, but only when photocatalysts with special designs are used. The other factor that should be considered before the practical application of SCPs is their selectivity to the reaction path. Selectivity is determined by equation (14.20) as the ratio of the quantum yield of *i*-product formation to the quantum yield of *r*-reactant decomposition:

$$S = \frac{\varnothing_i}{\varnothing_r} \tag{14.20}$$

Changes in selectivity depend on the correlation between the surface concentration (number) of electrons (e_s) and the surface concentration of holes (h^+_s). If the concentration of e_s is larger than the concentration of holes h^+_s, the reaction of reduction on SCPs is dominant (reduction of H^+); in the reverse case, the oxidation reaction is preferred. This factor is important when photo water splitting is carried out in solutions containing other redox systems.

More than a hundred SC materials, mostly oxides, were tested as potential photocatalysts in water splitting. Various syntheses, doping, cocatalysts, and specially organized nanostructures were applied, but still the efficiency of photocatalytic water splitting and H_2 formation is not satisfactory. Very rich data about applied photocatalysts, their efficiency, and special treatments can be found in reviews [17, 21, 23] and references therein. Note that some of these SC materials and strategies are also used in the photoelectrolysis of water (Section 14.3.2). Here, we briefly describe some approaches more characteristic for SCPs.

As in the case of SCEs, it is convenient to divide the photocatalysts into two groups: the photocatalyst absorbing the photons from the UV range of the solar spectrum and photocatalysts absorbing photons in the visible light range. The first group involves oxides such as TiO_2, Nb_2O_3, Ta_2O_3, Ga_2O_3, titanates, niobates, tantalates, a material with perovskite structure, and others. These SC materials are stable during illumination; however, their bandgap energy E_g is higher than 3 eV and they can only absorb photons in the UV range of the spectrum, which accounts for a small fraction of solar spectra (ca. 4%). The other group includes the photocatalysts possessing bandgap energy E_g lower than 3 eV, such as metal oxides (Fe_2O_3 and WO_3), sulfides, selenides (CdS and CdSe), perovskites ($BiFeO_3$ and $LaFeO_3$), and oxynitrides (TaON and T_3N_5). Unfortunately, some of them catalyze only half of the reaction of photocatalytic water splitting, mainly O_2 evolution (WO_3, In_2O_3, $SnNb_2O_3$, and Bi_2WO_3). The photocatalysts with $E_g < 3$ eV are often applied in the architecture of the Z-scheme system (see further). All these photocatalysts contain metal ions in their structures. Recently, the applications of metal-free photocatalysts are in progress. Plenty of such materials were studied, including elemental catalysts such as P, Si, and Se, and binary compounds such as carbon nitrides C_xN_y. The intensively studied material is g-C_3N_4, a graphite carbon-nitride-based polymer. It is a stable, visible light-active material with a bandgap energy of 2.7 eV. It was shown that this photocatalyst causes the evolution of H_2 from water at wavelength >420 nm, even without the cocatalyst used. Much more about metal-free photocatalysts and their application in hydrogen production can be found in the review [24].

Generally, three main steps can be distinguished during photocatalysis of water: the photon absorption by SCPs, separation of photogenerated charge carriers e–h^+, and reactions of H_2 and O_2 evolution on SCPs. The efficiency of water photocatalysis is determined by all of them and is low. Different strategies are developed to improve the efficiency of water splitting and the effectiveness of these steps. As in the case of photoelectrodes (Section 14.3.2), to improve photon absorption range, the particles of photocatalyst were doped with metals (Ag, Au, and Pt), nonmetals (N, P, and S), and metal ions (Cr^{3+}, Ni^{2+}, and Rh^{3+}) or were sensitized by using visible light sensitizer – noble metals (Ag, Au, and Pt) or organic dyes. Doping of SC materials changes their electronic structure causing a decrease in bandgap energy, or the formation of new energy levels in a bandgap. These effects shift the edge of photon absorption to the Vis range of the solar spectrum. However, the additional energy levels created in the bandgap can work as recombination centers decreasing the amount of photogenerated charge carriers. Metal nanoparticles deposited on photocatalyst surface can implement several functions. They are not only sensitizers but also can work as cocatalysts, and as surface traps protecting the charge carriers (electrons and holes) against recombination. Particularly interesting is the application of Ag and Au nanoparticles loaded on SCPs. These nano-MePs can absorb visible light via surface plasmon resonance, enhancing photocatalytic activity. The application of dyes is limited because during illumination they are oxidized by O_2 evolved at particles and decompose.

The other problems that should be solved are connected with the separation of electron–hole pairs generated during the illumination of SCPs and the separation of H_2 and O_2 evolved, before the backward reaction of water formation takes place. Few approaches are developed to cope with these problems. One of them is the application of cocatalysts, the other is the design of the systems containing two-photon absorbers (SCPs) referred to as Z-scheme and cascade system.

Many different cocatalysts were tested in the photocatalysis of water. Noble metals (Pt, Ru, Rh, Au, and Ag) were widely applied to improve the efficiency of hydrogen evolution. However, they are too expensive. Therefore, other nonnoble metal catalysts have been developed such as MoS_2, NiS, NiO, and CoO. For improvement of the OER, oxides IrO_2 and RhO_2 were applied, but also other oxides were tested such as NiO_x, CoO_x, and MnO_x. The cocatalysts can be loaded on SCPs in the form of nanoparticles by various methods. These methods are mainly divided into two groups. In one of them, the cocatalysts are grown directly on the surface of photocatalyst particles using a precursor. In other, the cocatalysts are synthesized and after that loaded on the surface of photocatalysts. The precipitation and photodeposition are used for the direct formation of cocatalysts, while for loading of cocatalysts impregnation, grinding, and calcination are applied.

How does this system cocatalyst/photocatalyst work? Figure 14.10 presents the processes between metal cocatalysts (MeC) and host photocatalysts (SCP). Let us suppose that the cocatalyst has Fermi energy $E_{F,Me}$ lower than the SC photocatalyst $E_{F,SC}$. After the deposition of Me nanoparticles on the photocatalyst, the equilibrium is established between electrons of both materials. This state is characterized by equality $E_{F,Me} = E_{F,SC}$ and is reached by the flow of electrons from the CB of SCP to the metal. As a result, the metal part of the interface is charged negatively, and the SC part of the interface is charged positively. Note that it means the formation at the SC surface of a thin space charge layer in a depletion state. The barrier formed at the interface (junction) is referred to as the Schottky barrier. During illumination and absorption of UV photons, this barrier decreases as the energy of SC electrons and $E_{F,SC}$ increase (to $E^*_{F,SC}$). These photoexcited electrons flow from CB to the metal, also increasing $E_{F,Me}$, and participate in the reduction of H^+ ions at the metal/solution interface. At the same time, the photogenerated holes oxidized water at the SCP/solution interface. To improve the efficiency of O_2 evolution, the additional cocatalyst can be deposited on SCPs. These cocatalysts are mainly oxides, and plenty of them have SC properties; therefore, the type of junctions that are formed between cocatalysts and photocatalysts depends on the type of SC and band energy matching.

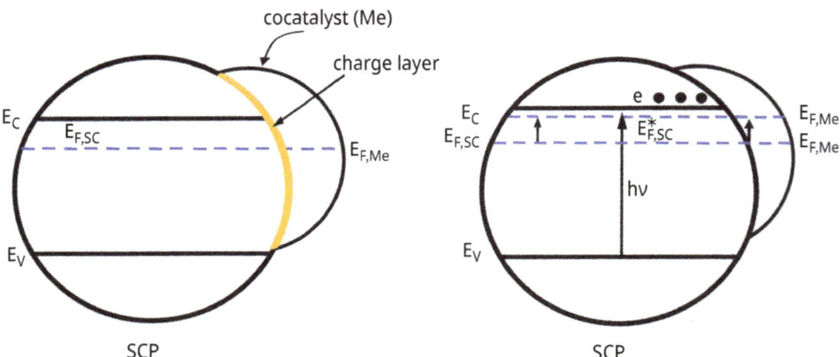

Fig. 14.10: Schematic illustration of cocatalyst/photocatalyst system. Principle of work. $E_{F,SC}$, $E^*{}_{F,SC}$, $E_{F,Me}$ – Fermi energies of an electron in semiconductor (SC) and metal (Me); * means illumination of SC.

The electric field of the junctions, formed at the interface of cocatalyst/photocatalyst, enhances significantly the separation of photogenerated electron–hole pairs, accelerates the motion of free charge carriers to the surface reaction sides, and prevents the backward reactions of water formation. The applied cocatalysts work also as ordinary catalysts, decreasing the activation energy of HER and OER. Much more about cocatalysts, their synthesis, and their application can be found in the review [25]. This cocatalyst/photocatalyst design can be classified as one photon absorber (SCP), one-stage system with four photons taking part in overall water splitting.

The design that is widely tested and applied is the so-called Z-scheme, a two-photon absorber two-stage system (Fig. 14.11). In the Z-scheme design, the reaction of photocatalytic water splitting is divided between two different photocatalysts – SCPs; hence, the evolved H_2 and O_2 are separated. These two stages are combined by a mediator–redox couple in the solution, or by a solid system. During illumination, in both particles (SCP1 and SCP2), charge carriers are photogenerated. At SCP1, electrons reduced H^+ ions to H_2 and holes oxidized the Red form of redox couple in the solution, while at SCP2 holes oxidized H_2O and electrons reduced the Ox form of redox couple (see also Fig. 11.7). If the solid mediator is used, the holes from SCP1 should recombine with electrons from SCP2.

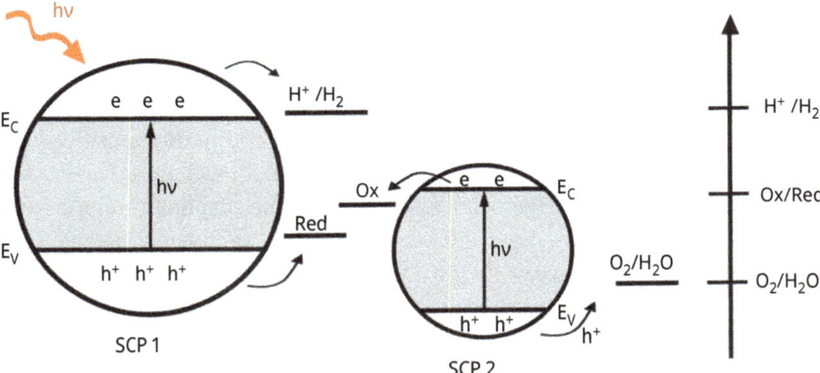

Fig. 14.11: Schematic energy diagram of photocatalytic water splitting Z-system (Z-scheme).

Plenty of SCPs were tested in the design of Z-scheme for water splitting. Among them $SrTiO_3$, TaON, $BaTaO_2N$, C_3N_4 and WO_3, TiO_2, and Ta_3N_5 are applied for H_2 and O_2 evolution, respectively. It is worth mentioning that the Z-scheme system containing $SrTiO_3$ (doped with Cr and Ta) and WO_3 coupled by mediator iodate/iodide ions can split water under visible light illumination. Z-designs applied for water splitting have some advantages:

- They enhance the separation of photogenerated charge carriers.
- They eliminate the backward reaction of water formation.
- They can utilize the visible part of the solar spectrum, as two particles with different E_g are used.

The other two absorber photon systems, which can utilize visible light and facilitate the separation of photogenerated electron–hole pairs, are cascade systems. Cascade is designed in the form of two different SCPs coupled together or SCP particles deposited on the nanosheet of SC. Systems such as $SrTiO_3/TiO_2$, WO_3/TiO_2, and CdS/TiO_2 were tested. They can be used for hydrogen production and degradation of organic compounds. However, there is no information about the ability for water splitting.

Plenty of efforts have been made to improve the efficiency of water photocatalysis. For many years, the rates of H_2 and O_2 evolutions were on the level μmol/h g not having practical meaning. The situation changed with the huge development of nanotechnologies. New strategies are developed, systems with nanoarchitecture are designed, and visible-light-driven photocatalysts are synthesized. It is not possible to describe all of them, but we should mention the one-step water splitting photocatalysts such as the plasmonic structure composites (Au–TiO_2 systems), solid solutions $[(Ga_{1-x}Zn_x)(N_{1-x}O_x)]$, and a new generation of cocatalyst design containing noble metal (core)/Cr_2O_3 (shell) nanoparticles. At present, the best rates of H_2 and O_2 evolutions were obtained on CoO one-step photocatalyst, under visible light, as 71.7 and 34.5 mmol/h g for H_2 and O_2 evolved, referred to as 5% STH [17, 23]. However, for most of photocatalysts, the rates of H_2 and O_2 evolutions are contained in the range from μmol/h g to few mmol/h g [17].

If we consider only H_2 evolution, the best efficiencies are obtained if the reaction of hydrogen generation is carried out in the aqueous solution containing sacrificial reagents. In these cases, the photocatalysts participate only in half-reaction of water splitting. The rates of H_2 evolution depend not only on the type of photocatalysts, and incident light used, but also on the type and concentration of sacrificial reagents. The rates of H_2 evolution are in the range of 10–150 mmol/h g. If the photocatalytic reaction is carried out in the presence of electron donors (hole scavengers), photogenerated holes oxidize the reducing reagent (electron donor) instead of water, while photogenerated electrons participate in H_2 evolution, enhancing the rate of reaction. If we want to enhance the rate of O_2 evolution, we utilize, as the sacrificial reagents, acceptors of electrons, which are reduced by photogenerated electrons, while photogenerated holes oxidize water. The reactions with sacrificial reagents are often used at the laboratory level for the evaluation of photocatalysts applied in water splitting. Alcohols, sulfide, sulfite ions, Ag^+, and Fe^{3+} are often used as electron and donor acceptors for such purposes. This half-reaction of H_2 evolution may be important for practical hydrogen production if, as reducing reagents, biomass or other abundant compounds will be used.

References

[1] Tashie Lewis BCh, Nnabuife SG. Hydrogen production, distribution, storage and power conversion in a hydrogen economy – A technology review. Chem. Eng. J. Adv. 2021, 8, 100172–100191.

[2] Smolinka T, Garche J. Electrochemical power sources. Fundamentals, systems and applications. Hydrogen production by water electrolysis. Amsterdam, Netherland, Elsevier, 2021.

[3] Bockris JO`M, Reddy AKN. Modern electrochemistry 2B. NY, USA, Kluwer Academic/Plenum Publisher, 2000.

[4] Wang S, Lu A, Zong CJ. Hydrogen production from water electrolysis: Role of catalysts. Nano Converg. 2021, 8, 4. doi: 10.1186/s40580-021-00254-x.

[5] Durovic M, Hnat J, Bouzek K. Electrocatalysts for the hydrogen evolution reaction in alkaline and neutral media. A comparative review. J. Power Sources 2021, 493, 229708.

[6] Song J, Wei C, Huang ZF, Liu C, Zeng I, Wang X, Hu ZJ. A review of fundamentals for designing oxygen evolution electrocatalysts. Chem. Soc. Rev. 2020, 49, 2196–2214.

[7] Li J. Oxygen evolution reaction in energy conversion and storage: Design strategies under and beyond the energy scaling relationship. Nano-Micro Lett. 2022, 14, 112.

[8] Zeng K, Zhang D. Recent progress in alkaline water electrolysis for hydrogen production and application. Prog. Energy Combust. Sci. 2010, 36, 307–326.

[9] Kecebas A, Kayfeci M, Bayat M. Electrochemical hydrogen generation. In: Calise F, D`Accadio M, Ferrevo D, eds. Solar hydrogen production. Processes, systems and technologies. Amsterdam, Netherland, Elsevier, 2019, 299–317.

[10] Kumar SS, Himabindu V. Hydrogen production by PEM water electrolysis – A review. Mater. Sci. Energy Technol. 2019, 2, 442–454.

[11] Tu H. Solid electrolytes. In: Bagotsky VS, ed. Fundamentals of electrochemistry. Hoboken, USA, Wiley & Sons Inc, 2006, 419–436.

[12] Laguna-Bercero MA. Recent advances in high temperature electrolysis using solid oxide fuel cells: A review. J. Power Sources 2012, 203, 4–16.

[13] Mohammed- Ibrahim J, Mousab H. Recent advances on hydrogen production through seawater electrolysis. Mater. Sci. Energy Technol. 2020, 3, 780–807.

[14] Rousseau R, Etcheverry L, Roumbaud E, Besseguy R, Delia M-L, Bergel A. Microbial electrolysis cell (MEC): Strengths, weaknesses and research need from electrochemical engineering standpoint. Appl. Energy 2020, 257, 113938.

[15] Hasany M, Mardanpour MM, Yaghmaei S. Biocatalyst in microbial electrolysis cells: A review. Int. J. Hydrog. Energy 2016, 41, 1477–1493.

[16] Walter MG, Warren EL, McKone JR, Boetcher SW, Mi Q, Santori EA, Lewis NS. Solar water splitting cells. Chem. Rev. 2010, 110, 6446–6473.

[17] Cao S, Piao L, Chen X. Emerging photocatalysts for hydrogen evolution. Trends Chem. 2020, 2, 57–70.

[18] Rajeshwar K. Hydrogen generation at irradiated oxide semiconductor – Solution interface. J. Appl. Electrochem. 2007, 37, 765–787.

[19] Diez-Garcia MI, Gomez R. Progress in ternary metal oxides as photocathodes for water splitting cells: Optimization strategies. Solar RRR 2022, 6, 2100871.

[20] Hu S, Levis NS, Ager JW, Yang J, McKone JR, Strandwitz NC. Thin film materials for protection of semiconducting photoelectrodes in solar fuel generators. J. Phys. Chem. C 2015, 119, 24201–24228.

[21] Wang Z, Li C, Domen K. Recent developments in heterogenous photocatalysts for solar –driven overall water splitting. Chem. Soc. Rev. 2019, 48, 2109–2125.

[22] Chowdhury F. Recent advances and demonstrated potentials for clean hydrogen via overall solar water splitting. MRS Adv. 2019, 4, 2771–2785.

[23] Ismail AA, Bahneman DW. Photochemical splitting of water for hydrogen production. Sol. Energy Mater. Sol. Cells 2014, 128, 85–101.

[24] Rahman MZ, Kibria G, Mullins Ch B. Metal-free photocatalysts for hydrogen evolution. Chem. Soc. Rev. 2020, 49, 1887–1931.

[25] Tian L, Guan X, Zong S, Dai A, Qu J. Cocatalysts for photocatalytic overall water splitting: A mini review. Catalyst 2023, 13, 355.

15 Electrochemical and photocatalytic methods in pollutant removal

15.1 General remarks

All living creatures need air and water for life. For centuries, the equilibrium has been established between the natural environment and human beings. From the nineteenth century with the Industrial Revolution, everything changed as a result of human activities, developments of technologies, and the increasing demand for goods and energy. We feel the climate changes, and our health is deteriorating as the amount of pollutants due to our activity increases in nature. The problems of detection and removal of pollutants are crucial in our days. The detection of some pollutants by using sensors based on electrochemical reactions is described in Chapter 13. The pollutants are everywhere. The main air pollutants are gases such as CO, SO_2, NO, NO_2, and NH_3, volatile organic compounds (VOCs), and particulates. They can be removed by applying physicochemical processes such as absorption, adsorption, and combustion. For instance, SO_2 is removed by a process called flue gas desulfurization, in which gas dissolves in/or reacts with absorbent, being trapped. Gas adsorption is used for the control and removal of VOCs. For such purposes also combustion is applied, converting organic pollutants into water and CO_2. Carbon dioxide for many years was treated as a natural air compound. At present, CO_2 is referred to as a greenhouse gas responsible for global warming and climate change. Among other greenhouse gases (e.g., H_2O, N_2O, CH_4, and O_3), CO_2 is not as effective in heat absorption and re-radiation as N_2O or CH_4; however, its emission is very high (82%) in comparison with CH_4 (10%) and N_2O (8%) and should be significantly reduced. The main source of CO_2 is human activity and increasing demand for energy, which is mostly produced by burning fossil fuels (coal, natural gas, petroleum, and gasoline). The best way to reduce the CO_2 level in the air is the application of alternative energy sources: solar energy (see Chapters 11 and 14), nuclear energy, and wind energy. Currently, the technology for the capture and storage of CO_2 is being developed. In this two-step technology, CO_2 is firstly separated from other gases and after that stored in geological beds In the presence of oxygen and in the humid environment, gases NO, NO_2, and SO_2 form nitrous, nitric, and sulfuric acids (Chapter 13) contributing to acidic rain precipitations. They fall to the ground, on surface waters, and at soil causing a pH decrease and other deteriorating effects. In the last decades, the acidification of oceans has been observed as a result of the huge dissolution of CO_2 and carbonic acid formation (H_2CO_3). Most of the gaseous pollutants have anthropogenic origins, but it is worth mentioning that NH_3 and H_2S can be produced in bacterial processes of biomass decomposition.

Water, an excellent solvent, can dissolve plenty of inorganic and organic compounds, and is therefore very susceptible to pollution. The ground and surface waters

https://doi.org/10.1515/9783111160986-016

(e.g., oceans, lakes, and rivers) are contaminated by air and soil pollutants, which can have natural or anthropogenic origins. The most polluted waters are wastewater which comes from domestic, industrial, and agricultural activities. These waters contain plenty of inorganic and organic contaminants. Among them are heavy metals, sulfides, nitrates, phosphates, organohalogens, organophosphorus, pharmaceutical compounds, and others. They exist in the form of dissolved species, colloids, and suspensions. Some of them are toxic and hazardous. The heavy metals are introduced into waters naturally, by erosion of minerals and soil, through volcanic eruptions and biogenic processes. They came also as a result of human activity from the metallurgy industry, factories of batteries, textiles, electroplating processes, and many others. In wastewaters, depending on the composition and solution pH, the heavy metals can exist as free ions, hydroxides, and stable complexes, formed with compounds such as chloride, citrate, cyanide, and ethylenediaminetetraacetic acid (EDTA). They are highly toxic, potentially carcinogenic, nonbiodegradable, and can accumulate in human beings and other biosystems. A variety of techniques have been developed and applied to remove heavy metals and other contaminants from water, such as adsorption, coagulation, reverse osmosis, and membrane filtration technology. Among them are electrochemical methods: electrodeposition (ED), electrodialysis (EDl), electrocoagulation (EC), and electroflotation (EF) [1]. The principles, some advantages, and disadvantages of these methods are briefly described in the next section.

In recent years, plenty of research has been carried out on the application of advanced oxidation processes (AOPs) for the purification of different wastewater. AOPs can be described as oxidative reactions in which input energy (chemical, light, and electrical) is used for the production of very reactive oxidizing species – hydroxyl radicals, enabling the in situ oxidation and decomposition of organic pollutants. It was shown that AOPs are very effective in the degradation of organic compounds and organic heavy metal complexes. AOPs can be realized by a variety of methods. Among them are Fenton oxidation, electro-Fenton oxidation, photocatalytic oxidation, and others. The application of photocatalytic oxidation of organic pollutants raised significant attention, as solar energy can be used for such purposes. Some principles of the application of electro-Fenton oxidation and photocatalysis for the removal of organic compounds are also described in this chapter.

15.2 Removal and recovery of heavy metals: electrochemical treatment

In the last few years, the fastest growing waste in the world is electronic waste (e-waste). Their production is estimated at about 50–60 million tons a year. Only a small amount of them (about 17%) is properly treated and recycled. Electronic waste (e-waste) means discarded electrical and electronic devices. All of them contain plenty of metals, divided mostly into three categories: (i) base metals which constitute about

30 wt% (e.g., Cu, Cd, Zn, Cr, Ni, and Co); (ii) precious metals constitute 0.1–1 wt% (e.g., Pd, Pt, Au, Ag, and Ir); and (iii) rare earth elements such as Gd, La, Sc, and Pr, and the amount of them is lower than 0.1 wt%. Since some of these metals are toxic, some are valuable, and the supply of some of them is not sufficient for further high-tech development, they have to be recovered. Different strategies are developed, but the first step is common – the extraction of metals by leaching the waste with acid, alkali, or even with complex compounds and conversion of solid metal-containing species into soluble ionic forms. An example is acid leaching of metal sulfides or hydroxides:

$$\mathrm{ZnS(s) + H_2SO_4(aq) + 1/2O_2(g) \rightarrow ZnSO_4(aq) + H_2O\ (l) + S(s)}$$

$$\mathrm{AgOH\ (s) + HNO_3(aq) \rightarrow AgNO_3(aq) + H_2O\ (l)}$$

Leachate solutions from acidic treatment (or others) of e-waste contain multiple metal ions, which differ from each other, posing a challenge to the selective recovery of pure metals. For such purposes, electrochemical methods can be used such as electrowinning, often referred to as electroextraction and electrorefining. In the electroextraction method, the metal ions dissolved in a leach solution are reduced onto the cathode and extracted in their metallic form. In the electrorefining method, an anode is made from impure metals, which have to be refined. During electrooxidation, the anode is partially dissolved in electrolytic media and the formed free metal ions migrate to the cathode, where they are reduced. In both methods, the main processes are ED processes. As pointed out above, metal contaminants exist in the ground and surface waters and in larger amounts in industrial wastewater. ED is applied for the removal of free metal ions from all kinds of water.

15.2.1 Electrodeposition and electrodialysis

Electrodeposition (ED) means electrolytic processes carried out in two-electrode cells at constant potential or current. They are governed by Faraday's law. During ED, new solid phases are formed. Much more about the principles, phase formations, and practical aspects can be found in Chapter 5. In the ED method, the free metal ions dissolved in wastewater or leaching solutions are reduced at the cathode. Their redox potential should be more positive than the redox potential of the H^+/H_2 redox couple. If not, the hydrogen is evolved at the electrode, disturbing the reduction of ions and decreasing the efficiency of the reaction:

$$\mathrm{Me}^{n+}(\mathrm{aq}) + ne \rightarrow \mathrm{Me\ (s)},\ E_{\mathrm{eq}} = E^0 + \frac{RT}{nF}\ln a_{\mathrm{Me}^{n+}}$$

$$2\mathrm{H}^+(\mathrm{aq}) + 2\mathrm{e} \rightarrow \mathrm{H_2(g)},\ E^0 = 0\ \mathrm{V}$$

The redox potential of H^+ ions depends on pH of the solution and is described by the equation:

$$E = -0.059\,\text{pH, and at pH 7 it is equal to} -0.413\,\text{V (vs. SHE)}$$

If we compare only the values of standard potentials of Me^{n+}/Me with this value, it is clearly seen that by ED, we can recover metals such as Ag, Cu, Pb, Ni, and Co, if their ions do not exist in the form of hydroxides or stable complexes. At the anode, the oxidation of some metallic contaminants takes place. This process is utilized in electrorefining:

$$\text{Me (s)} \rightarrow \text{Me}^{n+}(\text{aq}) + ne$$

If the potential of the anode is sufficient, the oxidation of water or hydroxide ions occurs, depending on the pH of the solution (see Section 14.2.1)

$$2\text{H}_2\text{O (l)} \rightarrow \text{O}_2(\text{g}) + 4\text{H}^+(\text{aq}) + 4\text{e}, \quad E^0 = 1.229\,\text{V (vs. SHE)}$$

$$4\text{OH}^-(\text{aq}) \rightarrow \text{O}_2(\text{g}) + 2\text{H}_2\text{O (l)} + 4\text{e}, \quad E^0 = 0.401\,\text{V (vs. SHE)}$$

In alkali media, most of the transition ions form insoluble hydroxides, which can be adsorbed on anode undergoing further oxidation to metal oxides:

$$\text{Me}^{n+}(\text{aq}) + n\text{OH}^-(\text{aq}) \rightarrow \text{Me(OH)}_{n,\text{ads}}$$

$$\text{Me(OH)}_{n,\text{ads}} \rightarrow \text{MeO}_n(\text{s}) + n\text{H}^+(\text{aq}) + ne$$

Effluents originating from electroplating baths or leaching solutions can contain a mixture of free heavy metal ions. In theory, they can be separated and selectively reduced by careful control of potential. However, in practice, the deposition reactions depend on the pH of the solution and impurities. It was shown that it is possible to deposit Cu from a solution containing Cu^{2+}, Cd^{2+}, or Zn^{2+} ions, but not Pb in the presence of In^{3+} ions, despite differences in the standard potentials (about 0.2 V). Selective recovery of Cu (95%) from e-waste was reached in an ammonia-based electrolyte, while selective recovery of Fe from the leaching of magnets (Nd–Fe–B) was obtained when ammonium sulfate and sodium citrate were used. The processes of selective recovery of metals from wastewater and leaching solutions are very complicated and require detailed analysis of solution composition. In EDs, applied for removal and recovery of metals from leaching solutions, aqueous media are mostly used. However, some metal ions have a very high redox potential, for example, $Al^{3+}/Al - 1.67$ V, $U^{3+}/U - 1.66$ V, $Pt^{2+}/Pt + 1.19$ V, and some react in aqueous solutions and cannot be removed in such a way. Therefore, often nonaqueous electrolytes are used. They include conventional organic solvents such as dimethylsulfoxide and dimethylformamide, but also ionic liquids (see Section 1.3.1). More details about the electrochemical recovery of metals from e-waste and wastewater can be found in reviews [2, 3] and references therein. The other problem that should be solved during the ED of metals from wastewaters is a low concentration of metal ions (not more than a few thousand ppm), resulting in the decrease of the ion transportation rate and the high values of concentration overpotential (see Section 1.1.3). A few methods are used to improve the transport of ions to the electrode and increase their concentration, such as extensive mixing of solution by using rotating electrodes, gas bubbling, concentrator electrochemical technique, or electrodialysis [1]. The selectivity and efficiency of metal recovery from wastewater may be improved when the EDl and ED are coupled together.

To date, EDl has been widely used in the desalinization of water and the brine industry. At present, the EDl method is also applied for the removal and recovery of heavy metal ions from the electroplating industry residues. The operating principles of EDl are simple. The scheme of the three-compartment EDl cell is shown in Fig. 15.1.

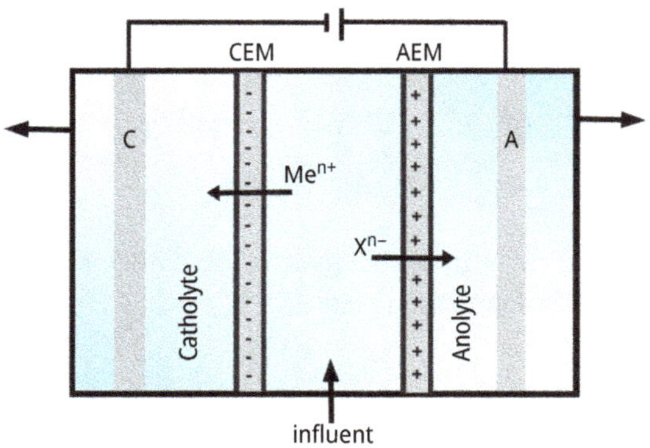

Fig. 15.1: Schematic illustration of an electrodialysis cell. A, anode; C, cathode; CEM, cation-exchange membrane; AEM, anion-exchange membrane.

Cell contains two electrodes (cathode and anode) and two ion-exchange membranes (IEM) (cation-exchange membrane (CEM) and anion-exchange membrane (AEM)). CEM is negatively charged and can attract positively charged metal ions, while AEM is positively charged and can attract negatively charged anions. The electric field, generated between two electrodes, works as a driving force causing the migration of ions toward electrodes. Migration of ions across the IEM allows an increase of cationic concentration in the catholyte part of the cell and anionic concentration in the anolyte part, respectively. The cell can be fed with industry effluents continuously and catholyte solutions may be removed for further treatments. In this case (Fig. 15.1), the cell contains two IEMs, but cells with only one or a stack of membranes are also used. More about membranes and membrane potential can be found in Section 1.4. The EDl may be carried out in two ways, one without separation of metal ions but with an increase of their concentrations, and the second with an increase of both separation and concentration. The second way is referred to as selective EDl and is very important for the recovery of metal from the industry effluents containing a mixture of metal ions. The selective separation can be achieved by pH matching, application of membranes selective to the charge of ions, and application of complexing compounds.

The specially prepared selective membranes are used for the separation of differently charged cations. For instance, the Nafion membrane covered with a thin layer of polyethyleneimine shows better selectivity in the separation of Cr^{3+} from a solution containing Na^{+} and Cr^{3+} ions. Lately, the heterogeneous and chelating membranes have been tested in the recovery of metals. The heterogeneous CEM consisting of acrylamide-2-methyl propane sulfonic acid was successfully used for the selective separation of Ni^{2+}, Pb^{2+}, and K^{+} ions. The other approach is the application of complexing agents such as chloride, citric acid, or widely used EDTA. They form complexes with metal ions changing their charge from positive to negative. For instance, Ni^{2+} and Co^{2+} ions can be separated in the presence of EDTA, since nickel ions form negatively charged complexes $Ni(EDTA)^{2-}$ transferred to the anolyte part of the chamber. The EDl technology is simple, and what is important is the removal of ions can be carried out in a continuous flow of industry effluent; therefore, the application of this method for recovery of ions is growing. However, some drawbacks have to be overcome such as the high demand for electrical energy caused by the resistance of EDl cells and the high cost of membranes. More detailed information about EDl and its application in the separation of metal ions can be found in review [4].

15.2.2 Electrocoagulation and electroflotation

The classical methods of the removal of organic pollutants and heavy metals from wastewater are coagulation/flocculation and flotation. Both methods have the same origin and are based on the precipitation of coagulated species. In both methods, coagulants such as Fe^{2+} and Al^{3+} are used. The sludge formed during the precipitation of species in coagulation/flocculation processes falls to the bottom of the reservoir. In the flotation, the precipitates or ions with surfactants are driven to the surface by gas bubbles. Coagulation processes are well-known in colloid science [5].

The wastewater contains plenty of colloid pollutants. They include organic species, metal oxides, sulfides, and biotic materials. Colloid systems consist of two phases: the dispersed phase containing particles with diameters ranging from 1 nm to 1 μm, and the continuous phase (dispersing phase). If this phase is liquid, we use the term "sol." Sols can be divided into two groups: lyophilic sols and lyophobic sols. In aqueous suspensions, they are referred to as hydrophilic and hydrophobic sols. Hydrophilic sols are formed mainly by large organic molecules (e.g., peptides) and their charge originates from ionized function groups. They are highly hydrated and stabilized by solvent (water) molecules. The hydrophobic sols contain insoluble particles of metal oxides, sulfides, chlorides, and others. Their charge comes from particles' ionized surface groups or chemically adsorbed ions (or both). Their stability is caused by the interaction between particles' charge and oppositely charged other ions existing in water. As a result, the electrical double layer (e.d.l.) is formed at the interface dispersed particle/solution. This e.d.l. consists of two parts: (i) compact layer referred to as the Helmholtz layer, with a linear potential drop in its space, due to the particle surface charge, water molecules, and solvated counter ions from solution, and (ii) diffusion layer, referred to as the Gouy–Chapman layer, where the distribution of ions is described by Maxwell's relation, and its potential changes exponentially with distance from the compact layer (see Section 1.1.1, Figs. 1.3–1.5, setting for the charge on a metal surface the charge on a particle surface). In an external electric field, the hydrophobic sol particles, together with their compact layer, move together as one species to the respective electrode, whereas the rest of their e.d.l. moves in the opposite direction. The "plane" that divides the two regions of the e.d.l. is called the plane (surface) of shear. The potential that exists on this surface is called the zeta potential (ζ) or electrokinetic potential and is an indirect measure of the charge of the colloid particles (Section 1.1.1., Fig. 1.9). It is worth stressing that this potential can be measured only when the colloidal particles are set in motion with respect to the dispersing phase by the presence of an external electric field. The formation of e.d.l. in the hydrophobic sols is responsible for all electrokinetic phenomena such as electroosmosis, electrophoresis, streaming potential, and sedimentation potential. The addition of ions that can adsorb at particle/solution interface, decreasing charge of e.d.l., and zeta potential destabilizes the colloid system. The particles become thermodynamically unstable and at zeta potential near zero begin to coagulate. The coagulation ability is dependent on the valence of coagulating ions: the greater the valence of ions, the better the ability for coagulation. The coagulation of hydrophilic colloids can be achieved by the application of other solvents than water, or ions with higher energy of hydration than energy of interaction of water molecules with colloid particles. Coagulation is connected to flocculation. Coagulation means the destabilization of colloids by neutralization of forces (mostly charges) which keep them separated, and results in the agglomeration on a molecular level and formation of microparticles. Flocculation means the agglomeration of destabilized colloid particles in flocks (large aggregates) by flocculants such as polyacrylamide (PAM), polyaluminum chloride, or polyferric sulfate. Note that very often both terms are used interchangeably. In coagulation, the coagulant agents Fe^{2+} and Al^{3+} come from ionized salts, mostly chlorides and sulfates. In electrocoagulation, Fe^{2+} and Al^{3+} ions are produced electrochemically, in situ in wastewaters, by anodic electrooxidation of Fe, steel, or Al. At the cathode, the reduction of water takes place with H_2 evolution. The reactions that occur at the anode and further in the solution are complicated and depend on the pH of the solution (wastewater). The Pourbaix diagrams of Fe/H_2O, Al/H_2O, equilibriums, and reactions in such systems are described elsewhere [6, 7]. Here we consider some of them. Let us start with the Fe anode. The reactions that occur in acidic media are

$$\text{Anode:}\quad \mathrm{Fe\,(s)} \longrightarrow \mathrm{Fe^{2+}(aq)} + 2\mathrm{e},\ E^0 = -0.441\,\mathrm{V\,(vs.\,SHE)}$$

$$\text{Cathode}\quad 2\mathrm{H^+(aq)} + 2\mathrm{e} \longrightarrow \mathrm{H_2(g)},\quad E^0 = 0\,\mathrm{V\,(vs.\,SHE)}$$

The reactions at the anode are multistep reactions with the formation of adsorbed species (FeOH and $FeOH^+$). Fe^{2+} ions are formed in the oxidation reaction of $FeOH^+_{ads}$:

$$FeOH^{+}{}_{ads} + H^{+}(aq) \rightarrow Fe^{2+}(aq) + H_2O\,(l)$$

In the presence of O_2 in acidic media, Fe^{2+} ions are oxidized to Fe^{3+} ions and insoluble $Fe(OH)_3$ is formed:

$$4Fe^{2+}(aq) + 10\,H_2O(l) + O_2(g) \rightarrow 4Fe(OH)_3(s) + 8H^{+}(aq)$$

In alkaline (basic) media, the ferrous ions produced in the anode oxidation react with hydroxide ions forming insoluble $Fe(OH)_2$. The reactions are

$$\text{Anode:}\quad Fe(s) \rightarrow Fe^{2+}(aq) + 2e,\; Fe^{2+}(aq) + 2OH^{-}(aq) \rightarrow Fe(OH)_2(s)$$

$$\text{Cathode:}\quad 2H_2O\,(l) + 2e \rightarrow H_2(g) + 2OH^{-}(aq),\;\; E^0 = -0.828\text{V (vs. SHE)}$$

Insoluble $Fe(OH)_2$ and $Fe(OH)_3$ hydroxides remain in wastewaters as suspensions, which can remove pollutants by the formation of surface complexes or by electrostatic interactions followed by coagulation. In complexes, the pollutants work as ligands to hydrous ions:

$$\text{Pollutant-H} + (HO)OFe(s) \rightarrow \text{pollutant-OFe}\,(s) + H_2O(l)$$

If in EC, Al anode is used, electrooxidation leads to the formation of Al^{3+} ions:

$$Al\,(s) \rightarrow Al^{3+}(aq) + 3e,\; E^0 = -1.676\text{ V (vs. SHE)}$$

Al^{3+} ions are not stable in aqueous solutions and undergo spontaneous hydrolysis. In these reactions, different species are generated depending on the pH of the solutions. In acidic media $Al(OH)_2{}^{+}$, $Al(OH)^{2+}$ soluble species dominate and insoluble aluminum hydroxide $Al(OH)_3$ is formed in reactions:

$$Al^{3+}(aq) + H_2O\,(l) \rightarrow Al(OH)^{2+}(aq) + H^{+}(aq)$$

$$Al(OH)^{2+}(aq) + H_2O\,(l) \rightarrow Al(OH)_2{}^{+}(aq) + H^{+}(aq)$$

$$Al(OH)_2{}^{+}(aq) + H_2O\,(l) \rightarrow Al(OH)_3(s) + H^{+}(aq).$$

In alkaline media, at $pH < 9$, insoluble $Al(OH)_3$ species are formed, while at $pH > 9$ soluble $Al(OH)_4{}^{-}$ ions dominate in solutions. At the same time, at the cathode, H_2 is evolved (see Fe reactions). All the species generated during and after the dissolution of the anode are charged. As in the case of Fe, they can remove the pollutants in two ways: by the neutralization of the negative charge of colloid particles followed by coagulation and by the formation of surface complexes with pollutants. The H_2 evolved during the reduction of hydrogen ions or water molecules at the cathode can also be applied for the removal of pollutants by the flotation process. EC technology has been widely used for the removal of suspended particles of many pollutants; among them are organic compounds, heavy metals (e.g., Zn, Ni, Co, Mn, and Mo), and also anions such as CN^-, F^-, and PO_4^{3-}. Note, that by using EC we can only remove pollutants from water, but we cannot separate them. Much more about EC applications and parameters influencing the efficiency of EC, an interested reader can find in references [1, 8]. Flotation and electroflotation are other methods widely used for the removal of suspended species: metal particles, colloids, flocks, and oil contaminants from wastewater. Removal is achieved by the bubbling gases, which adhere to the species and flow them to the surface. In EF [9], the gases are generated electrochemically, and these gases are used for the removal of pollutants. As in EC, in EF, the hydrogen is evolved at the cathode and the pollutants can flow to the surface. Additionally at proper potential or current density, at the anode not only Al is oxidized to Al^{3+} ions but also water is decomposed into O_2:

$$\text{Anode}\quad 2H_2O\,(l) \longrightarrow O_2(g) + 4H^+(aq) + 4e, \quad E^0 = 1.229\,\text{V (vs. SHE), acidic media}$$

$$\text{Anode}\quad 4OH^-(aq) \longrightarrow O_2(g) + 2H_2O\,(l) + 4e, \quad E^0 = 0.401\,\text{V (vs. SHE), basic media}$$

Both gases H_2 and O_2 are used in flotation processes. EF can be applied separately, but very often the combination of EC and EF is used. In Fig. 15.2, the processes of coagulation, flotation, and sedimentation, which proceed during EC and EF, are schematically presented.

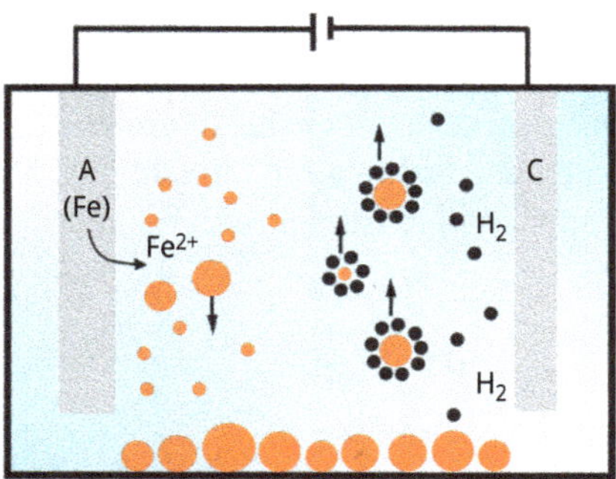

Fig. 15.2: Schematic illustration of electrocoagulation, electroflotation, and sedimentation in an electrochemical reactor.

All these methods, described above, can be used for metal removal and recovery. However, the combination of ED and EDI has a major advantage, over EC and EF processes: the recovery is selective. Therefore, these methods are often used for recovering valuable metals needed for high-tech devices. During the application of EC and EF methods, the heavy metals and other metals precipitate together with organic pollutants forming flocks, sludge, or surface suspensions. Such a mixture needs further treatment, so the cost of recovery of metals increases. By using EC and EF, we can remove from wastewater organic pollutants, but many of them are toxic and should be degraded. For such purposes, other methods are used.

15.3 Degradation and removal of organic pollutants: electrochemical treatment

For the removal and degradation of organic pollutants, oxidizing processes are used. Among them, we can distinguish the special group referred to as AOPs. These processes are widely used for the purification of wastewater from biologically toxic, non-degradable organic materials such as pesticides, volatile organics, or detergents. During AOPs, all these pollutants and others can be converted to CO_2 and H_2O, and inorganic compounds (the mineralization of organics). It was shown that AOPs are very efficient and decrease the concentration of some pollutants to very low levels from ppm to ppb. Generally, AOPs mean the oxidation reactions of organic compounds by very strong hydroxyl radicals $^{\bullet}OH$ generated chemically, electrochemically, or by light, mainly from H_2O_2. It is worth mentioning that $^{\bullet}OH$ radicals have one of the highest redox standard potential $E^0 = 2.8$ V (vs. SHE) for the $^{\bullet}OH/H_2O$ system [10]. Processes where hydroxyl radicals are generated electrochemically are referred to as electrochemical AOPs (EAOPs). Alternative to AOPs, EAOPs, and technologies connected with them, other oxidative electrochemical technologies are developed. They are based on common anodic oxidation processes. Below, we briefly describe Fenton and electro-Fenton reactions, where the source of hydroxyl radicals $^{\bullet}OH$ is hydrogen peroxide (H_2O_2) and some of the anodic processes utilized in the degradation of organic pollutants.

15.3.1 Fenton and electro-Fenton oxidation

Fenton processes (FPs) refer to the reactions where reagents Fe^{2+} and H_2O_2, often called Fenton reagents, are involved in the generation of hydroxyl radicals $^{\bullet}OH$. Plenty of studies were carried out to describe the mechanism of these reactions. Generally, we can distinguish a few crucial reactions in which radicals ($^{\bullet}OH$ and $HO_2^{\bullet}$) are formed and Fe^{2+} ions are recovered:

$$Fe^{2+}(aq) + H_2O_2(aq) \rightarrow Fe^{3+}(aq) + {}^{\bullet}OH + OH^-(aq)$$

$$^{\bullet}OH + H_2O_2(aq) \rightarrow HO_2^{\bullet} + H_2O\,(l)$$

$$Fe^{3+}(aq) + HO_2^{\bullet} \rightarrow Fe^{2+}(aq) + H^+(aq) + O_2(g)$$

$$Fe^{3+}(aq) + H_2O_2(aq) \rightarrow Fe^{2+}(aq) + HO_2^{\bullet} + H^+(aq)$$

FPs are much more complicated. The interested reader is referred to reviews [10, 11] and references therein. It was mentioned above that Fenton reactions are widely applied in the degradation of organic species. The degradation is caused by $^{\bullet}OH$ and organic radicals generated during oxidation and fragmentation of organic compounds:

$$RH + {}^{\bullet}OH \rightarrow R^{\bullet} + H_2O$$

Organic radicals are very reactive and in further reactions may produce various radical species ($RO^{\bullet}$ and $ROO^{\bullet}$). All these generated radicals, ferrous and ferric ions, react further in chain reactions resulting in the mineralization of organic compounds. The crucial parameter in the Fenton reactions is the pH of the wastewater. The reactions should be carried out at pH 3, which means that the pH of neutral or slightly basic wastewater should be adjusted. The cost of application of the Fenton reaction to purification is fairly high, as expensive chemical reagents are required. The cost may decrease when electro-FPs (EFPs) are applied.

In EFPs, the Fenton reagents (Fe^{2+} and H_2O_2) are formed electrochemically in situ. The wastewater and other waters flow continuously through an electrochemical reactor, where Fenton reagents are produced at electrodes. The reagents react further generating the radicals and initiating the chain of reactions shown partially above. The EFPs can also be realized in other ways. In one way, only one reagent is produced electrochemically (mostly H_2O_2) and iron (II) salts are externally added, in another way both reagents are supplied externally. Let us concentrate on the first case, where Fenton reagents are produced electrochemically. At a proper current density (voltage) in an electrochemical reactor, an anode (Fe or steel) dissolves with the formation of ferrous ions:

$$Fe\,(s) \rightarrow Fe^{2+}(aq) + 2e, \; E^0 = -0.441\,V\,(vs.\,SHE)$$

while at the cathode in an acidic solution, peroxide (H_2O_2) is formed in the reduction of O_2 dissolved in solution, or coming from the air. Generally, in acidic solutions, O_2 can be reduced in two pathways: direct four-electron reduction or two-electron reduction which involves the formation of H_2O_2:

$$O_2(g) + 4H^+(aq) + 4e \rightarrow 2H_2O\,(l), \quad E^0 = 1.229\,V\,(vs.\,SHE)$$

or

$$O_2(g) + 2H^+(aq) + 2e \rightarrow H_2O_2(aq), \; E^0 = 0.695\,V\,(vs.\,SHE)$$

Both reactions are sluggish, and the exchange current of these reactions ranges from 10^{-10} to 10^{-11} A/cm^2. The catalysts have to be applied in both reactions. Plenty of studies were carried out to determine the mechanism and to find the proper catalysts. It was shown that direct reaction is preferable (water formation) when, in the initial step of the reaction, O_2 is strongly adsorbed and O–O bonds are broken (see Section 4.5.1). The catalysts that are used in the reaction of H_2O_2 formation should weakly adsorb O_2 molecules, have a low value of activation overpotential for O_2 reduction, and also release H_2O_2 before further reduction or decomposition occurs:

$$H_2O_2(aq) + 2H^+(aq) + 2e \rightarrow 2\,H_2O\,(l),\ E^0 = 1.76\,V\,(vs.\,SHE)$$

$$2H_2O_2(aq) \rightarrow O_2(g) + 2H_2O\,(l)$$

It was found that carbon materials have such properties toward H_2O_2 formation. Plenty of them are used and tested in the electrochemical reactors applied in the H_2O_2 industry and the treatment of wastewater. Among them are graphite, carbon foam, mesoporous carbon, graphene composites, carbon nanotubes with metal particles, and others (see Section 4.3.2). When H_2O_2 is produced at the cathode, but Fe^{2+} ions come from added Fe(II) salts, the inert anode is applied, made mainly from oxides such as IrO_2/RuO_2, TiO_2, and Nb_2O_5. In this case at the anode, water is oxidized with O_2 evolution. Much more information about this reaction and the catalyst applied, an interested reader can find in Section 14.2.1. The organic pollutants from wastewater can be removed by EFPs in electrochemical reactors. Their design may be different, but a few parameters are important and common:

- The pH of effluents should be kept at 3. In other cases, Fe^{3+} ions created in Fenton reactions precipitate as $Fe(OH)_3$; moreover, H_2O_2 is not stable in neutral and alkaline solutions.
- In many cases, the conductivity of wastewater is low; therefore, some electrolytes should be added.
- An essential parameter is current density determined by operating conditions.
- It was established that the optimal temperature for Fenton reactions is 30 °C.

If we compare EFPs with classical ones (FPs), we clearly see the advantages of EFPs:

- The EFPs are less expensive. H_2O_2 can be produced in situ in wastewater; therefore, the transport of reagent is eliminated.
- By using proper current density or potential (voltage), the processes are more controllable.
- By using the proper pH of reaction media, the production of sludge is limited or fully eliminated.

Hydroxyl radicals generated during both Fenton methods are very strong oxidants able to nonselectively react with organic pollutants yielding their complete mineralization.

15.3.2 Anodic oxidation

Electrochemical oxidation processes (anodic oxidation processes – AOx) are widely applied in the degradation and removal of organic pollutants. They are friendly to the environment and able to mineralize nonbiodegradable organic compounds. These

AOx processes can undergo two routes: (i) direct oxidation occurring at the anode surface and (ii) indirect oxidation where electrochemically generated mediators (oxidant) oxidize the pollutant in solution. In direct AOx, the organic molecules adsorbed on the anode surface are oxidized in direct charge (electron) transfer without involvement of any other substances:

$R_{ads}(M) \rightarrow R^+ + e$, M is the active site of the anode surface and R is organic compounds

This process may be carried out at potentials lower than the potential of water oxidation and oxygen evolution. However, very often, direct transfer reaction resulted in the formation of insoluble, not fully oxidized product fouling and poisoning electrode surface. This effect can be avoided by leading the oxidation at potentials higher than the potential of water decomposition, or by using indirect oxidation. If the oxidation is carried out according to the first way, the processes can be catalyzed by intermediate, oxygen-containing radicals ($^\bullet OH$, $O^\bullet$) generated during water oxidation and oxygen evolution (see Section 14.2.1). They adsorb on the anode active side M and may participate in the oxidation of organic molecules:

$$M + H_2O\,(l) \rightarrow M[^\bullet OH]_{ads} + H^+(aq) + e$$

$$M[^\bullet OH]_{ads} \rightarrow M[O^\bullet]_{ads} + H^+(aq) + e$$

$$R + M[^\bullet OH]_{ads} \rightarrow RO + H^+(aq) + M + e$$

$R + M[O^\bullet]_{ads} \rightarrow RO + M$, RO is partially oxidized organic compounds

The most important is the hydroxyl radical $^\bullet OH$, which can be chemisorbed or weakly adsorbed on the anode surface depending on the anode nature. If hydroxyl radicals are weakly adsorbed (physical adsorption), the radicals are freer to react with pollutant molecules at electrodes and assist in the complete mineralization of organics:

$$R + M[^\bullet OH]_{ads} \rightarrow CO_2(g) + H_2O\,(l) + H^+(aq) + M + e$$

Chemisorbed radicals can cause partial oxidation of organic compounds (RO).

A different mechanism is proposed for the oxidation of organics at some metal oxide anode (MO_x). In the first step, the $^\bullet OH$ radicals are also formed during water decomposition and are chemisorbed on/in active sites of anode, followed by oxidation and formation of so-called higher oxide ($MO_{(x+1)}$). This higher oxide acts as a mediator in the selective oxidation of organic pollutants:

$$MO_x(s) + H_2O\,(l) \rightarrow MO_x(^\bullet OH)_{ads} + H^+(aq) + e$$

$$MO_x(^\bullet OH)_{ads} \rightarrow MO_{(x+1)}(s) + H^+(aq) + e$$

$$MO_{(x+1)} + R \rightarrow MO_x(s) + RO$$

If the adsorption of $^{\bullet}OH$ radicals on metal oxide anode is weak, the hydroxyl radical may assist in the nonselective complete oxidation of organic pollutants:

$$MO_x(^{\bullet}OH)_{ads} + R \rightarrow MO_x + CO_2 + H_2O$$

Note the possibility of competitive reaction of O_2 evolution; therefore, O_2 can also participate in the oxidation of pollutants. Since in the case of direct anodic oxidation processes, the $^{\bullet}OH$ radicals also participate, and the direct AOx reactions belong to EAO processes.

In indirect anodic oxidation, the pollutants are oxidized by strong oxidizing reagents generated electrochemically in situ in wastewater at anode or cathode (H_2O_2). These reagents work as mediators in charge transfer between the electrode and organics in solution. They may be divided into two groups: the metallic redox couple such as Ag^+/Ag^{2+} ($E^0 = 1.98$ V) and Co^{2+}/Co^{3+} ($E^0 = 1.82$ V) and strong oxidizing reagents such as chlorine (Cl^-/Cl_2, $E^0 = 1.358$), ozone (O_2/O_3, $E^0 = 2.075$ V), peroxide (H_2O/H_2O_2, $E^0 = 1.763$ V), and persulfate ($SO_4^{2-}/S_2O_8^{2-}$, $E^0 = 2.01$ V) [10]. All these compounds possess the high values of standard redox potentials pointing to their oxidation power. The processes where metallic redox couples take place are referred to as mediated electrochemical oxidation (MEO). They are preferably used for the treatment of solid waste and concentrated solutions containing more than 20% of waste. The metal ions in an acidic solution are anodically oxidized to the higher oxidation state, in which they react with organic pollutants inducing their degradation:

$$Me^{n+} \rightarrow Me^{(n+1)+} + e, \qquad Me^{(n+1)+} + R \rightarrow Me^{n+} + CO_2 + H_2O$$

Such reactions were successfully applied for the degradation of cellulose, oil, resins, urea, organic acid, and others. When wastewaters contain low concentrations of organic pollutants (lower than 20%), the effluents are treated by other oxidizing reactants electrogenerated in situ. They originate from the oxidation of water and ions present in wastewater effluents such as chlorides, sulfates, and carbonates. Among them, chlorine is most widely applied in wastewater treatment. Chlorides are oxidized in the reaction:

$$2Cl^-(aq) \rightarrow Cl_2(g) + 2e, \; E^0 = 1.358\,\text{V (vs. SHE)}$$

Chlorine is not stable undergoing disproportionation in water solution:

$$Cl_2(g) + H_2O\,(l) \rightarrow HClO\,(aq) + H^+(aq) + Cl^-(aq)$$

$$Cl_2(g) + 2\,OH^-(aq) \rightarrow ClO^-(aq) + H_2O\,(l) + Cl^-(aq)$$

Hypochlorous acid and hypochlorite ions can undergo further oxidation to ClO_2^- chlorite ions and ClO_3^- chlorate ions in an alkaline medium. All these chlorine-containing oxidants are strongly reactive and can cause the mineralization of pollutant molecules, but they also can produce some harmful, unwanted products. Chlorine was used for the degradation of phenols, glucose, textile dyes, pesticides, and others. Much higher oxidant power than chlorine has other reagents: ozone, persulfate, and peroxide (H_2O_2/H_2O, $E^0 = 1.765$ V) generated at the cathode (see Section 15.3.1):

$$3H_2O\,(l) \rightarrow O_3(g) + 6H^+(aq) + 6e, \quad E^0 + 1.51\,V\,(vs.\,SHE)$$

$$2SO_4{}^{2-}(aq) \rightarrow S_2O_8{}^{2-}(aq) + 2e, \quad E^0 = 2.01\,V\,(vs.\,SHE)$$

All these oxidants described above and others were tested in wastewater treatment for removal of very hard degradable toxic organic pollutants.

The anodic oxidation of pollutants is carried out in a typical electrochemical reactor equipped or not with a system for the continuous flow of wastewater. The main parts of reactors are electrodes. The cathode is conventionally made from steel or graphite. The heart of the reactor is the anode, its material, catalytic properties, and adsorption ability to decide on the reaction pathway and the selectivity and efficiency of mineralization of pollutants and their removal. Anodes applied in the oxidation of organic pollutants differ in their efficiency and selectivity. On some of them, the reaction of oxidation of organics is completed with the formation of CO_2 and H_2O, and at others, partial oxidation occurs. It was found that anodes such as SnO_2, PbO_2, and BDD (boron-doped diamond) at which the potential of O_2 evolution is higher (1.9 V, 1.9 V, 2.3 V in H_2SO_4 solution) favor the complete oxidation of organic pollutants. Others such as RuO_2, IrO_2, and Pt, characterized by lower potential evolution of O_2 (1.47 V, 1.52 V, 1.6 V in H_2SO_4 solution) participate in partial oxidation of organic pollutants.

In practice, most anodes can participate in both AOx pathways. Applying high enough potential to the anode, all reactions with direct charge transfer and mediator generation may occur simultaneously. Depending on the pollutants that have to be removed, different anodes are applied in AOx processes such as Pt, metal oxides, mixed metal oxides, or graphite. Pt is considered a good anode for many electrooxidation reactions, for example, phenol, dyes, and herbicides. However, as the cost of Pt is too high, it should be replaced with other materials. Metal oxides such as IrO_2, RuO_2, Nb_2O_5, Ta_2O_5, SnO_2, and mixed oxides are very effective in promoting Cl_2 and hypochlorite production. Some of them, IrO_2 and RuO_2, are widely used as catalysts in water oxidation and O_2 evolution. It was shown that at high potential, the strong O_3 oxidant may be produced at PbO_2. In recent years, the application of BDD (boron doped diamond) electrodes in AOx processes has aroused interest. BDD electrodes are chemically stable, are resistive against corrosion even in harmful media, and have a very large potential for oxygen evolution (2.3 V vs. SHE). Therefore, BDD anodes are widely used for near complete removal of very hazardous pollutants from wastewater such as textile dyes, pharmaceutical contaminants, or surfactants. The other materials widely used for AOx processes are carbon-based and graphite electrodes, for example, graphite rods, activated carbon, carbon papers, carbon brushes, and carbon nanostructures. Carbon electrodes cannot work at high potentials; however, they are often used, since they are cheap and may be fabricated as large porous assemblies with large surfaces. The interested readers can find much more about the application of AOx processes in the degradation of pollutants in reviews [11–13] and references therein.

15.4 Solar energy: pollutant degradation and removal

Sunlight is a renewable and sustainable source of natural energy that can be utilized for the degradation and removal of pollutants from our environment. For such purposes, solar energy may be used in two ways: indirect and direct. We have above-described processes and electrochemical methods applied in the degradation and removal of inorganic and organic pollutants from industrial and domestic wastewater. Some of these methods such as electrodeposition (ED), electrocoagulation (EC), and anodic oxidation electrolysis require a large amount of electrical energy delivered by conventional energy sources. These electrochemical reactors can also be fed with electrical energy coming from photovoltaic panels converting solar energy into electricity. This can be treated as an indirect application of solar energy in pollutant removal. The direct way of the utilization of solar energy in the purification of wastewater and air is the application of photocatalytic processes. Photocatalytic processes convert solar energy into chemical energy, making it possible to mineralize organic pollutants. They can be divided into homogeneous and heterogeneous processes. Typical examples of homogeneous photoprocesses are photo-Fenton reactions, where reagents and catalysts are in the same phase. The heterogeneous photocatalysis is carried out on the solid phase: suspension of powdered catalyst or catalyst particles, nanoparticles embedded in membranes, or other materials. The compounds used in heterophotocatalysis are semiconductors (SC). Much more information about SC materials, their electrochemical, and photoelectrochemical properties, and their application in photoelectrochemical cells, an interested reader can find in Chapters 10 and 11. In Chapter 12 and Section 14.3.3, the basis of photocatalytic processes on SC particles (SCPs), their application in water photocatalysis, and the efforts to improve the light absorption and quantum efficiency of SCPs are described. Here, we briefly described the application of photocatalytic processes in the degradation and removal of some pollutants pointing also to the possibility of using pollutants as electron donors or acceptors in photocatalytic reactions, yielding the formation of valuable products, for example, H_2, CH_3OH, and HOOH. Before going further, we have to bring to mind how these systems work. Three steps can be distinguished during irradiation of SCPs by photons with energy $h\nu \geq E_g$: (i) photogeneration of charge carriers: electrons in the conduction band (CB) and holes (h^+) in the valence band (VB); (ii) separation of charge carriers: electrons and holes; and (iii) participation of charge carriers in reactions at SCP/solution interface. Electrons generated in CB reduce electron acceptors (A), whereas holes from VB oxidize the electron donors (D). Processes are illustrated in Fig. 15.3.

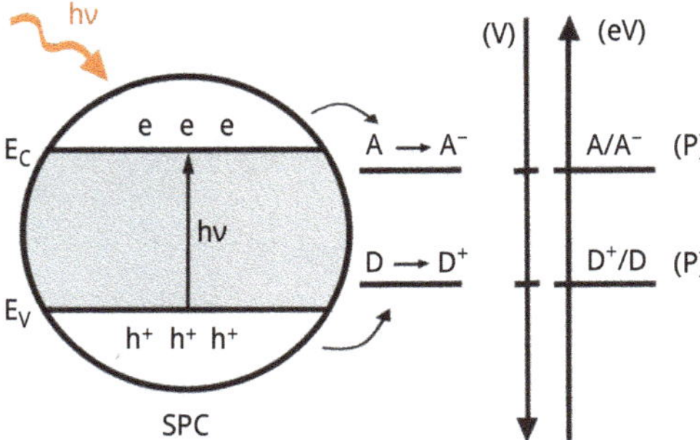

Fig. 15.3: Schematic illustration of photocatalytic reaction. A, acceptor; D, donor; P, pollutant.

Reactions occur when the thermodynamics requirements are fulfilled. Generally, the energy of the CB edge, E_C, has to be higher than the energy level of the A/A^- couple, that is, it is located at a more negative potential relative to the redox potential of A/A^- in solution. The energy of the VB edge, E_V, should be lower than the energy level of the D/D^+ couple, that is, located at a more positive potential than the redox potential of the D/D^+ couple in solution (see Fig. 15.3). The roles of electron acceptors or electron donors (known also as sacrificial agents) can play pollutant molecules. Contaminant molecules adsorbed on SCP surfaces are irreversibly reduced or oxidized. During their degradation, sometimes, valuable products are formed. Let us consider some examples of pollutant degradation.

15.4.1 Inorganic pollutants

For illustrative purposes, we choose carbon dioxide CO_2 (greenhouse gas) and H_2S pollutants existing in natural gas, petroleum, and wastewaters, being also products of biomass fermentation. Their degradation differs; carbon dioxide adsorbed on the SCP surface is reduced by electrons from CB during illumination of SCPs, while H_2S is oxidized as a gas, or sulfide ions in solution by photogenerated holes from VB. The process of reduction of CO_2 is difficult as the energy of the C = O bond is high (850 kJ/mol). During the reduction of CO_2 in aqueous solutions, different products can be formed, including CO, HCOOH, HCHO, CH_3OH, CH_4, and C_2H_5OH. These reactions are multielectron reactions involving from 2 to 12 electrons. For instance, 2 electrons are involved in the formation of carbon monoxide and formic acid, 4 electrons participate in formaldehyde formation, 6 in methanol formation, and 8 and 12 in methane and ethanol formation. Some of these reactions are quoted in Section 11.3.4. The values of redox potentials of these reactions are close to each other, so during CO_2 reduction, a mixture of products can be formed. For instance, the redox potentials of CO_2/HCOOH, CO_2/CO, CO_2/HCOH, and CO_2/CH_3OH, in an aqueous solution, at pH 7 are as follows: −0.58 V,

−0.52 V, −0.46 V, −0.37 V vs. NHE, respectively. They are not only close to themselves but also close to the equilibrium potential of H^+/H_2 at pH 7 equal to −0.413 V; therefore, during CO_2 reduction, hydrogen may also be evolved. Plenty of researches were carried out on the photocatalytic reduction of CO_2 on different kinds of SCPs. In Fig. 12.4 (Section 12.3), you find the energetic diagram pointing to the correlation between some popular oxide photocatalysts and redox potentials of CO_2 reduction. Oxides such as TiO_2, ZnO, and NiO may be used for such purposes.

However, these oxides absorb the UV part of the solar spectrum; therefore, plenty of efforts are made for their modification, especially doping with metals (Pt), nonmetals (N), structural changes by using materials in the form of nanorods, nanotubes, and so on. Other photocatalysts were also tested for CO_2 reduction such as sulfides (CdS and Bi_2S_3), metal-organic compounds, and photocatalysts modified with enzymes. Much more about the photocatalytic reduction of CO_2 and applied photocatalysts, one can find in reviews [14, 15] and references therein. However, the main problem is still not solved: how to selectively reduce CO_2, for instance, to CH_3OH, or more generally, how to carry out the synthetic reactions

$$n\mathrm{CO_2} + n\mathrm{H_2O} + h\nu \rightarrow (\mathrm{CH_2O})_n + n\mathrm{O_2}$$

in which the absorbed greenhouse gas will be utilized and converted into valuable products.

During the reduction of CO_2 by electrons from CB, the photogenerated holes in VB react with electron donors (D) specially added to the solution or oxidize water molecules producing a radical intermediate or O_2. Other processes occur during the illumination of SCPs in the presence of H_2S. H_2S is a very interesting compound, in which the O molecule atom is replaced by S, and the molecule may be split with H_2 evolution, according to the decomposition reaction of H_2S (g), or H_2S (aq):

$$\mathrm{H_2S} \rightarrow \mathrm{H_2(g)} + \mathrm{S(s)}$$

This reaction requires much less energy than in the case of water decomposition. Standard Gibbs free energy of reactions $\Delta_r G^0$ are 33.56 kJ/mol and 27.89 kJ/mol, respectively, while $\Delta_r G^0$ for H_2O (l) is equal to 237.13 kJ/mol. In consequence, there is a large difference in the standard redox potential of reactions of H_2O (l) and H_2S (aq) splitting (1.229 V and 0.144 V vs. SHE, respectively). In solution, H_2S is a weak aprotic acid that ionizes in two steps:

$$\mathrm{H_2S} \leftrightarrow \mathrm{H^+} + \mathrm{HS^-},\ \mathrm{p}K = 7, \text{and further } \mathrm{HS^-} \leftrightarrow \mathrm{H^+} + \mathrm{S^{2-}},\ \mathrm{p}K = 11.96$$

During photocatalytic decomposition of H_2S in solution, photogenerated electrons and holes reduce proton ions and oxidize sulfide ions, respectively:

$$\mathrm{SC} + h\nu \rightarrow \mathrm{e\,(CB)} + \mathrm{h^+(VB)}$$

$$\mathrm{2H^+(aq)} + \mathrm{2e\,(CB)} \rightarrow \mathrm{H_2(g)}$$

$$2HS^{-}(aq) + 2h^{+}(VB) \rightarrow 2H^{+}(aq) + S_2{}^{2-}(aq)$$

$$2S^{2-}(aq) + 2h^{+}(VB) \rightarrow S_2{}^{2-}$$

Photocatalytic decomposition of H_2S can be also carried out in the gas phase. In this case, H_2 and S are produced. The decomposition potential of H_2S is much lower than the potential of water splitting; therefore, SC materials possessing low bandgap energy E_g are widely used. They can absorb the photons from the visible part of the solar spectrum by increasing the efficiency of decomposition of H_2S and hydrogen evolution. Plenty of SC materials were tested; among them are metal sulfides CdS and ZnS. To enhance their activity, some efforts have been made to combine them with other SC, forming heterojunctions, for example, CdS/TiO_2, CdS/ZnO, and ZnS/CdS, to dope them with ions such as Cu^{2+} and Ni^{2+} (ZnS), or to use them in nanostructures. Much more about photocatalytic H_2S decomposition and applied photocatalysts, you can find in review [16]. Note that most of the results are obtained by researchers using different sources (UV–Vis lamps) for illumination; however, you can also find some results where systems with simulated sunlight were used. The photocatalytic processes can be used for the removal of heavy metal ions in wastewater [17]. They are reduced and the electrons photogenerated in CB of SC photocatalyst participate in the reduction. It was shown that such photocatalysts as TiO_2–ZrO_2 can easily reduce Cu^{2+} and Pb^{2+} ions to Cu, Pb, and Cr(VI) ions to Cr(III) in reaction:

$$Cr_2O_7{}^{2-} + 14H^{+} + 6e\,(CB) \rightarrow 2Cr^{3+} + 7H_2O$$

The photocatalysts and photocatalytic processes also have substantial significance in the degradation and removal of organic pollutants.

15.4.2 Organic pollutants

In Section 15.3, we described the processes (AOPs) that cause the oxidation of organic pollutants to a high degree, even to mineralization. A strong oxidant – hydroxyl radical $^{\bullet}OH$ – participates in these reactions; however, other radicals such as $HO_2{}^{\bullet}$ can also be involved. As was pointed out, the radicals are produced during catalytic processes in Fenton reactions between Fe^{2+} and H_2O_2, or during anodic oxidation of water, where radicals are intermediates in oxygen evolution. Such radicals can also be formed during photocatalytic oxidation of water or photolysis of solution containing Fe^{2+} ions and H_2O_2 (Fenton system). It was shown that illumination of the Fenton system with UV/Vis light in the range of 400–200 nm (approximately) accelerated the degradation rate of pollutants. It results from the additional formation of $^{\bullet}OH$ radicals, and they are not only created in elementary Fenton reaction:

$$Fe^{2+}(aq) + H_2O_2(aq) \rightarrow Fe^{3+}(aq) + {}^{\bullet}OH + OH^-(aq)$$

but also in reactions:

$$Fe^{3+}(aq) + H_2O\ (l) + h\nu \rightarrow Fe^{2+}(aq) + {}^{\bullet}OH + H^+(aq)$$

$$H_2O_2(aq) + h\nu \rightarrow 2{}^{\bullet}OH$$

Thus, illumination not only regenerates Fe^{2+} ions crucial in catalytic reactions but also increases the amount of $^{\bullet}OH$ radicals participating in the reactions. Such Fenton reactions during illumination are referred to as photo-FPs and belong to the group of homogeneous photocatalysis as Fe^{2+} ions, H_2O_2, and pollutants are in the same liquid phase. Recent works have shown that instead of the application of expensive illumination with a UV–Vis lamp, the sunlight energy may be used in FPs and remediation of wastewaters. Such reactions are referred to as solar-FP. Much more about different FPs and their applications in the degradation of the pollutants, the interested readers can find in review [10] and references therein.

The FPs may be realized also in another way, using solid photocatalysts – iron-based compounds such as iron oxides, for example, α-Fe_2O_3 (hematite), γ-Fe_2O_3 (maghemite), Fe_3O_4 (magnetite), iron (hydroxyl) oxides FeOOH, or spinel ferrites $MeFe_2O_4$ (Me – Mn, Fe, Co, and Ni). They can be used in the form of particle suspensions, particles embedded in membranes, or different nanostructures. These compounds have large adsorption ability toward pollutant molecules. Some of them possess the bandgap energy E_g lower than 2.8 eV (α-Fe_2O_3 and γ-Fe_2O_3 are 2.2 eV and 1.83 eV, respectively), so they can absorb the photons from the visible part of the solar spectrum. They are also partially dissolved in dark and under light, being the source of Fe^{2+} and Fe^{3+} ions at the particle surface. Ferric ions can react further with H_2O_2 produced during photoreduction of O_2:

$$SC + h\nu \rightarrow e\,(CB) + h^+(VB)$$

$$e\,(CB) + O_2(g) + 2H^+(aq) \rightarrow H_2O_2$$

$$Fe^{2+} + H_2O_2 \rightarrow Fe^{3+} + 2{}^{\bullet}OH_{ads}$$

The hydroxyl radicals oxidize adsorbed pollutant molecules causing their degradation:

$$R_{ads} + {}^{\bullet}OH_{ads} \rightarrow RO_{ads}\ (\text{partially oxidized intermediate}) \rightarrow nCO_2 + nH_2O$$

Plenty of iron-based materials were tested in photocatalytic degradation of different pollutants, among them were pharmaceutical drugs. For instance, $CoFe_2O_4$ was used for the degradation of doxycycline, Fe_2O_3-modified graphene nanoplatelets for the removal of amoxicillin, and the maghemite nanoparticles for the degradation of tetracycline. More detailed information, interested readers can find in the review [18] and references therein.

The degradation of pollutants is caused not only by $^{\bullet}OH$ and $HO_2^{\bullet}$ radicals generated during different Fenton systems or anodic oxidation, these radicals are also formed during photocatalytic reactions of oxygen reduction and water oxidation. During illumination of SCPs, photogenerated CB and VB electrons and holes can participate in reactions:

$$SC + h\nu \rightarrow e\,(CB) + h^+(VB)$$

$$h^+(VB) + H_2O_{ads} \rightarrow H^+ + {}^{\bullet}OH$$

$$h^+(VB) + OH^- \rightarrow {}^{\bullet}OH$$

$$e\,(CB) + O_2 \rightarrow O_2^{\bullet -}$$

$$O_2^{\bullet -} + H^+ \rightarrow HO_2^{\bullet},\ HO_2^{\bullet}\ \text{notation is equal to}\ HOO^{\bullet}$$

Radicals $HO_2^{\bullet}$ and $O_2^{\bullet -}$ undergo further reactions

The interested reader can find the sequence of reactions and diagram presenting the correlation between energy of CB, VB edges, and redox potentials of oxygen radicals in Section 12.3 (Fig. 12.3). There are plenty of organic pollutants in wastewater such as phenolic compounds, pesticides, hydrocarbons, dyes, and drugs. For their degradation and removal, mostly binary metal oxides such as TiO_2, ZnO, SnO_2, and Fe_2O_3, and materials with oxide structures such as perovskites (ABO_3), that is, $SrTiO_3$ and $NaTaO_3$ are used, since they are stable in a hazardous environment and can participate in production of highly oxidant radicals. They degrade a wide range of organic pollutants in biodegradable nontoxic molecules, which can be further mineralized to CO_2 and H_2O. Detailed information about the application of photocatalysts in the degradation of different classes of pollutants can be found in reviews [15, 17–19]. Photocatalytic degradations and removal of pollutants are carried out in photocatalytic reactors. They differ in configurations and design, and their construction depends on the kind of pollutants. Their description is beyond the scope of this book. Interested readers can find such information in [20].

References

[1] Yasri NG, Gunasekaran S. Electrochemical technologies for environmental remediation. In: Anjum NA, Gill SS, Tuleja Z, eds. Enhancing cleanup of environmental pollutants. Non biological approaches. Dordrecht, Germany, Springer, 2017, vol. 2, 5–73.

[2] Rai V, Liu D, Xia D, Jayaraman Y, Gabriel JCP. Electrochemical approaches for the recovery of metals from electronic waste: A critical review. Recycling 2021, 6, 53.

[3] Qasem NAA, Mohammed RH, Lawal DU. Removal of heavy metal ions from wastewater: A comprehensive and critical review. Npj Clean Water 2021, 4, 36. doi: 10.1038/s41545-021-00127-0.

[4] Juve J-MA, Christensen FMS, Wang Y, Wei Z. Electrodialysis for metal removal and recovery. A review. Chem. Eng. 2022, 435, 134857.

[5] Hiemenz PC, Rajagoplan R. Principles of colloid and surface chemistry. NY, USA, Marcel Dekker Inc, 1997.

[6] Bard AJ. Encyclopedia of electrochemistry of elements. V9, Part A: Hg, Fe, H. NY, USA, Marcel Dekker Inc, 1979.

[7] Bard AJ. Encyclopedia of electrochemistry of elements. NY, USA, Marcel Dekker Inc, 1976.

[8] Mickova IL. Advanced electrochemical technologies in wastewater treatment. Part I Electrocoagulation. Am. Sci. Res. J. Eng. Technol. Sci. 2015, 14, 233–257.

[9] Mickova IL. Advanced electrochemical technologies in wastewater treatment. Part II Electro – Flocculation and Electro-Flotation. Am. Sci. Res. J. Eng. Technol. Sci. 2015, 14, 273–294.

[10] Brillas E, Sires I, Oturan MA. Electro-Fenton processes and related electrochemical technologies based on Fenton`s reaction chemistry. Chem. Rev. 2009, 109, 6570–6631.

[11] Vasudevan S, Oturan MA. Electrochemistry and water pollution. In: Lichtfous E, Schwarzbauer J, Robert D, eds. Green materials for energy, products and depollution. Environmental Chemistry for a sustainable world. Dordrecht, Germany, Springer, 2013, vol. 3, 27–68.

[12] Panizza M, Cerisola G. Direct and mediated anodic oxidation of organic pollutants. Chem. Rev. 2009, 109, 6541–6569.

[13] Sires I, Brillas E, Oturan MA, Rodrigo MA, Panizza M. Electrochemical advanced oxidation processes: Today and tomorrow. A review. Environ. Sci. Pollut. Res. 2014, 21, 8336–8367.

[14] Wang W-N, Soulis J, Yang JY, Biswas P. Comparison of CO_2. Photoreduction systems. A review. Aerosol Air Qual. Res. 2014, 14, 533–549.

[15] Guo W, Guo T, Zhang Y, Yin L, Dai Y. Progress on simultaneous photocatalytic degradation of pollutants and production of clean energy: A review. Chemosphere 2023, 333, 139486.

[16] Preethi Y, Kanmani S. Photocatalytic hydrogen production. Mater. Sci. Semicond. Process 2013, 16, 561–575.

[17] Ren G, Han H, Wang Y, Liu S, Zhan J, Meng X, Li Z. Recent advances of photocatalytic application in water treatment: A review. Nanomater 2021, 11, 1804.

[18] Olusegun SJ, Souza TGF, Souza GO, Osial M, Mohallen NDS, Ciminelli VST, Krysinski P. Iron-based materials for the adsorption and photocatalytical degradation of pharmaceutical drugs. A comprehensive review of the mechanism pathway. J. Water Process Eng. 2023, 51, 103457.

[19] Al-Nuaim MA, Alvasiti AA, Shnain ZY. The photocatalytic process in the treatment of polluted water. Chem. Pap. 2023, 77, 677–701.

[20] Shanghagi M, Sargazi H, Bazargan A, Bellardita M. Photocatalytic reactor types and configuration. In: Bazagran A, ed. Photocatalytic water and wastewater treatment. London, United Kingdom, Iwa Publishing, 2022, 73–110.

List of abbreviations

A	Anode, acceptor
AC	Acetonitrile
AC	Alternating current
AEM	Anion-exchange membrane
AFC	Alkaline fuel cell
AFM	Atomic force microscopy
AOx	Anodic oxidation
AOP	Advanced oxidation process
AWE	Alkaline water electrolysis
BET	Adsorption isotherm, Brunauer S, Emmet PH, and Teller EJ
BFC	Biofuel cell
C	Cathode
CB	Conduction band
CCS	Carbon capture and storage
CE	Counter electrode
CEM	Cation-exchange membrane
CNFs	Carbon nanofibers
CNTs	Carbon nanotubes
CP	Conducting polymer
CPE	Constant phase element; composite polymer electrode
CV	Cyclic voltammetry, cyclic voltammogram
D	Donor; diffusivity of charges; diffusion coefficient
*x*D	Dimensional (*x* – 0, 1, 2, 3)
DC	Direct current
DET	Direct electron transport
DMF	Dimethylformamide
DMFC	Direct methanol fuel cell
DMSO	Dimethylsulfoxide
DSSC	Dye-sensitized solar cell
E	Enzyme
E_a	Activation energy
EAOP	Electrochemical advanced oxidation process
EC	Electrocoagulation, ethylene carbonate
ECALE	Electrochemical atomic layer epitaxy
e.d.l	Electrical double layer
EDTA	Ethylenediaminetetraacetic acid
ED	Electrodeposition
EDI	Electrodialysis
EDLC	Electrical double layer capacitor
EF	Electroflotation
EFP	Electro-Fenton process
EIS	Electrochemical impedance spectroscopy
EMF	Electromotive force
EPS	Electrochemical power sources
EQE	Quantum conversion efficiency
EQCM	Electrochemical quartz-crystal microbalance
ES	Enzyme–substrate complex

https://doi.org/10.1515/9783111160986-017

FA	Formic acid
FP	Fenton process
G	Graphene
GC	Glassy carbon
GDH	Glucose dehydrogenase
GO	Graphene oxide
$(GO)_x$	Glucose oxidase
HER	Hydrogen evolution reaction
HHV	High heating value of hydrogen
HOR	Hydrogen oxidation reaction
HOPG	Highly oriented pyrolytic graphite
HRP	Horseradish peroxidase
HTE	High-temperature electrolysis
IPCE	Incident photon to current efficiency
IL	Ionic liquid
ISE	Inorganic solid electrolyte; ion-selective electrode
ITIES	Interface between two immiscible electrolyte solutions
Lac	Laccase
LHE	Light-harvesting efficiency
LHV	Low heating value of hydrogen
LTE	Low-temperature electrolysis
M	Active side of electrode/catalyst
MCFC	Molten carbonate fuel cell
MCO	Multicopper oxidase
MEA	Membrane electrode assembly
MEC	Microbial electrolysis
MEO	Mediated electrochemical oxidation
MET	Mediated electron transport
MFC	Membrane fuel cell
MHLD	Marcus, Hush, Levich, Dogonadze theory
NHE	Normal hydrogen electrode
OCP	Open-circuit potential
OER	Oxygen evolution reaction
OHP	Outer Helmholtz plane
OPD	Overpotential deposition
ORR	Oxygen reduction reaction
Ox	Oxidized form of Redox system
PAA	Porous anodic alumina
PAFC	Phosphoric acid fuel cell
PAM	Polyacrylamide
PC	Propylene carbonate
PEC	Photoelectrochemical cell
PEM	Proton membrane electrolysis
PEMFC	Proton-exchange membrane fuel cell
PV	Photovoltaic cell
QC	Quartz crystal
QCM	Quartz-crystal microbalance
QDSSC	Quantum dot-sensitized solar cell
QS	Quantum size

RE, Ref	Reference electrode
rds	Rate-determining step
Red	Reduced form of Redox system
RGO	Reduced graphene oxide
RHE	Reference hydrogen electrode
S	Substrate
SC	Semiconductor
SCC	Stress corrosion cracking
SCE	Saturated calomel electrode, semiconductor electrode
SCP	Semiconductor particle
SERR	Surface-limited replacement reaction
SHE	Standard hydrogen electrode
SMR	Steam methane reforming
SOE	Solid oxide electrolysis
SOFC	Solid oxide fuel cell
SPE	Solid polymer electrolyte
SSE	Solid-state electrolyte
STE	Energy conversion of solar to electricity
STF	Energy conversion of solar to fuel
STH	Energy conversion of solar to hydrogen
STM	Scanning tunneling microscopy
SRH	Shockley–Read–Hall (recombination)
t_i	Ionic transference number of "i"
U_i	Electrolytic mobility
UHV	Ultra-high vacuum
UPCD	Underpotential codeposition
UPD	Underpotential deposition
WE	Working electrode
VB	Valence band
VOC	Volatile organic compound

List of symbols

Symbol	Description	Unit
A	Surface area	cm^2
a_i	Activity of species "i"	mol/dm^3
$C_{e.d.l}$	Differential capacitance of the electrical double layer	$\mu F/cm^2$
$C_{G\text{-}Ch}$, C_H	Differential capacitance of Gouy–Chapman and Helmholtz layers, respectively	$\mu F/cm^2$
C_F	Faradaic capacitance	$\mu F/cm^2$
D	Diffusion coefficient	cm^2/s
E	Electrical potential, electrode potential	V
E^f	Formal potential of an electrode	V
E^0	Standard potential of an electrode	V
E_{eq}	Equilibrium (Nernst) potential of an electrode	V
ΔE_{eq}	Difference of equilibrium potentials of electrodes in an electrochemical cell, electromotive force (emf), and zero-current cell potential	V
ΔE^0_{eq}	Difference of standard potentials of electrodes in an electrochemical cell	V
ΔE	Potential (voltage) applied to the electrolysis cell	V
ΔE^0_{tn}	Thermoneutral potential of electrolysis cell	V
E_{ML}	Potential of UPD monolayer formation	V
ΔE_{UPD}	Difference between the reversible and the peak potentials during UPD formation	V
f_i	Activity coefficient	Dimensionless
f_0	Resonance frequency of a quartz crystal microbalance	Hz, s^{-1}
Δf	Change of resonance frequency of a quartz crystal microbalance	s^{-1}
G	Thermodynamic potential, Gibbs free energy	J/mol
$\Delta G^{0\#}$	Standard Gibbs free energy of activation	J/mol
ΔG^0_{ads}	Standard Gibbs free energy of adsorption	J/mol
ΔG_c	Gibbs free energy of formation of critical size nuclei	J/mol
ΔG^0_{mix}	Standard Gibbs free energy of the mixing (alloy)	J/mol
$\Delta_r G$	Gibbs free energy of reaction	J/mol
$\Delta_r G^0$	Standard Gibbs free energy of reaction	J/mol
H	Enthalpy	J/mol
$\Delta_r H$	Enthalpy of reaction	J/mol
$\Delta_r H^0$	Standard enthalpy of reaction	J/mol
$\Delta H^{0\#}_a$	Standard activation enthalpy	J/mol
ΔH^0_{ads}	Standard enthalpy of adsorption	J/mol
I_{ch}	Charging current of the cell	A
I_d	Discharge current of the cell	A
i, I	Current	A
j	Current density	A/cm^2
j_0	Exchange current density	A/cm^2
$j_{0,A}$, $j_{0,C}$	Exchange current density of reaction at anode (A) and cathode (C)	A/cm^2
j_L	Limiting current density	A/cm^2
$j_{L,A}$, $j_{L,C}$	Limiting current density of reaction at anode (A) and cathode (C)	A/cm^2
K	Equilibrium constant	Dimensionless
K_{ads}	Adsorption equilibrium constant	Dimensionless
k	Rate constant for homogeneous reaction	Depend on order
$k_{f,b}$	Rate constant for heterogeneous forward and backward reactions	cm/s

https://doi.org/10.1515/9783111160986-018

m, M	Mass and molecular weight	g, g/mol
n	Number of electrons transferred per molecule	Dimensionless
n_i	Number of moles of species	"i"
nF	Charge transferred per mol	C/mol
N_0	Initial number of active nucleation sites	Dimensionless
P	Power	W
p	Pressure	Pa
Q, q	Charge	C
R	Resistance	Ω
r, h	Radius and height of disk shape nuclei	nm
R_{in}, R_{ex}	Internal and external resistance of the cell	Ω
S	Entropy	J/mol•K
$\Delta_r S$	Entropy of reaction	J/mol•K
$\Delta_r S^0$	Standard entropy of reaction	J/mol•K
T	Temperature	K, C^0
U	Internal energy	J/mol
U_{ch}	Charge potential of the cell	V
U_{d}	Discharge potential of the cell	V
V	Volume	m^{-3}
W	Warburg impedance	Ohm
W_{a}	Wagner number	Dimensionless
w_{rev}, w_{max}	Reversible work and maximal work	J
x_A, x_B	Molar fraction of components	Dimensionless
z_i	Valency of ion "i"	Dimensionless

Greek symbols

α	Cathodic transfer coefficient	Dimensionless
B	Anodic transfer coefficient	Dimensionless
γ	Molar surface free energy, interfacial tension, N/m; activity coefficient	Dimensionless
δ	Thickness of diffuse layer	m
ε	Dielectric permittivity	Dimensionless
ε_0	Dielectric permittivity of vacuum	F/m
ζ	Electrokinetic potential and zeta potential	V
η	Overpotential	V
η_{a}	Activation overpotential	V
η_{c}	Concentration overpotential	V
η_{Ohm}	Ohmic overpotential	V
$\eta_{\text{a,A}}$, $\eta_{\text{a,C}}$	Activation overpotential of reaction at anode (A) and cathode (C)	V
$\eta_{\text{c,A}}$, $\eta_{\text{c,C}}$	Concentration overpotential of reaction at anode (A) and cathode (C)	V
η_{F}	Faradaic efficiency	Dimensionless
Θ	Degree of surface coverage	Dimensionless
γ	Activity coefficient of an ion	Dimensionless
κ	Ionic conductivity	Ohm^{-1} m^{-1}
Λ	Molar conductivity	Ohm^{-1} m^{2} mol^{-1}

λ	Debye screening length, m; molar conductivity of an ion at infinite dilution	$(\text{Ohm}^{-1}\ \text{m}^2\ \text{mol}^{-1})$
μ_i	Chemical potential of species "*i*"	J/mol
$\widetilde{\mu}_i$	Electrochemical potential of species "*i*"	J/mol
ν_i	Stoichiometric coefficient of species "*i*"	Dimensionless
ν	Rate of potential sweep	mV/s
ρ	Density, g/cm^3, charge density, C/cm^3; resistivity	Ohm m
σ	Surface charge density, C/cm^2; conductivity	$(\text{Ohm}^{-1}\ \text{m}^{-1})$
τ	Time constant	s
Φ	Electron work function	eV
φ	Internal potential	V
χ	Surface potential	V
ψ	External potential	V
ω	Angular frequency	rad/s
ν	Reaction rate, mol/dm^3 s	mol/m^2 s

Symbols applied in photoelectrochemistry of SC (Chapters 10–12, 14, and 15)

C_H	Capacity of Helmholtz layer	$\mu F/cm^2$
C_{SC}	Capacity of space charge layer in SC	$\mu F/cm^2$
C_{SS}	Capacity of surface states	$\mu F/cm^2$
c	Bulk concentration	mol/dm^3
c_{Ox}, c_{Red}	Concentration of oxidized and reduced forms of the redox couple in solution	mol/dm^3
D_n, D_p	Diffusion coefficient of electrons and holes	cm^2/s
D_{Redox}	Density of electron states in redox system	$J^{-1}\ cm^{-3}$
D_{ox}	Density of unoccupied electron states in redox system	$J^{-1}\ cm^{-3}$
D_{Red}	Density of occupied electron states in redox system	$J^{-1}\ cm^{-3}$
E	Energy, potential	eV, V
E_F	Fermi energy level of electron in SC	eV
$E_{F,\,Redox}$	Fermi energy level of electron in redox system, in equilibrium	eV
$E_{F,\,dec,\,n}$, $E_{F,\,dec,\,p}$	Fermi energy (electrochemical potential) for the decomposition reaction of SC electrode with participation of electrons and holes	eV
E_C, E_V	Energy of the edges of conduction and valence bands (CB and VB) in the SC bulk	eV
$E_{C,\,S}$, $E_{V,\,S}$	Energy of the edges of conduction and valence bands at the surface of SC	eV
E_g	Energy of bandgap	eV
E_{SS}	Energy level of the surface state	eV
${}_nE_F^*$, ${}_pE_F^*$	Energy of quasi-Fermi levels for electrons and holes	eV
E (V)	Electrode potential	V
E^0	Standard electrode potential	V
E_{fb}	Flat band potential	V
E_{ph}^{OC}	Open-circuit photopotential	V
E_d^{oc}	Open-circuit potential, in dark	V
E_{ph}	Photopotential, energy of photon	V, eV
E_{Redox}, E_{eq}	Equilibrium potential of redox system	V

E^0_{Redox}	Standard potential of redox system	V
$E^0_{dec,n}$	Standard potential of SC decomposition with participation of electrons	V
$E^0_{dec,p}$	Standard potential of SC decomposition with participation of holes	V
ff	Fill factor	Dimensionless
$\Delta_r G$	Gibbs free energy of reaction	J/mol
$\Delta_r G^0$	Standard Gibbs free energy of reaction	J/mol.
G_n, G_p	Rate of generation of electrons and holes under external stimulus	$cm^{-3}\,s^{-1}$
I_0	Light incident intensity	W/cm^2
I_{ab}	Absorbed light (photon) flux	W/cm^2
j_n, j_p	Current density of electrons and holes	A/cm^2
$j_{n,diff}$	Diffusion current density of electrons	A/cm^2
$j_{p,diff}$	Diffusion current density of holes	A/cm^2
$j^{ph}_{sc,c}$	Short-circuit photocurrent density	A/cm^2
$j_{CB}, j_{0,CB}$	Current density, exchange current density, and currents that flow across conduction band	A/cm^2
$j_{VB}, j_{0,VB}$	Current density, exchange current density, and currents that flow across valence band	A/cm^2
j_{corr}	Corrosion current density	A/cm^2
j_{ph}	Photocurrent density (SC electrode)	A/cm^2
k	Rate constant of chemical reaction (first order)	s^{-1}
K_{ads}	Equilibrium adsorption constant	Dimensionless
L_D	Debye length	μm
L_{diff}	Diffusion length	μm
L_n, L_p	Diffusion length of electrons and holes	μm
L_H	Thickness of Helmholtz layer	nm
L_{SC}	Thickness of space charge layer in SC	nm, μm
N_c, N_V	Density of electron states in conduction and valence bands of SC	$J^{-1}\,cm^{-3}$
N_D, N_A	Concentration of donors and acceptors in the SC bulk	cm^{-3}
N_{SS}	Density of surface states	J/cm^2
n, p	Concentration of electrons in conduction band and holes in valence band	cm^{-3}
n_0, p_0	Concentration of electrons in conduction band and holes in valence band in the equilibrium state	cm^{-3}
P_{in}	Input power of solar irradiation	W/m^2
P_{out}	Output power of irradiated system	W/m^2
Q_{SC}	Charge of space charge layer in SC	C
Q_{SS}	Charge of surface states	C
R	Rate of hydrogen generation	mol/s
R_n, R_p	Recombination rates of electrons and holes	$cm^{-3}\,s^{-1}$
S	Selectivity	Dimensionless
t	Time	s
U	External potential applied to the SC electrode	V.

Greek symbols

ε	Power efficiency	Dimensionless
ε_{STH}	Solar to hydrogen conversion efficiency	Dimensionless
ε_{thresh}	Threshold efficiency	Dimensionless
ϵ_{stor}^{ch}	Energy storage efficiency	Dimensionless
ϵ_{SC}	Relative dielectric permittivity of SC	Dimensionless
α	Absorption coefficient	m^{-1}
λ	Wavelength	m
τ_n, τ_p	Lifetime of electrons and holes	s
μ_n, μ_p	Mobility of electrons and holes	$m^2/V\ s$
$\tilde{\mu}_e(SC)$	Electrochemical potential of electrons in SC	eV
$\tilde{\mu}_e$ (Redox)	Electrochemical potential of electrons in redox system	eV
φ_G	Galvani potential	V
φ_{G-Ch}	Potential drop in Gouy–Chapman layer	V
φ_H	Potential drop in the Helmholtz layer	V
φ_s, φ_b	Potential at the surface, in the bulk of SC	V
φ_{SC}	Potential drop in space charge layer in SC	V
φ_{SC}^0	Potential drop in space charge layer; the SC is in equilibrium with redox system	V
$\varnothing$	Quantum yield	Dimensionless
$\varnothing_i$	Quantum yield of *i*-product formation	Dimensionless
$\varnothing_r$	Quantum yield of *r*-reactant decomposition	Dimensionless
v_r	Rate of chemical reaction	$mol/dm^3\ s$
v	Frequency	s^{-1}
η_A, η_C	Overpotential of reactions at anode (A) and cathode (C)	V

Constants

e	Elementary charge and absolute value of electrical charge	1.602×10^{-19} C
ϵ_0	Dielectric permittivity of vacuum	$8.854 \times 10^{-12}\ J^{-1}\ C^2\ m^{-1}$, F/m
k_B	Boltzmann's constant	1.380×10^{-23} J/K
h	Planck's constant	6.626×10^{-34} Js
R	Gas constant	8.314 J/K mol
F	Faraday's constant	9.648×10^4 C/mol

Index

The index lists only those page numbers when a given term is used or explained for the first time, or where another aspect of the same term is described

https://doi.org/10.1515/9783111160986-019

www.ingramcontent.com/pod-product-compliance
Lightning Source LLC
Chambersburg PA
CBHW080657220326
41598CB00033B/5244

* 9 7 8 3 1 1 1 1 6 0 3 4 4 *